Advances in Air Pollution Modeling for Environmental Security

NATO Science Series

A Series presenting the results of scientific meetings supported under the NATO Science Programme.

The Series is published by IOS Press, Amsterdam, and Springer (formerly Kluwer Academic Publishers) in conjunction with the NATO Public Diplomacy Division.

Sub-Series

I. Life and Behavioural Sciences	IOS Press
II. Mathematics, Physics and Chemistry	Springer (formerly Kluwer Academic Publishers)
III. Computer and Systems Science	IOS Press
IV. Earth and Environmental Sciences	Springer (formerly Kluwer Academic Publishers)

The NATO Science Series continues the series of books published formerly as the NATO ASI Series.

The NATO Science Programme offers support for collaboration in civil science between scientists of countries of the Euro-Atlantic Partnership Council. The types of scientific meeting generally supported are "Advanced Study Institutes" and "Advanced Research Workshops", and the NATO Science Series collects together the results of these meetings. The meetings are co-organized by scientists from NATO countries and scientists from NATO's Partner countries – countries of the CIS and Central and Eastern Europe.

Advanced Study Institutes are high-level tutorial courses offering in-depth study of latest advances in a field.
Advanced Research Workshops are expert meetings aimed at critical assessment of a field, and identification of directions for future action.

As a consequence of the restructuring of the NATO Science Programme in 1999, the NATO Science Series was re-organized to the four sub-series noted above. Please consult the following web sites for information on previous volumes published in the Series.

http://www.nato.int/science
http://www.springeronline.com
http://www.iospress.nl

Series IV: Earth and Environmental Series – Vol. 54

Advances in Air Pollution Modeling for Environmental Security

edited by

István Faragó
Eötvös Loránd University,
Budapest, Hungary

Krassimir Georgiev
Bulgarian Academy of Sciences,
Sofia, Bulgaria

and

Ágnes Havasi
Eötvös Loránd University,
Budapest, Hungary

Springer

Published in cooperation with NATO Public Diplomacy Division

Proceedings of the NATO Advanced Research Workshop on
Advances in Air Pollution Modeling for Environmental Security
Borovetz, Bulgaria
8–12 May 2004

A C.I.P. Catalogue record for this book is available from the Library of Congress.

ISBN 10 1-4020-3350-8 (PB)
ISBN 13 978-1-4020-3350-6 (PB)
ISBN 10 1-4020-3349-4 (HB)
ISBN 10 1-4020-3351-6 (e-book)
ISBN 13 978-1-4020-3349-0 (HB)
ISBN 13 978-1-4020-3351-3 (e-book)

Published by Springer,
P.O. Box 17, 3300 AA Dordrecht, The Netherlands.

www.springeronline.com

Printed on acid-free paper

TABLE OF CONTENTS

PREFACE

The protection of our environment is one of the major problems in the society. More and more important physical and chemical mechanisms are to be added to the air pollution models. Moreover, new reliable and robust control strategies for keeping the pollution caused by harmful compounds under certain safe levels have to be developed and used in a routine way. Well based and correctly analyzed large mathematical models can successfully be used to solve this task. The use of such models leads to the treatment of huge computational tasks. The efficient solution of such problems requires combined research from specialists working in different fields.

The aim of the NATO Advanced Research Workshop (NATO ARW) entitled "Advances in Air Pollution Modeling for Environmental Security" was to invite specialists from all areas related to large-scale air pollution modeling and to exchange information and plans for future actions towards improving the reliability and the scope of application of the existing air pollution models and tools. This ARW was planned to be an interdisciplinary event, which provided a forum for discussions between physicists, meteorologists, chemists, computer scientists and specialists in numerical analysis about different ways for improving the performance and the quality of the results of different air pollution models.

The NATO ARW was held at Borovetz (Bulgaria), in the period 8-12 June, 2004. The participants were partly outstanding specialists with international reputation, partly very talented young researchers who will once belong to the first category. About 46 delegates from 17 NATO member countries and partner countries actively participated in the workshop. (In addition to the above NATO participants, there were ten further participants, supported by BULAIR.)

The main objectives of this meeting were the following:
(a) improving the abilities of air pollution models to calculate reliable predictions of the pollution levels in a given domain and in real time by using adequate description of the physical and chemical processes,
(b) implementation of advanced numerical methods and algorithms in the models,
(c) efficient utilization of up-to-date computer architectures,

(d) development of mechanisms for studying particles (including here fine and ultra-fine particles), biogenic emissions, etc.
(e) optimization techniques in the study of the pollution levels, etc.

Plans for developing more advanced and more reliable air pollution models were also discussed. The adaptation of the existing and new models to the new generation of computers was one of the major topics of this meeting.

There were 45 plenary talks given. In the first working day we organized a discussion "Running comprehensive air pollution models on different kinds of computers architectures", while on the next day we had a discussion on "Emission control in air pollution problems". On the third day we organized a discussion about "Data assimilation and solving big optimization problems".

We hope that the participants of the meeting have got new motivation to further applications of the existing results with possible cooperation with specialists from different NATO countries. The exchange of the experience and knowledge of the specialists in air pollution modeling, numerical mathematics, optimal control and parallel computing could facilitate very much the further activities on this area. The participants could present their latest results in the area of air pollution modeling and – as long-term benefits – they will do cooperative research towards reducing the trans-boundary transport of air pollution. It is expected that new collaborations between the NATO and Partner countries institutions and new teams for joint research will be established. The exchange of knowledge, ideas and tools for treatment of the air pollution models for environmental security and the optimal control of emissions can be used for an improvement of the existing models and the development of new ones in the near future.

The editors

ACKNOWLEDGEMENTS

The NATO ARW entitled "Advances in Air Pollution Modeling for Environmental Security" was fully supported by the grant ARW.EST.980503 from the NATO Scientific Program. This support allowed us (Istvan Farago from the Eotvos Lorand University, Hungary and Krassimir Georgiev from the Bulgarian Academy of Sciences, Bulgaria) to organize this interdisciplinary meeting. The expenses of ten further participants were fully covered by the project EC 5FP Contract EVK2-CT-2002-80024 BULAIR.

The editors are indebted to the people involved in the organization work. Our special thanks to

- all other members of the Organizing Committee:
  Artash Aloyan (Russian Academy of Sciences, Russia)
  Laszlo Bozo (Hungarian Meteorological Service, Hungary)
  Peter Builtjes (TNO-MEP, The Netherlands)
  Ivan Dimov (Bulgarian Academy of Sciences, Bulgaria)
  Adolf Ebel (University of Cologne, Germany)
  Clemens Mensink (VITO, Belgium)
  Roberto San Jose (Technical University of Madrid, Spain)
  Dimiter Syrakov (Bulgarian Academy of Sciences, Bulgaria)
  Emanuel Vavalis (University of Crete, Greece)
  Zahari Zlatev (NERI, Denmark);

- Company for International Meetings (CIM Ltd), Sofia for its help in the technical organization of the meeting and accommodation of the participants;
- LogiCom Kft and Gabor Horvath for their help in informatics.

We acknowledge very much the help provided in the process of preparing the Proceedings, mainly, to

- Annelies Kersbergen (Springer NATO Publishing Unit) for the successful collaboration;
- Katalin F.Ható for her very valuable work in the text edition process of the manuscript.

We would like to express our gratitude to Dr. Alain Jubier (Programme Director) and Lynne Nolan (Programme Secretary) from Environmental and Earth Science & Technology of the NATO, who helped us in any moment of our work.

The workshop could not have been successful without the active contribution of all participants.

The editors

MATHEMATICAL MODELING OF THE REGIONAL-SCALE VARIABILITY OF GASEOUS SPECIES AND AEROSOLS IN THE ATMOSPHERE

Artash Aloyan and Vardan Arutyunyan

Institute of Numerical Mathematics, RAS, 8 Gubkin str., 119991, Moscow, Russia

Abstract: A three-dimensional numerical model of atmospheric hydrodynamics, transport and photochemical transformation of gas-phase pollutants and aerosol dynamics is considered. New particle formation occurs via binary homogeneous nucleation of sulphuric acid and water vapor, proceeding under the conditions of temperature and humidity fluctuations. The kinetic processes are described by multidimensional equations of condensation and coagulation, where the size-distribution function is given explicitly. Wind flow fields and turbulent characteristics are calculated from the mesoscale nonhydrostatic model of atmospheric hydrodynamics over complex topography. A series of numerical experiments are performed aimed at modeling the photochemical air pollution and aerosol dynamics in two specific regions as well as formation of sulfate aerosol particles in the northern hemisphere. A comparison is given for calculated and measured ozone concentration data.

Key words: Atmosphere, photochemical air pollution, aerosol, nucleation, condensation, coagulation, numerical modeling.

1. INTRODUCTION

Sulfate aerosol plays an important role in different regions: free atmosphere, marine boundary layer, Arctic, urban polluted areas, etc. Tropospheric ozone pollution is another problem of primary interest. Resulting from the interaction between nitrogen oxides and volatile organic compounds, this secondary pollutant is a key indicator for photochemical air pollution. For aerosol modeling, photochemical transformation products are

I. Farago et al. (eds.), Advances in Air Pollution Modeling for Environmental Security, 1–10.

of primary importance due to the fact that their vapor condensation leads to primary clusters initiating aerosol formation. Then these clusters interact with background atmospheric nuclei.

At present, there are a number of models for photochemical transport (Hass et al., 1993; Stockwell et al., 1990; Zlatev et al., 1992) and aerosol dynamics including chemistry (Ackermann et al., 1998; Meng et al., 1998; Wexler et al., 1994). Some aerosol models employ aerosol size distribution as a simple function (lognormal, gamma distribution, monodisperse spectrum, etc.). Another approach is used in the aerosol models where the aerosol size distribution is described by non-equilibrium distribution functions (Aloyan et al., 1997; Aloyan, 2000; Meng et al., 1998). In Binkowski and Shankar (1995), a regional aerosol model is described including primary emission, three-dimensional transport and chemistry of sulfuric aerosol particles, based on the oxidation of sulfur dioxide.

In this paper, we describe the structure of the numerical models designed for regional-scale transport and transformation of gaseous pollutants and aerosols in the atmosphere (Aloyan, 1998; 2000; Aloyan et al., 1995; 1997; 2004) as well as present some results of the numerical calculations performed on the basis of these models.

2. HYDRODYNAMIC MODEL

The mesoscale nonhydrostatic hydrodynamic model is based on the equations of atmospheric thermo-hydrodynamics, written in a divergent form in the Cartesian coordinate system and with a terrain-following transformation (Aloyan et al., 1995; Penenko and Aloyan, 1985). The horizontal turbulent exchange coefficients for momentum, heat, and moisture are determined using the Reynolds stress tensor, and the vertical turbulent exchange coefficients are obtained from a (k-ε)-model. The structure of the lower atmospheric layer is described by the Monin–Obukhov similarity theory and the Businger empirical functions. The earth-surface temperature is calculated from the balance equation and the equation of heat and moisture distribution in soil.

3. TRANSPORT MODEL

The basic equations for the concentration change rates of multicomponent gas species and aerosols are represented as (Aloyan, 2000):

$$\frac{\partial C_i}{\partial t} + u_j \frac{\partial C_i}{\partial x_j} = F_i^{gas} - P_i^{nucl} - P_i^{cond} + P_i^{phot}$$
$$+ \frac{\partial}{\partial x_1} K_{11} \frac{\partial c_i}{\partial x_1} + \frac{\partial}{\partial x_2} K_{22} \frac{\partial c_i}{\partial x_2} + \frac{\partial}{\partial x_3} K_{33} \frac{\partial c_i}{\partial x_3}, \quad (j = \overline{1,3}) \tag{1}$$

$$\frac{\partial \varphi_k}{\partial t} + (u_j - \delta_{j3} w_g) \frac{\partial \varphi_k}{\partial x_j} = F_k^{aer} + P_k^{nucl} + P_k^{cond} + P_k^{coag}$$
$$+ \frac{\partial}{\partial x_1} K_{11} \frac{\partial \varphi_k}{\partial x_1} + \frac{\partial}{\partial x_2} K_{22} \frac{\partial \varphi_k}{\partial x_2} + \frac{\partial}{\partial x_3} K_{33} \frac{\partial \varphi_k}{\partial x_3}, \quad (j = \overline{1,3}) \tag{2}$$

Here C_i, $i = 1, \ldots, N_g$ and φ_k, $k = 1, \ldots, N_a$, are the concentrations of gaseous species and aerosols, respectively; N_g and N_a are the numbers of gaseous components and aerosol fractions, respectively; w_g is the gravitational settling; F^{gas} and F^{aer} stand for emissions of gaseous species and aerosols; P^{nucl}, P^{cond}, P^{coag}, and P^{phot} are the operators of nucleation, condensation, coagulation, and photochemical transformation, respectively.

Usually, in photochemical pollution modeling, Eq. (2) is not used, and the kinetic terms P^{nucl} and P^{cond} are omitted from the right-hand side of Eq. (1). Unlike this, our study of sulfate aerosol formation makes use of both Eqs. (1)–(2), but of all the species only SO_2 is considered, and hence the photochemical term P^{phot} is omitted.

Eqs. (1)–(2) are considered in the domain $D_t = D \times [0, T]$, where $D = \{(x,y,\sigma); x \in [-X, X]; y \in [-Y, Y]; \sigma \in [0, H]\}$ is the upper boundary of the model domain. Initial conditions at $t = 0$ are assumed to be given. At the lateral boundaries, the meteorological fields, gaseous admixtures and aerosols are supposed to match smoothly with their respective background values. At the upper boundary of the air mass ($z = H$), the perturbations due to surface roughness are assumed to decay. To set the boundary condition at $z = h$ (where h is the surface layer height), the pollutant surface resistance is represented as a sum of the following three terms: aerodynamic resistance r_a, quasi-laminar resistance of the surface layer r_b, which is due to molecular diffusion near the surface of the Earth, and the surface resistance r_c, which accounts for processes at the deposition surface. Hence, the pollutant fluxes in the surface layer take the following form:

$$\nu \frac{\partial \varphi}{\partial z} = \frac{\varphi - \varphi_{surf}}{r_a + r_b + r_c}, \text{ where } r_b = \frac{2}{\kappa u_*} \left(\frac{Sc}{Pr} \right)^{2/3}. \tag{3}$$

Here $Sc = 1.5 \times 10^{-5}$ m^2s^{-1} is the Schmidt number, $Pr = 0.71$ is the Prandtl number, and u_* is the dynamical speed.

4. CHEMICAL MODEL

We assume that in the model the gaseous species NO, NO_2, CO, CH_4, H_2CO, and other organic compounds are emitted into the atmosphere. The chemical mechanism used in the model contains typical nitrogen, sulfur, and carbon cycles of reactive species (a total of 44 chemical species and 204 chemical reactions). The complete list of chemical species is as follows: $O(^1D)$, $O(^3P)$, O_3, OH, HO_2, H, NO_3, NO, NO_2, SO_2, H_2CO, HCO, H_2O_2, CO, CH_4, HNO_2, HNO_3, SO_3, SO_2^*, N_2O_5, CH_3, CH_3O_2, CH_3O, CH_3OH, CO_2, $CH_3O_2NO_2$, HCOOH, $HOCH_2O$, $HOCH_2O_2$, CH_3OOH, HO_2NO_2, CH_3ONO_2, ISOP, XO2, ALD2, C2O3, PAN, TOL, TO2, PAR, OLE, ROR, XO2N, ETH. Here the Carbon-Bond Mechanism (Gery et al., 1989) nomenclature is partially used. More details on the chemical model can be found in (Aloyan et al., 1995; 2004).

5. AEROSOL MODEL

New nuclei are formed from vapor mainly by the binary homogeneous nucleation of sulfuric acid and water drops. Suppose that there is a binary cluster in the supersaturated vapor atmosphere at temperature T and at pressure Pv; this cluster consists of n_w molecules of the substance w and n_a molecules of the substance a with the mole fractions x_{iv}, $i = w,a$. We assume that the particles have a spherical shape and are in liquid phase. Then the free energy for the generation of a liquid particle in the binary mixture can be represented as:

$$W = \Delta G = n_w \Delta\mu_w + n_a \Delta\mu_a + A\sigma \quad (4)$$

where ΔG is the variation of Gibbs free energy, A is the surface area, σ is the surface tension, $\Delta\mu_i = \mu_{il}(T, P_v, x_{il}) - \mu_{iv}(T, P_v, x_{iv})$, where μ_{il} and μ_{iv} are chemical potentials in the liquid and vapor phases, respectively. The function ΔG for supersaturated vapor has a maximum at the point of critical radius r^*. Vapor is supersaturated for all drops with $r > r^*$.

The parameterization used in this work can be found in Kulmala and Laaksonen (1998), Vehkamaki et al. (2000).

The critical radius (r^*) and the energy for critical-cluster formation (w^*) are determined as

$$r^* = \frac{2\sigma(x^*)v_i(x^*)}{kT\ln\left(\rho_i^{free} / \rho_{i,s}^{free}(x^*)\right)}; \quad w^* = \frac{4}{3}\pi r^{*2}\sigma(x^*). \quad (5)$$

Finally, the nucleation rate has the form

$$J = Z\rho(1,2)\exp\left[-\frac{w^* - w(1,2)}{kT}\right], \quad (6)$$

where $\rho(1,2)$ and $w(1,2)$ are the number concentrations of the formation energy of sulfuric acid dihydrate, and Z is a kinetic pre-factor.

The equation of the particle mass distribution for outer particles through attachment of vapor molecules and new-particle formation is

$$\frac{\partial \varphi_g}{\partial t} + \frac{\partial v_g \varphi_g}{\partial g} = J_{nucl}(g,t) +$$

$$\frac{1}{2}\int_0^g K(g,g_1)\varphi_{g-g_1}\varphi_{g_1}dg_1 - \int_0^\infty K(g,g_1)\varphi_{g_1}dg_1 \quad (7)$$

where g_1 is the mass of particle, φ_g is the concentration of particles with masses between g and $g + dg$, $K(g, g_1)$ is the collision frequency of particles with masses of g and g_1, $J_{nucl}(\varphi_g, t)$ is the nucleation rate. This equation describes also the condensation growth and particle merging due to coagulation. Eq. (7) corresponds to the P^{coag} and P^{cond} operators in Eqs. (1)–(2).

6. NUMERICAL RESULTS

Numerical experiments were conducted to simulate the photochemical pollution in the Houston–Galveston area (USA), using a database of the TexAQS-2000 field campaign (August 22 to September 1, 2000). Area- and point-source emission data on the following species were used: NO, NO_2, OLE, PAR, TOL, XYL, FORM, ALD2, ETH, MEOH, ETOH, ISOP, SO_2, and CO. The values of point-source emissions were given on two grid domains: one with X = 1620 km, Y = 1656 km, and horizontal steps of 36 km in both directions, and the other with X = 332 km, Y = 260 km, and horizontal steps of 4 km. The lateral and upper (at z = 2.3 km) boundary conditions were given by the hourly values of these species. For each time step, the concentrations of given species at lateral boundaries of the smaller domain were determined by interpolating the solution obtained for the larger domain. In Figure 1, time series plots of calculated and measured hourly-mean ozone concentrations are shown at 4 points located in the model domain. In general, the diurnal change of ozone concentrations is well reproduced by the model.

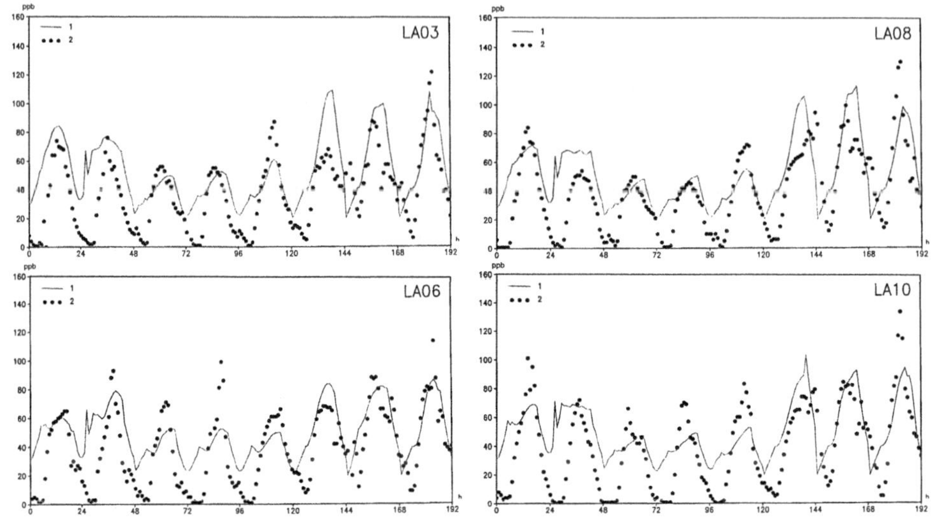

Figure 1: Calculated and measured ozone concentrations at 4 points.

Numerical experiments were also performed for the Baikal lake region, taking into account the following interconnected processes: atmospheric dynamics (photochemical and aerosol transport), binary homogeneous nucleation, condensation/evaporation, and coagulation. First, the atmospheric circulation parameters were obtained from the hydrodynamic model and, then, the problem of gas–aerosol transport was solved. In Figures 2–5, the five local emission (SO_2, NO_x, and CO) sources are shown in circles. In Figure 2, the concentration of H_2SO_4 is demonstrated for t = 14 h. A total of 30 particle size intervals were considered in the range from 0.1 to 5 μm. Figure 3 gives the nucleation rate (in units $cm^{-3}s^{-1}$) and Figure 4 gives the threshold concentration of H_2SO_4. In Figure 5, the particle-size variation due to coagulation and condensation (the distribution of particles with radius 0.1 μm) is presented.

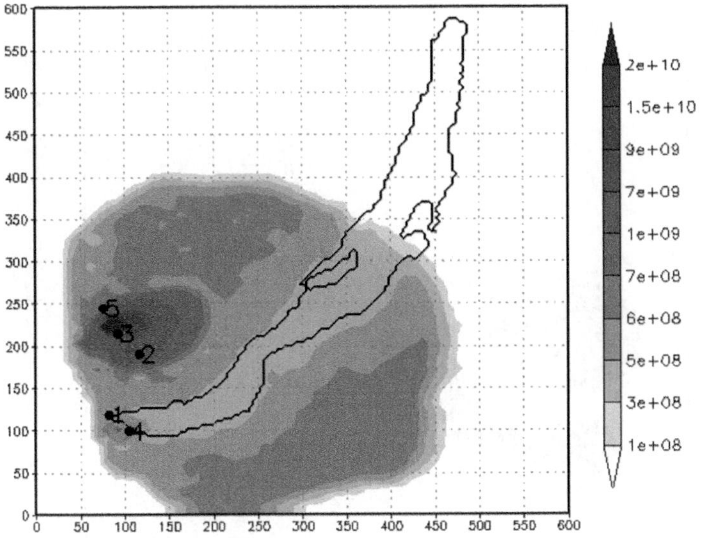

Figure 2: Concentration of H_2SO_4, t = 14 h.

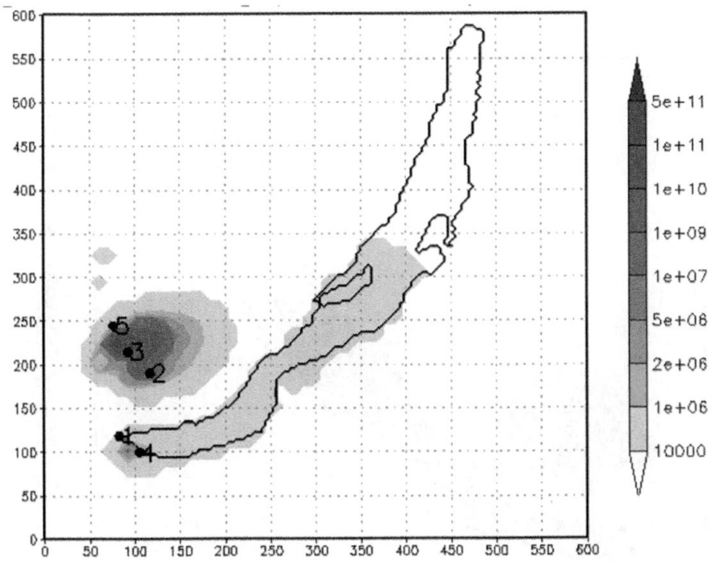

Figure 3: Nucleation rate, t = 14 h.

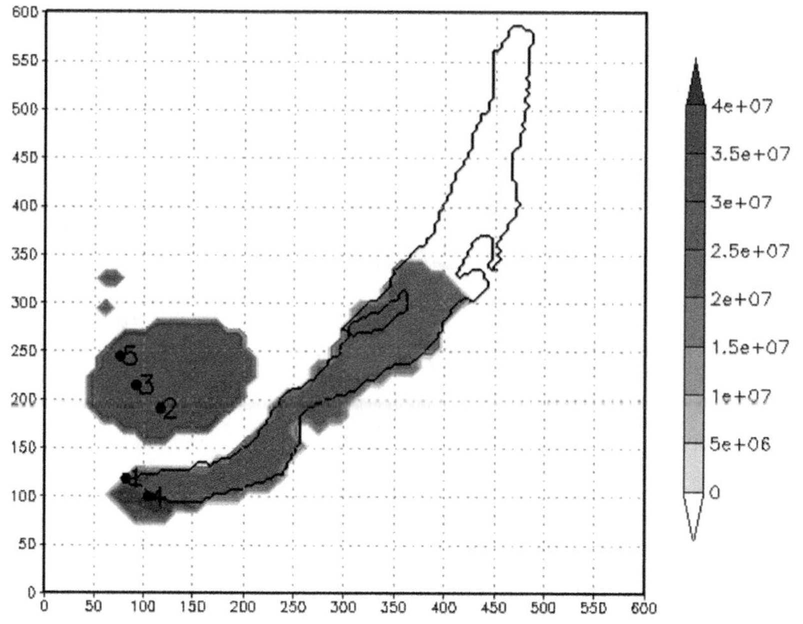

Figure 4: Threshold concentration of H_2SO_4, t = 14 h.

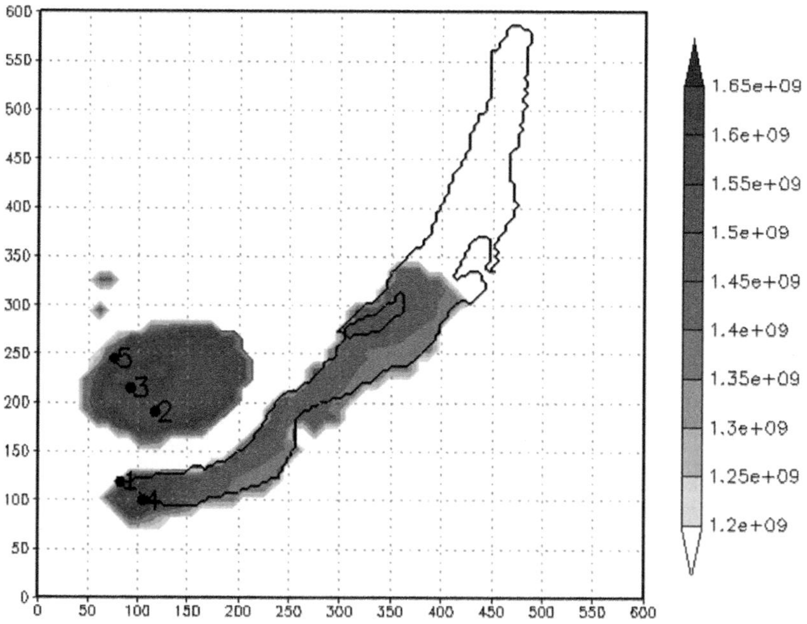

Figure 5: Particle-size spectrum, r = 0.1 μm, t = 14 h.

Also, using the kinetic equations, the problem of sulfate aerosol formation was considered in the northern hemisphere. These numerical experiments are based on Eqs. (1)–(2) written in the spherical coordinate system (Aloyan, 2000), using a 144 x 72 x 15 grid with the upper boundary of the model domain at a height of 9.2 km. As an illustration, in Figure 6, the number concentration is presented for particles with radius r = 0.11 μm, after 30 days of simulation.

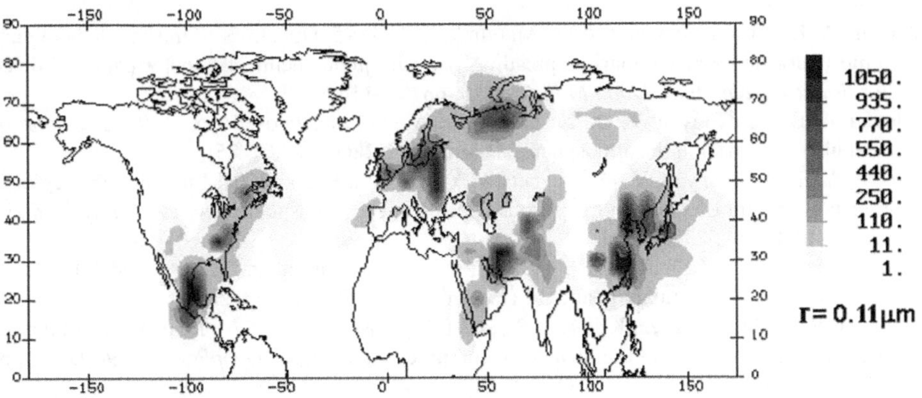

Figure 6: Aerosol concentration, t = 30 days, z = 6800 m.

7. CONCLUSION

- The model can be used to study the variability of gaseous species and aerosols in the lower troposphere.
- New-particle formation is conditioned by binary nucleation both in the regional and global scales.
- The formation of sulfate aerosol depends on temperature and moisture characteristics of the atmosphere and has an essential diurnal variation pattern for urban areas and a seasonal pattern for the upper troposphere.
- The difference between measured and calculated concentrations can be explained by uncertainties in meteorological data and turbulence parameters in night-time stable-stratification conditions as well as the fact that no account of gas–aerosol interaction was given.

Acknowledgements: This work was supported by the Russian Foundation for Basic Research, project nos. 03-05-64484 and 03-01-96-187 (r2003yug), and by the ISTC project no. 1908.

REFERENCES

Ackermann, I. J., Hass, H., Memmesheimer, M., Ebel, A., Binkowski, F. S., and Shankar, U., 1998, Modal aerosol dynamics model for Europe: Development and first applications. *Atmos. Environ.* **32**, pp. 2981–2999.

Aloyan, A. E., 1998, Mesoscale modelling of wet convection and gas-aerosol interaction. *Russ. J. Num. Anal. Math. Modelling* **13(6)**, pp. 447–453.

Aloyan, A. E., 2000, Mathematical modelling of the interaction of gas species and aerosols in atmospheric dispersive systems. *Russ. J. Num. Anal. Math. Modelling* **15(1–4)**, pp. 211–224.

Aloyan, A. E., Arutyunyan, and V. O., Marchuk, G. I., 1995, Dynamics of mesoscale boundary atmospheric layer and impurity spreading with the photochemical transformation allowed for. *Russ. J. Num. Anal. Math. Modelling* **10**, pp. 93–114.

Aloyan, A. E., Arutyunyan, V. O., Lushnikov, A. A., and Zagaynov, V. A., 1997, Transport of coagulating aerosol in the atmosphere. *J. Aeros. Sci.* **28(1)**, pp. 67–85.

Aloyan A. E., Arutyunyan, V. O., Kuznetsov, Yu. A., and He, J., 2004, Modeling the regional transport and transformation of gaseous species in the atmosphere. *Izv. RAN: Phys. Atmos. Ocean* **40**(4), pp. 523–534.

Binkowski, F. S., and Shankar, U., 1995, The Regional Particulate Matter Model 1. Model description and preliminary results. *J. Geophys. Res.* **100**(D13), pp. 26191–26209.

Gery, M. W., Whitten, G. Z., Killus, J. P., and Dodge, M. C., 1989, A photochemical kinetics mechanism for urban and regional-scale computer modeling. *J. Geophys. Res.* **94**(D10), pp. 12925–12956.

Hass, H., Ebel, A., Feldmann, H., Jakobs H. J., and Memmesheimer, M., 1993, Evaluation studies with a regional chemical transport model (EURAD) using air quality data from the EMEP monitoring network. *Atmos. Environ.* **27A**, pp. 867–887.

Kulmala, M., Laaksonen, A., and Pirjola, L., 1998, Parameterization for sulfuric acid/water nucleation rates. *J. Geophys. Res.* **103**, pp. 8301–8307.

Meng, Z., Dabdub, D., and Seinfeld, J. H., 1998, Size-resolved and chemically resolved model of atmospheric aerosol dynamics. *J. Geophys. Res.* **103**, pp. 3419–3435.

Penenko, V. V., and Aloyan, A. E., 1985, Models and Methods for Environment Control Problems. Nauka, Moscow (in Russian).

Stockwell W. R., Middleton P., Chang J. S., and Tang X., 1990, The second-generation Regional Acid Deposition Model chemical mechanism for regional air quality modeling. *J. Geophys. Res.* **95**, pp. 16343–16367.

Vehkamaki, H., Kulmala, M., Napari, I., Lehtinen, K. E. J., Timmreck, C., Noppel, M., Laaksonen, A., 2002, An improved parameterization for sulfuric acid-water nucleation rates for tropospheric and stratospheric conditions. *J. Geophys. Res.* **107**(D22), p. 4622.

Wexler, A. S., Lurmann, J. H. and Seinfeld, J. H., 1994, Modeling urban and regional aerosols – 1. Model development. *Atmos. Environ.* **28,** pp. 531–546.

Zlatev, Z., Christensen, J., and Hov, O., 1992, A Eulerian air pollution model for Europe with non-linear chemistry. *J. Atmos. Chem.* **15**, pp. 1–37.

AIR POLLUTION MODELING IN ACTION

Simulation in action

Katalin Balla, Sándor Márton and Tamás Rapcsák

Computer and Automation Research Institute, Hungarian Academy of Sciences, 1111 Budapest, Kende utca 13-17, Hungary

Abstract: In the paper we describe an air pollution model as applied to a composting plant. Two technologies of the composting process are compared. The model is based upon the Hungarian National Standards. The region is analyzed from meteorological and geomorphologic points of view. For illustration we add grayscale maps of relative concentration for three cases.

Key words: Air pollution modeling, Hungarian National Standards, composting plant, surface sources, point sources.

1. INTRODUCTION

Recently, there is a growing interest in environmental problems. The possible outbreak of the pollution from an industrial plant is just one of the sources that may cause the ruin of the environment and undermine the population's health. Typically, all media – the ground water, the surface water and the air – may be concerned; thus an overall environmental assessment study must cover simulation of the possible events in each of the media.

In general, these studies set several targets. The first one is the prediction of the processes running in and around the plants when they operate normally and when any malfunction occurs. To prevent hard and/or long lasting consequences a monitoring system is to be built up. A monitoring system can be located at properly chosen points and can register the dangerous pollution level with proper timing if it is known in advance when and where the deviation from the normal level can be observed in due course. As usual, the environmental authorities give license for operation of the plant if the operation is "safe". On one hand, it means that during the normal operation the pollution remains much below the fixed threshold. The monitoring system

I. Farago et al. (eds.), Advances in Air Pollution Modeling for Environmental Security, 11–22.

must be capable to measure the actual environmental load. On the other hand, the hazard of malfunctioning must be still acceptable. By the help of proper monitoring, mishaps must be recognized in due course in order to stop them before the fatal consequences become inevitable.

The results of the simulation have financial consequences, too. In some cases, the technology must be improved to ensure that the admissible levels be not exceeded. In other cases, either a more complex or a more frequent monitoring is required to control the processes. Obviously, the investors are interested in a low-cost production that includes the low cost spent for the environmental investments. For this reason, it is not excluded that some times even the generation or release of hazardous pollutants is underestimated. On the other hand, especially in the last two decades, the population exhibits high sensitivity to environmental issues. Both the environmental awareness is continuously raising and there is a fear - solid or unfounded - of improperly modeled environmental impacts. Nowadays, the confrontation with the civil society in environmental issues seems to be a regular event.

The environmental authorities must reflect all these concerns. Their decisions are based upon a unified process of evaluation. An application for a license for operation must contain an overall environmental impact assessment. Environmental laws fix the targets. The methods of evaluation are described in the Hungarian National Standards (HNS).

2. AIR POLLUTION MODELING

The air pollution process – if it does not violate some restrictions – must be simulated by a model fixed in the HNS. It is obligatory, no other models are allowed. Basically, it is a bi-Gaussian plume model. A detailed explanation of the particular modeling aspects and corresponding formulas may be found, e.g., in Szepesi, 1981. On the connection to the simpler Pasquill's model see Mészáros et al., 1999. The HNS decompose the model up to algorithmic steps. It is also fixed what situations must be simulated and what kind of averages must be computed (Hungarian Standards, 1980, 1981a, 1981b, 1982, 1985). The threshold values are contained in governmental and departmental order.

Even with these restrictions, the simulations require cooperation and synergy of specialists of different fields. The problem solution consists of three phases.

- Evaluation of the technological processes, the geomorphologic and the meteorological situation at the location of the plant.
- Data processing. This is partially a software-engineering problem of medium size. It includes the organization and check of the input data, the

computation as prescribed by the algorithm and organization of the output into tables, figures and maps.

- Evaluation of the results. As a first step the results must undergo a check whether they do not indicate the inapplicability of the model. An additional sensitivity analysis prevents the misuse of incorrect or estimated data and uncertainties of measurements. Provided the results are meaningful, the evaluation of the environmental impacts follows. Suggestions if necessary for the change of the technology are given. Proposals to the operation of the monitoring system are formulated.

In the next parts of the paper we demonstrate some points of the above process as they were carried out within a case study. It concerns the transmission of chemical substances emitted by a composting plant. The question is whether a technological renewal changes the environmental impacts.

3. MODELS FOR THE COMPOSTING PLANT

The former technology runs an open-air composting process (Figure 1). Not surprisingly, the inhabitants of the nearby settlement have been complained of the offensive effects of the bad smell. Due to technical development, a significant part of the process can be relocated indoors (Figure 2) and the air is to be released through a 36-meter high stack. The other part of the process that remains outdoors is planned to emit a reduced amount of chemical substances generating bad smell effects. This study examines and compares the effects of the old and new technologies.

The old technology is modeled by a purely surface source while the new one is a combination of a surface source and a point source. Practically, these two models are variants of a plume model. The difference is in the definition of the turbulent dispersions.

As regards the smell substances, there are no limit values in the HNS. The lack of limit values raises difficulties. Neither the modern sampling nor the analytic methods can detect the extremely low concentration of sulfides in a distance of 1 km. There was no successful detection in case of known compounds, either. At stacks and at their vicinity, the dynamic olfactometric method works. However, even its application is useless at a distance of 1 km or more.

Figure 1: Open-air composting.

Figure 2: Indoor composting.

The computations were carried out for a 1 kg/h emission and the concentration was measured in $\mu g/m^3$. For the heaviest polluting component, the ammonia, the emission values were given. The corresponding concentration values are multiples of that obtained for the unit emission.

The basic (military) map with scale 1:10 000 was transformed into a digital map following the configurations of the terrain, as well. The software goes through the points of the map. The concentration is computed point-by-point taking the input data and the HNS into account pointwise. Finally, the isoconcentration curves are drawn in the basic map.

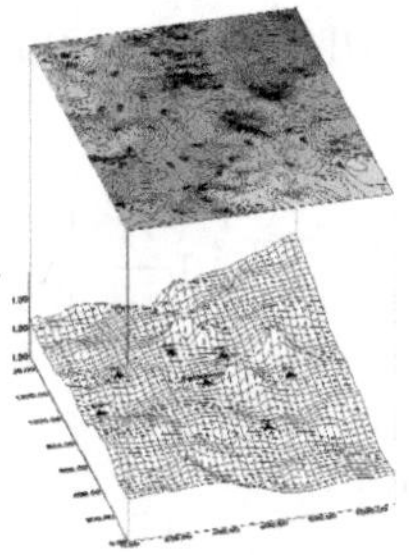

Figure 3: Terrain configuration.

4. HNS AND THE PARAMETERS

In the computations we used HNS, 1981a for point sources with the effective height of the emission computed by HNS, 1985. Surface and line sources are described in HNS, 1981b. Formulas for the turbulent dispersion are taken from HNS, 1980. The definition of the data is given in HNS 1982.

The parameters needed for computations:

1. Based on HNS, 1980, exponents of the wind profile are computed by the use of Pasquill's stability indicator. Its values are computed from the data obtained from the closest monitoring point of the Hungarian Meteorological Service. With the help of the same HNS and Tables 3, 4 and 5 of its Appendix M1, we obtained the atmospheric stability values tabulated in Table 1. The corresponding wind profile exponents are given in Table 2.

	Daytime			Night	
Wind speed [m/s]	2	5	8	1	2
spring	B	D	D	F	E
summer	B	C	D	F	E
autumn	C	D	D	F	E
winter	D	D	D	F	E

Table 1: Atmospheric stability values. Pasquill's stability indicator.

Pasquill's stability indicator	A	B	C	D	E	F,F*
Windprofile exponent (p)	0.079	0.143	0.196	0.270	0.363	0.440

Table 2: Pasquill's stability indicator vs. wind profile exponent.

2. HNS, 1980 defines the turbulent dispersion coefficients for point sources. In the horizontal direction, orthogonal to the wind, the coefficient is

$$\sigma_y = 0.08\left(6p^{-0.3} + 1 - \ln\frac{H}{z_0}\right)x^{0.367(2.5-p)},$$

while the vertical coefficient is

$$\sigma_z = 0.38\left(8.71 - \ln\frac{H}{z_0}\right)x^{1.55\exp(-2.35p)},$$

where p is the windprofile exponent, H is the effective emission height, z_0 is the roughness parameter and x is the distance from the source. The modification for the surface sources is given by formulas

$$\sigma_{sy} = \left(\sigma_{0y}^{\ 2} + \sigma_y^{\ 2}\right)^{1/2}, \sigma_{sz} = \left(\sigma_{0z}^{\ 2} + \sigma_z^{\ 2}\right)^{1/2}$$

where σ_{0y} is the initial horizontal dispersion coefficient. It is equal to the width of the surface source divided by 4.3. σ_{0z} is the initial vertical dispersion coefficient. It is equal to the height of the surface source divided by 2.15. For a wind speed lower than 2 m/s, the horizontal coefficient is multiplied by 4.

3. For elevated points a correction factor to the dispersion coefficients is given, see Table 3. For the missing Pasquill's category C, we worked with an interpolated estimated factor.

Pasquill's stability indicator		B	D	F
correction factor	strongly dissected terrain	1.4	4.0	15.0
	medium sectiondis	1.2	2.5	8.0

Table 3: Correction factors for elevated terrain.

4. The wind speed characteristic for the plume was computed as an average of the speed in the atmospheric layer between heights H-2σ_z and H+2σ_z. Corrections w.r.to the data obtained from the meteorological service took the geomorphologic analysis into account.
5. The values of the surface roughness are given in HNS, 1980, too. For the area in question we made computation both with 0.1 (flat area covered by vegetation) and 0.3 (forest).

Finally, the model requires the following data:

- height of the surface source: 2 m, or

- for the point source
  - stack: height: 36 m, cross-section: 0.875 m^2,
  - speed of the emitted air: 15m/s,
  - temperature of the emitted gas: 50 °C,
  - temperature of the air, see Table 4.

When a point source continuously emits gaseous polluting materials, the relative concentration at the receptor point (*x,y,z*) for short average time (1 h) is computed by the following formula:

$$c = \frac{1}{2\pi\sigma_y\sigma_z u_m}\exp\left[-\frac{1}{2}\left(\frac{y}{\sigma_y}\right)^2\right]\left\{\exp\left[-\frac{1}{2}\left(\frac{z-H}{\sigma_z}\right)^2\right]+\exp\left[-\frac{1}{2}\left(\frac{z+H}{\sigma_z}\right)^2\right]\right\}$$

$$\cdot\exp\left(-\frac{0.693x}{u_m T_{1/2}^{SZ}}\right)\exp\left(-\frac{0.693x}{u_m T_{1/2}^{A}}\right)\exp\left(-\frac{0.693x}{u_m T_{1/2}^{N}}\right).$$

Here u_m is the wind speed (for short average time), $T_{1/2}^{SZ}, T_{1/2}^{A}, T_{1/2}^{N}$ are the characteristic halving time of chemical reactions of the gaseous substances, that of the wet sedimentation of the gaseous substances and that of the wet sedimentation of solid polluting materials, respectively. The HNS suggest to take the halving times equal 1, in general.

In case of surface source, the relative concentration formula is the same as for a point source, just the turbulent dispersions σ_y and σ_z are replaced by σ_{sy} and σ_{sz}. .

Season	Daytime temperature (°C)	Night temperature (°C)
spring	15	7
summer	26	14
autumn	14.5	6.5
winter	3	-3

Table 4: Average temperature.

5. EVALUATION OF THE METEOROLOGICAL AND THE GEOMORPHOLOGICAL SITUATION

The closest village Perbál is located to the east of the plant. A significant smell effect may occur at Perbál, if a west wind blows. However, the data show that the most frequent directions are north, north-northwest, northwest. There is also a relatively frequent direction, the east. The northeast-east-southeast winds are not blowing towards Perbál. There is a negligible amount of wind, cca. 5% of all that blows to Perbál. When one tabulates the seasonal frequencies vs. the wind directions and speed, then it turns out that the north and northwest winds are fresh and strong. Thus, the smell substances are mixed with a high-speed air and they are diluted. It means that either the smell effects decrease or they cannot be observed at all. These phenomena are confirmed by modeling. With the same stability value, the higher the wind speed is, the lower the concentration is. On the other hand, the stability situation corresponding to night period – Pasquill's category F – gives rise to the highest concentration.

Above the purely meteorological data, the geomorphologic analysis should be taken into account. At the west skirts of Perbál there is a valley in the northwest-southeast direction. It redirects the west wind into the direction of the valley. The valley plays the role of a partial windbreak for the whole Perbál's region.

These circumstances define the range of wind directions for which the modeling must be carried out. These are the following.

- West wind: this wind may move the smell substances from the plant to the village;
- West-northwest wind: due to the redirection of the wind, it is more realistic than the previous one. However, it touches essentially only the northwest part of the village.

The west-southwest wind would move the smell substances into the east-northeast parts of Perbál. However, this wind direction is very rare and the valley decreases its impact even more. Practically, this case can be neglected.

The west wind is very rare in each season. Its most frequent speed is 2 m/s. The west-northwest wind is of relatively low speed; the most frequent values are 2 m/s in spring and autumn, 2-3 m/s in summer and 2-5 m/s in winter. While the north-northwest wind direction is of the highest frequency, the secondary maximum is attained at the northwest direction. These winds, however, pass by the side of Perbál. Moreover, the northwest winds are mostly of 5 m/s and 7-8 m/s speed. They help rather to eliminate the "gathered" smell effects at the valley located at the northwest corner of Perbál than to strengthen them. To show that a higher wind speed results in a lower concentration, the simulation was done both for 5 m/s and 8 m/s. The almost wind-free situation was modeled by 0.3 m/s wind speed.

For the night periods, Pasquill's stability indicator has no seasonal differences, it is equally F. Therefore, in the case of surface sources there was no need to model by seasons. The point source behaves differently: the released air lifts into height depending on the outer temperature. As to the wind speed in daytime, it is the highest at midday hours and it is the lowest at nights and in early morning hours.

6. CONCLUSIONS AND MAPS

In the previous sections we explained how the models are built and what are the simulations to be done. Altogether, 192 simulations were carried out for daytime and 36 for night periods.

6.1 General conclusions

When the stability category is fixed, the higher the wind speed is, the lower the concentration is. It means that the plume is shorter. When the wind speed is the same, the increasing stability category (A→F) causes the increase of the concentration.

However, Pasquill's category is not independent of the wind speed. When the wind speed is increasing, the stability category either remains unchanged or increases. A difference of 3 m/s in the wind speed has a greater impact on the concentration than a change in the stability category.

When the wind speed is very low (0.3 m/s) the plum is much larger both in length and in width than in the case of higher speed values.

Surface sources:

From a meteorological point of view the spring and the autumn is very similar. The wind speeds 0.3 m/s, 5 m/s and 8 m/s are accompanied by the same stability categories B, D and D. Thus, the concentration is similar. There is a small difference for 2 m/s. The winds of 8 m/s speed are accompanied by the same stability categories in all seasons. In this case the concentration is the lowest: the 0.1 $\mu g/m^3$ isoconcentration curve is just touching the skirts of Perbál. In spring, in autumn and in winter, if 5 m/s wind blows, then the stability is D and the 0.17 $\mu g/m^3$ isoconcentration curve goes there. The highest values are 0.6-0.7 $\mu g/m^3$ at 0.3 m/s in all seasons with the exception of summer.

With the new technology, the surface emission is decreased essentially. Although the surface becomes smaller, it is far to compensate the decrease of the emitted amount.

Point sources:

The larger plumes are observable in summer and in winter. In summer it is due to the higher outer temperature that does not allow the air to lift too high. In winter, the explanation for the larger plum is the stability category D.

Computations justified the geomorphologic analysis. The higher concentration values were obtained where the inhabitants complained of the bad smell effects. And vice versa, no bad effects were reported from locations where we got low concentration values.

The computation confirms also that if a forest would cover the region between the plant and the village, then there would be a smaller smell effect.

6.2 Conclusions for the ammonium and the dimetil-sulfid

Even with the old technology and a surface emission of 28.1 kg/h, the concentration of the ammonium is diluted to the $1/10^{th}$ of the threshold value at the skirts of Perbál. The new technology yields a release of 20.2 kg/h at the stack. At the same location, the concentration is practically zero. This conclusion holds for the worst meteorological conditions. The ammonium was taken for comparison since its surface emission value is the highest with respect to the surface emission threshold value. The same holds for the ratio of the air quality threshold value and the emitted value. The results demonstrate that even in the case of the ammonium the concentration is negligible with respect to the air quality threshold value. It is less than 10% of the limit.

The highest emission with the new technology belongs to dimetil-sulfid, it is 0.0061 kg/h. However, even with the old technology, when 0.021 kg/h is released, the concentration with respect to the limit value (80 $\mu g/m^3$) is negligible even at the plant itself.

6.3 Maps

For visualization of the general results, a large number of colored maps were drawn with isoconcentration curves for unit loads (relative concentration curves) and for true ones. Here we enclose only one grayscale map of the relative concentrations for each technology.

Acknowledgement: The work was supported by OTKA (Hungarian National Scientific Found) under Grant No. T043276.

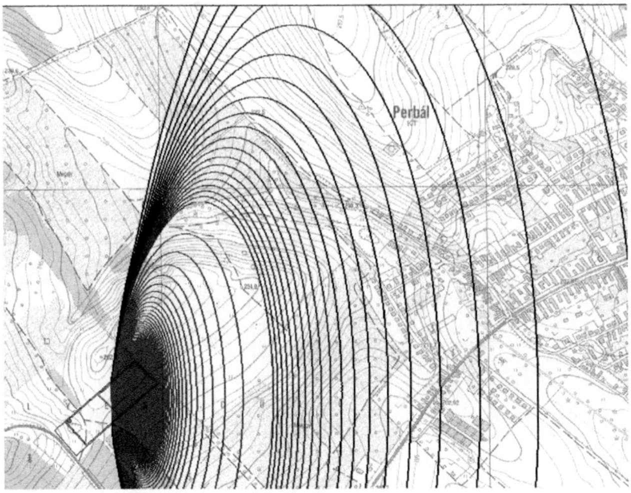

Figure 4: Old technology, surface source, west wind, 0.3 m/s, Pasquill's category B, daytime, spring, autumn, flat, without forest.

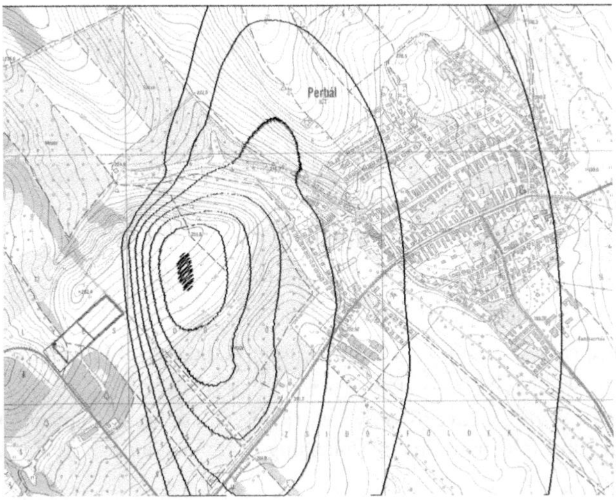

Figure 5: New technology, stack, west wind, 0.3 m/s, Pasquill's category A, daytime, temperature 26 °C, summer, flat, without forest.

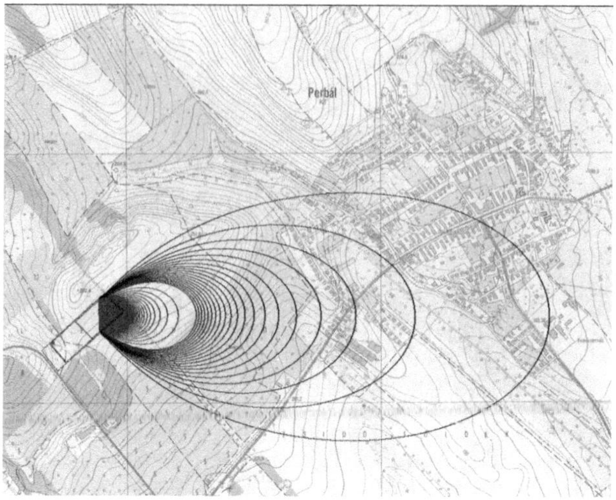

Figure 6: New technology, surface source, west wind, 2 m/s, Pasquill's category B, daytime, spring, summer, flat, without forest.

REFERENCES

Hungarian National Standards, 1980, MSZ 21457/4-80.

Hungarian National Standards, 1981a, MSZ 21459/1-81.

Hungarian National Standards, 1981b, MSZ 21459/2-81.

Hungarian National Standards, 1982, MSZ 21459/4-82.

Hungarian National Standards, 1985, MSZ 21459/5-85.

Mészáros, Cs., Rapcsák, T., Sági, Z., 1999, Pollution in the air. In: Large-scale computations in air pollution modelling. Kluwer, pp. 235-247.

Szepesi, D. (ed.), 1981, A levegőkörnyezet (levegőminőség és humánkomfort) tervezése. Műszaki Könyvkiadó. (Environmental planning of the air (air quality and human comfort), in Hungarian.)

ADVANCES IN URBAN METEOROLOGY MODELLING

Ekaterina Batchvarova[(1)] and Sven-Erik Gryning[(2)]
[(1)]National Institute of Meteorology and Hydrology, 66 Blvd. Tzarigradsko Chaussee, BG 1784 Sofia, Bulgaria; [(2)]Risø National Laboratory, DK-4000 Roskilde, Denmark

Abstract: In this study we focus on the lowest part of the urban boundary layer, which is connected to a horizontal scale of 2-3 km and a vertical scale covering the surface boundary layer. The structure of the urban surface boundary layer is discussed with an emphasis on the turbulence parameters controlling dispersion processes in it.

Key words: Urban boundary layer, atmospheric turbulence, atmospheric dispersion, applied modeling, meteorological experiments.

1. INTRODUCTION

Atmospheric motions take place at an immense variety of scales - from large weather systems of thousand kilometers in size down to eddies on a scale of millimeters responsible for the ultimate dissipation into heat. Therefore it is important to define the limits of models and parameterisations in time and space identifying the processes that are dominant and those which can be neglected.

The urban boundary layer is one of the challenging areas of parameterisation in meteorology, because of its spatial and time variability. Typical scale for the urban area is of the order of 20 km, but the urban area is not homogeneous on this scale. It is subdivided into a large number of areas, such as the central, residential, recreation and industrial parts. These are referred to as neighborhoods. It can be noted that the typical neighborhood scale for an urban area is rather similar to the typical landscape patches of agriculture and forest in Europe. When the area is characterized by narrow

I. Farago et al. (eds.), Advances in Air Pollution Modeling for Environmental Security, 23–32.

irregular streets and houses of varying height the momentum transport is controlled by the street configuration and the dissipative stress is expected to be important. When a rectangular street system of broader streets and high buildings is considered, momentum transport is dominated by the drag forces of the high buildings. An area with high buildings and vast space among them is a separate case.

2. TURBULENCE CHARACTERISTICS

The spatial and temporal variability is a distinguished feature of the turbulent flow over urban areas. This variability can be simplified by introducing neighborhoods that are complexly interacting with the flow and forming internal boundary layers. On the level of street canyons each part of the city at each moment is characterized by different flow characteristics. This flow regime is called the roughness sub-layer (Roth, 2000). It is highly inhomogeneous both in its vertical and horizontal structure. Turbulent fluxes and variances can be expected to be functions of height in it. At a level of 3-5 times the average building height the structure of the flow over a neighborhood becomes more uniform and is considered to be in equilibrium with the underlying surface area, this regime is known as inertial sub-layer, and the Monin-Obukhov similarity theory applies. This means that turbulent fluxes and variances are near constant with height. Higher up the differences in meteorological fields introduced by surface characteristics are blended and the boundary layer in unstable conditions is forced by area aggregated features.

Understanding the behaviour of the crosswind and vertical fluctuations of the wind velocity, σ_v and σ_w is important because they are controlling parameters for the dispersion of plumes in the atmosphere.

Here we compare commonly used parameterisations of σ_w and σ_v (Gryning et al., 1987) to observations from 3 urban experiments; details of the experiments are given in Section 3. In the report of the COST Action 710 (Cenedese et al., 1998), these parameterisations were validated on a large number of data sets and found to perform well. They read:

$$\sigma_w^2 = u_*^2\left[1.5\left(\frac{z}{z_i}\right)^{2/3}\left(\frac{w_*}{u_*}\right)^2 \exp\left(-2\left(\frac{z}{z_i}\right)\right) + \left(1.7-\left(\frac{z}{z_i}\right)\right)\right] \tag{1}$$

where the convective velocity scale is $w_* = \left((g/T)\overline{w'T'}\, z_i\right)^{1/3}$, with g for the acceleration due to gravity, T for temperature and z_i for mixed layer height, and

$$\sigma_v^2 = 0.35 w_*^2 + (2 - z / z_i) u_*^2 \ . \qquad (2)$$

The parameterisations are developed for flat homogeneous terrain and therefore are expected not to be valid in the roughness sub-layer but in the inertial sub-layer.

2.1 BUBBLE experiment

From the BUBBLE experiment (Rotach et al., 2004 and Gryning et al., 2004) we use the observations from Sperrstrasse (Figure 7). The local building height is 14.6 m. An aerosol Lidar located within Basel 5 km from the experimental area gave information on the height of the mixed layer.

Comparisons were performed between parameterisations and the measurements of σ_w and σ_v at the heights of 17.9 and 31.7 m at the Sperrstrasse mast. At both heights the parameterisation gave higher values than actually measured in such a way that the agreement between parameterised and measured values improved with height.

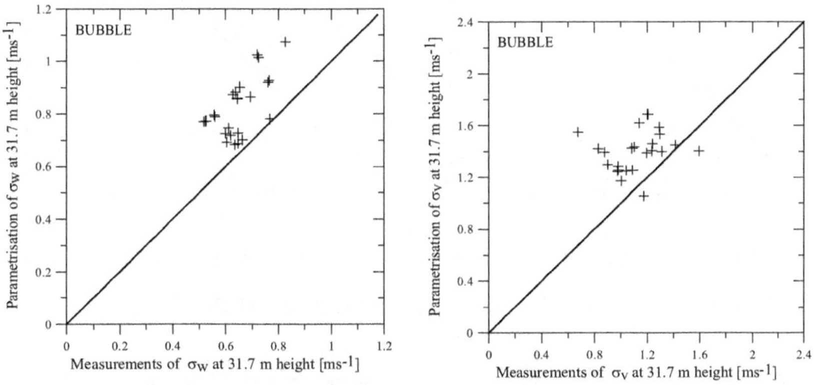

Figure 1: Observed half-hourly averaged values versus parameterisations of σ_w (left panel) and σ_v (right panel) at a height of 31.7 m at the Sperrstrasse tower, Basel (BUBBLE experiment, June-July 2002).

At the level of 17.9 m the parameterised values of σ_w and σ_v were 25% and 40% larger than the measured ones. At 31.7 m the difference was reduced to 10% and 20%, respectively, see Figure 1. The strong height dependence of both σ_w and σ_v is illustrated in Figure 2, showing simultaneous values of σ_w and σ_v at 17.9 m plotted against the 31.7 m level. For σ_w the measured values at 31.7 m are 10% higher as compared to the 17.9 m level; the percentage is 20% for σ_v.

The strong vertical variability of both the vertical and lateral variances, σ_w and σ_v, indicates that this layer is not part of the inertial sub-layer, where fluxes should be near constant with height. It rather suggests that the layer belongs to the roughness sub-layer, where the flow has a considerable spatial and vertical variability.

The good performance of the parameterisations for σ_w and σ_v when compared to the measurements at the highest level of observation (31.7 m) is a very promising result, indicating that the upper level is close to the transition to the inertial sub-layer.

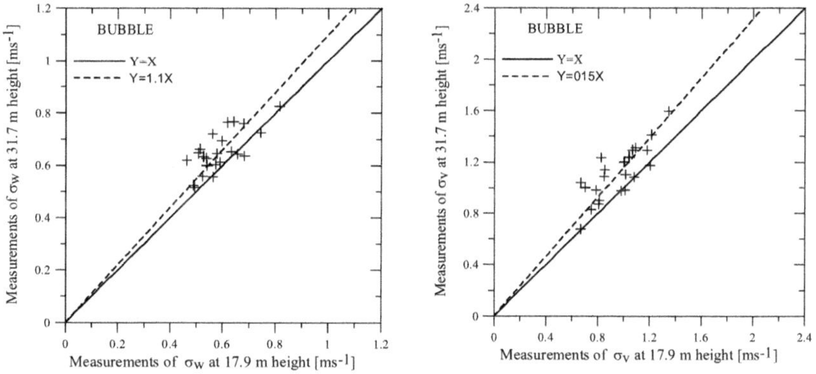

Figure 2: Observed half-hourly averaged values at the 17.9 m and 31.7 m levels of σ_w (left panel) and σ_v (right panel) at Sperrstrasse tower, Basel (BUBBLE experiment, June-July 2002).

2.2 Sofia 2003 experiment

The parameterisations were further tested on data from the Sofia 2003 experiment (Batchvarova et al., 2004) at 20 and 40 m above ground on the meteorological tower of the National Institute of Meteorology and Hydrology. The site is typical for an Eastern European suburban area, see Figure 8. The parameterisation of both the vertical and lateral variances, σ_w and σ_v is compared to measurements at a height of 40 m in Figure 3. The agreement suggests that at that height the transition between the roughness sub-layer and the inertial sub-layer has occurred.

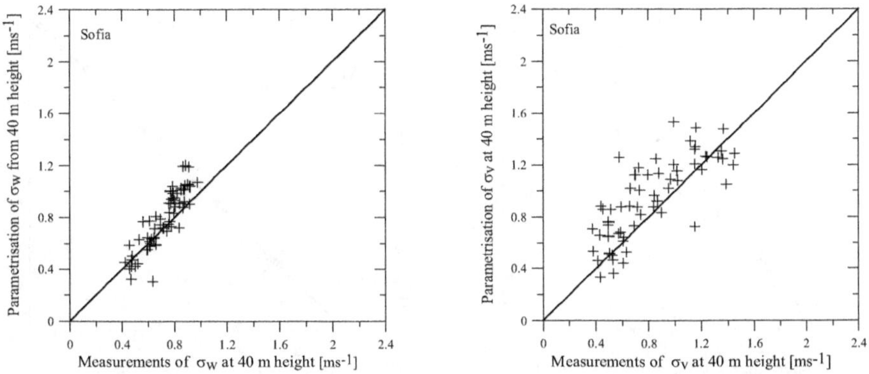

Figure 3: Observed half-hourly averaged values versus parameterisations of σ_w (left panel) and σ_v (right panel) at 40 m at Sofia tower, NIMH (Sofia experiment, Sept. - Oct. 2003).

The direct comparison between 20 m and 40 m measured variances shows that the depth of the roughness sub-layer is more than 20 m at this neighborhood, Figure 4.

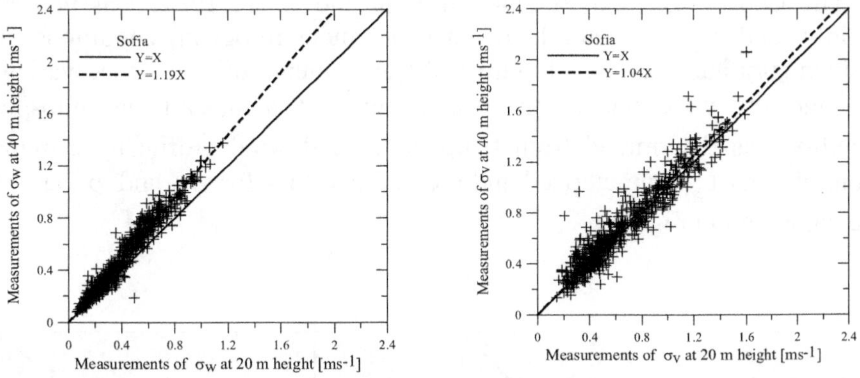

Figure 4: Observed half-hourly averaged values at the 20 m and 40 m levels of σ_w (left panel) and σ_v (right panel) at Sofia tower, NIMH (Sofia experiment, Sept. - Oct. 2003).

In Figure 5 only the measurements of the variances under convective conditions are shown for 5 days during which high resolution radiosoundings were performed. The planetary boundary layer height used in the parameterisation of variances was estimated from the radiosonde temperature profiles. The separation of convective cases from all (comparing Figures 4 and 5) results in minor differences only.

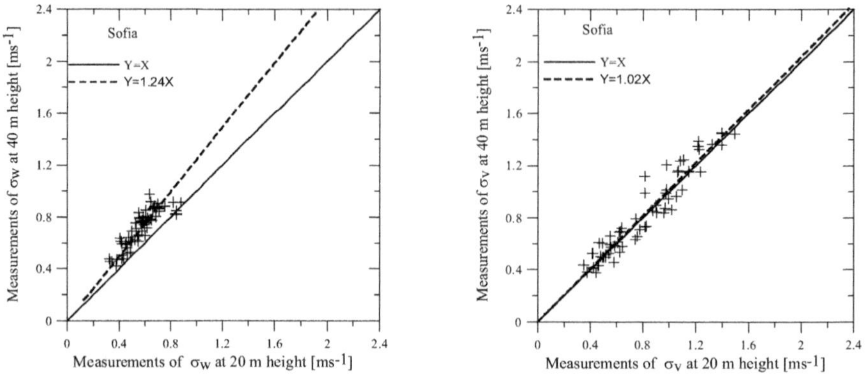

Figure 5: Observed half-hourly averaged values at the 20 m and 40 m levels of σ_w (left panel) and σ_v (right panel) at Sofia tower, NIMH (Sofia experiment, Sept. - Oct. 2003) for days with PBL measurements.

2.3 Copenhagen experiment

The Copenhagen experiment (Gryning and Lyck, 1984) was performed mainly under near neutral to unstable meteorological conditions in a residential/urban area, see Figure 9. Measurements of turbulence variances, σ_w and σ_v, were carried out at a height of 115 m, and the atmospheric stability was determined from temperature and wind profile measurements along the mast. Parameterised and measured values for σ_w and σ_v at 115 m are compared in Figure 6.

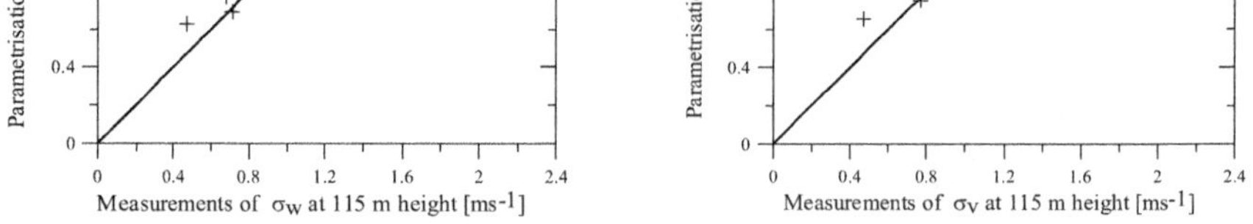

Figure 6: Observed half-hourly averaged values versus parameterisations of σ_w (left panel) and σ_v (right panel) at 115 m at the Gladsaxe tower, Copenhagen.

3. EXPERIMENTS

3.1 BUBBLE experiment

From the BUBBLE experiment in Basel, Switzerland, (Rotach et al., 2004; Gryning et al., 2004) meteorological measurements from a mast in central Basel are used, where turbulence measurements were performed half-hourly within and above the street canyon. METEK ultra sonic anemometers were mounted at 6 levels on the tower, namely at 3.6, 11.3, 14.7, 17.9, 22.4 and 31.7 m above ground. The local building height amounts to 14.6 m, see Figure 7. The measurements are taken from the days with tracer experiments (Gryning and Batchvarova, 2005) which are mostly convective, with low wind and directions generally perpendicular to the street. An aerosol LIDAR located within the city, 5 km from the experimental area gave the height of the mixing layer.

Figure 7: The meteorological tower at Sperrstrasse in Klein Basel, BUBBLE experiment.

3.2 Sofia 2003 experiment

The turbulence regime in the surface boundary layer in the vicinity of the National Institute of Meteorology and Hydrology was investigated during an experimental campaign in September 2003 (Batchvarova et al., 2004). Two ultra-sonic METEK anemometers and a fast Krypton hygrometer were mounted on a tower at 20 and 40 m above ground level, Figure 8. The convective boundary layer development was studied through high resolution low ascent velocity radiosoundings released every 2 hours starting at 7 and ending at 19 local summer time.

Figure 8: The meteorological tower at the National Institute of Meteorology and Hydrology (NIMH) in Sofia, sonic anemometers were mounted near the platforms (upper panel) and a view from the tower to the west (lower panel) with a sonic anemometer in the foreground.

3.3 Copenhagen experiment

The experimental site is mainly residential (Figure 9) in both upwind and downwind directions (Gryning and Lyck, 1984). The meteorological measurements included continuous recording of the three-dimensional wind velocity fluctuations at a height of 115 m, performed solely during tracer releases, and time averaged profile measurements of wind and temperature along a 200 m tower. From these measurements hourly values of σ_w and σ_v at 115 m were determined. The friction velocity and sensible heat flux (composed to the Monin-Obukhov length) were calculated from the profile measurements. The mixing height was determined from the standard routine radiosoundings that were launched 4 km northeast of the mast. All tracer experiments were performed during daytime in neutral to slightly convective atmospheric conditions.

Figure 9: Cup-anemometer, wind-vane and vertical propeller used in the Copenhagen experiment, mounted at 115 m on the Gladsaxe TV mast. The signals were recorded continuously during the tracer experiments and later digitized with 1 Hertz. These measurements were used to determine hourly values of σ_w and σ_v.

4. RESULTS AND DISCUSSION

The ability to predict the variances, σ_w and σ_v has been investigated for three urban experiments.

Observations from the BUBBLE experiment show a marked increase in both σ_w and σ_v between the 17.9 m and 31.7 m levels. It is also noted that the performance of the parameterization of σ_w and σ_v increases with height, being a 10% overestimate for σ_w and 20% for σ_v at 31.7 m. It can be noted that even the 31.7 meter level might still not be within the inertial sub-layer.

Observations from Sofia (20 and 40 m levels) indicate a somewhat different structure. As for BUBBLE, an increase in σ_w and σ_v is observed between the lower level (20 metres) and upper level (40 metres). For σ_w the change is about the same level as observed for BUBBLE, but for σ_v it is smaller. Furthermore, the parameterizations of σ_w and σ_v performs rather well at the upper level.

Copenhagen measurements are available only at one height (115 m). There are rather few observations and the parameterizations are found to perform fairly well, especially for σ_v. But the limited number of measurements precludes any conclusions.

The result of the analysis suggests that the use of parameterizations for σ_w and σ_v is feasible within the urban inertial sub-layer. This is of considerable interest for dispersion modelling in the urban boundary layer, because σ_w and σ_v are controlling parameters for spreading of plumes in vertical and lateral directions, respectively. These aspects are discussed for the BUBBLE and Copenhagen experiments in Gryning and Batchvarova (2005).

Acknowledgement: The studies presented here are supported via a NATO-CLG 979863; a Swiss-Bulgarian Institutional Partnership 7IP 065650 related to COST715 and the BULAIR project EVK2-CT-2002-80024.

REFERENCES

Batchvarova E., Gryning, S.-E., Rotach, M. W., Christen, A., 2004, Modelled aggregated heat fluxes compared to turbulence measurements at different heights. Proc. 9th Int. Conference on Harmonisation within atmospheric dispersion modeling for regulatory purposes, 1-4 June 2004, Garmisch Partenkirchen, Germany, vol 2, pp 7-12.

Cenedese, A., Cosemans, G., Erbrink, H., Stubi, R., 1998, Vertical profiles of wind, temperature and turbulence. In: Harmonisation of the pre-processing of meteorological data for atmospheric dispersion models. COST action 710 Final report. Office for Official Publications of the European Communities.

Gryning, S.-E. and Batchvarova, E., 2005, Advances in urban dispersion modeling. NATO advanced research workshop: Advances in air pollution modelling for environmental security, Borovetz (BG), 8-12 May 2004., *This volume.*

Gryning, S.-E.; Lyck, E., 1984, Atmospheric Dispersion from Elevated Sources in an Urban Area: Comparison between Tracer Experiments and Model Calculations. *J. Climate Appl. Meteorol.* **23**, pp 651-660.

Gryning, S.-E., Holtslag, A. A. M., Irwin, J. S., Sivertsen, B., 1987, Applied Dispersion Modelling Based on Meteorological Scaling Parameters. *Atmos. Environ.* **21**, pp 79-89.

Gryning, S.-E., Batchvarova, E., Rotach, M. W., Christen, A., Vogt, R., 2004, Roof-level SF_6 tracer experiments in the city of Basel - Report published in the institute series of IAC-ETH.

Rotach, M. W., Gryning S.-E., Batchvarova, E., Christen, A., Vogt, R., 2004, Pollutant dispersion close to an urban surface - BUBBLE Tracer experiment. Accepted for publication in *Journal of Meteorology and Atmospheric Physics.*

Roth, M., 2000, Review of atmospheric turbulence over cities. *Quart. J. R. Meteorol. Soc.* **26**, pp. 941-990.

MODELLING STUDIES ON THE CONCENTRATION AND DEPOSITION OF AIR POLLUTANTS IN EAST-CENTRAL EUROPE

László Bozó
Hungarian Meteorological Service, 1675 Budapest, P.O. B. 39

Abstract: Model estimations are presented on the annual average atmospheric concentrations and depositions of sulfur and nitrogen species, as well as on toxic elements in Hungary and its surrounding regions. Spatial and temporal variations of the distributions are discussed in the paper. The temporal variation of atmospheric sulfur budget was calculated based on a simple box model. In the case of Pb, the cumulative atmospheric depositions were also estimated for the period of 1955-2015. It was found that Hungary belongs to the moderately polluted regions in Europe. It is also demonstrated that significant part of tropospheric ozone and its precursors are imported to Hungary through long-range atmospheric transport processes.

Key words: Long-range transport models, trace elements, atmospheric deposition.

1. INTRODUCTION

For the past decades, energy generation, industrial production and transportation have caused serious environmental contamination in East-Central Europe. The rate of contamination can vary from place to place as a function of source densities and intensities of pollutant fluxes as well as meteorological conditions. The pattern of pollution may be characterized not only by local, highly concentrated sites such as densely populated urban areas, but also by lower concentrations of pollution widely dispersed over the landscape including agricultural regions, forests and surface waters. Trace gases and aerosol particles can be transported far away from their sources before being deposited on the surface. Due to the deep economic changes in this region during the past 15 years, its energy and industry structures were

I. Farago et al. (eds.), Advances in Air Pollution Modeling for Environmental Security, 33–40.

reorganized, which resulted in a significant decrease of pollutant emission. Out-of-date industrial technologies are being replaced by less energy consumer and more environment friendly ones. Thermal power plants are being equipped with efficient sulfur and dust filters. Leaded gasoline has been phased out of use. The rates of changes in pollutant emission, however, were different in the countries of the region. As a result of these changes, now the region is coping with less serious atmospheric environmental problems than 10-20 years ago.

2. EMISSION OF POLLUTANTS IN HUNGARY

Due to the geographical size and location of the country, local air quality is highly influenced by both Hungarian and transboundary sources. At the same time, Hungarian sources have a strong effect on the air quality and rates of atmospheric deposition in its surrounding regions.

In Hungary, the SO_2 emission decreased significantly during the past decade in all important source categories: energy, industry, transportation, agriculture, home heating and other services. The release of 1,020 ktonnes estimated for 1990 was reduced to 480 ktonnes in 2000. In the case of nitrogen-oxides, the annual emission of 240 ktonnes in 1990 was reduced to 180 ktonnes within 10 years. It should be noted, however, that the relative contribution of transportation to the total emission increased during the same period.

The emission rates of heavy metals also decreased between 1990-2000. The most significant decrease was detected in the case of Pb (from 695,000 kg to 37,000 kg). The corresponding annual emission rates in 1990 and 2000 are 5,600 kg and 2,700 kg for Cd, 43,000 kg and 37,000 kg for Ni, 6,800 kg and 4,200 kg for Hg, 150,000 kg and 110,000 kg for V.

3. SULFUR AND NITROGEN SPECIES

Due to the significant reduction of SO_2 emission during the past decade, the annual average concentrations have also decreased. Based on EMEP model calculations (EMEP, 2003), local maxima could be detected in the northern part of West Hungary and in the northern part of East Hungary (where still lignite based power plants are in operation, see Figure 1). The model also shows that Hungarian sulfur emitters contribute to the total atmospheric sulfur deposition in Hungary at a rate of 43%, the remaining part is transported from abroad. The atmospheric sulfur budget of Hungary, however, is still negative, so higher amount of sulfur is emitted than deposited by dry and wet atmospheric processes in the country. The rate of Hungarian

SO_2 emission in 2000 was 0.24 Tg (S) yr^{-1}, while the rates of dry and wet deposition were 0.07 Tg (S) yr^{-1} and 0.04 Tg (S) yr^{-1}, respectively, thus the total atmospheric sulfur budget was –0.13 Tg (S) yr^{-1}. The value of the sulfur budget in 1990 was –0.38 Tg (S) yr^{-1}, so a significant improvement could be estimated for the last decade of the past century.

The annual average NO_2 concentration under regional background conditions is in the range of 1.5-3.0 μg (N) m^{-3} in Hungary. It should be noted, however, that in the vicinity of major roads and within urban areas, where traffic is the dominating NO_2 source, the average concentrations are much higher. Total atmospheric nitrogen deposition is also calculated by the EMEP model. It includes the dry and wet deposition of gases and aerosol particles containing nitrogen. Its range is between 400-1500 mg N m^{-2} yr^{-1} (Figure 2). Local maxima were detected in the northern and western part of the country. Concerning the atmospheric nitrogen budget of Hungary, it can be stated that it is more or less balanced for both reduced and oxidized nitrogen compounds.

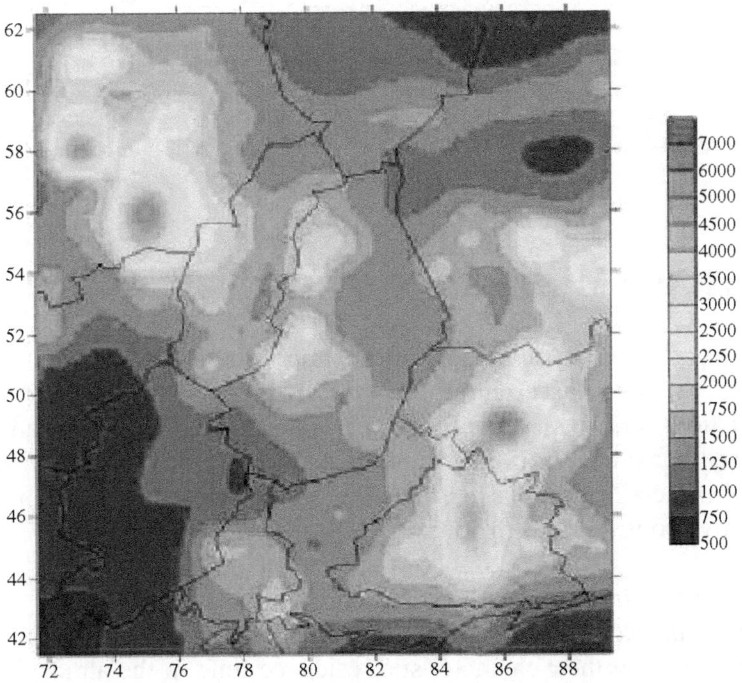

Figure 1: Atmospheric deposition of sulfur in East-Central Europe (mg S m^{-2} yr^{-1}).

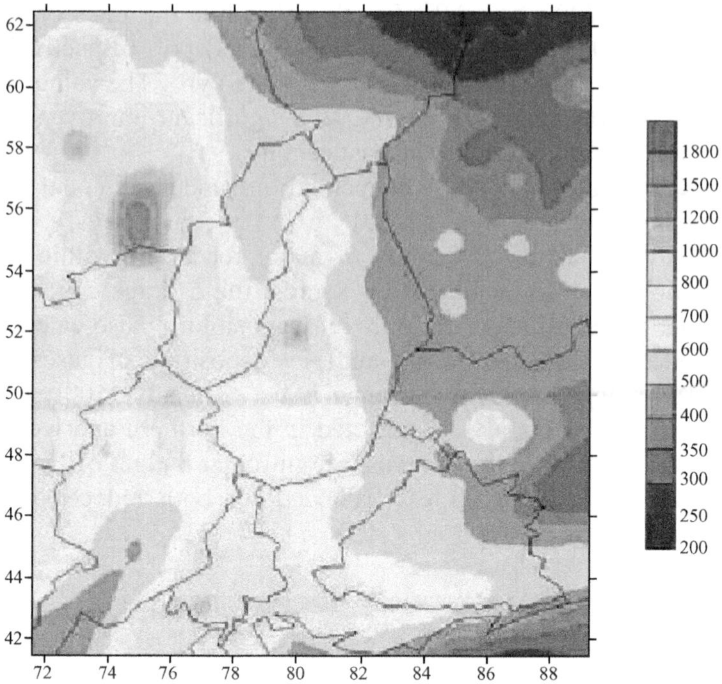

Figure 2: Atmospheric deposition of nitrogen in East-Central Europe (mg N m^{-2} yr^{-1}).

4. TRACE METALS

The most significant changes in the atmospheric environment of the past decades in the East-Central European region is connected with the introduction and widespread use of unleaded gasoline. Lead emission decreased considerably in all countries of the region, resulting in decreasing atmospheric concentrations and total deposition rates, as well as in re-ranking the relative contributions of source categories.

Long-range transport of atmospheric trace metals is not simply a clean air protection problem in Europe but a major part of environmental policy as well: considering their characteristic residence time in the atmosphere (a few days) and the sizes of the countries in Europe, international co-operation is needed to control the transboundary flux and deposition of these pollutants.

Model computations were performed based on the application of the Lagrangian-type long-range transport model, TRACE (Bozó, 2000) on the total deposition of lead and cadmium. Even relatively low fluxes across the Earth's surface can result in accumulation of toxic trace metals in various soils. From the soils they can be taken up by the plants and may be leached out into the ground water.

The total deposition of lead is shown in Figure 3. The rate of annual deposition in 2000 over Hungary was in the range of 1000-4000 g km^{-2} yr^{-1}, much below the ecological threshold in Europe (250,000 g km^{-2} yr^{-1}). Figure 4 shows the total deposition of cadmium in the East-Central European region computed by the TRACE model for the year of 2000: one can see that ecological threshold (5,000 g km^{-2} yr^{-1}) was not exceeded in any parts of the region: it varies in the range of 20-300 g km^{-2} yr^{-1}. Based on the comparisons with regional background measurements, it was found that our model simulations underestimate the rate of annual deposition by 10-20%. The reason for this could be the underestimations in emission data of trace elements and the uncertainty in the parameterization of deposition processes in the atmosphere.

The rate of lead deposition varied significantly during the past decades. Due to the cumulative characteristics of lead in our environment, it is advisable to estimate the cumulative lead deposition in Hungary for the past 45 years and to provide some quantitative estimates for the next decade. This type of simulation was also done by means of TRACE model computations. Historical emission data were taken from Olendrzynski et al. (1995), while future scenarios are based on the calculations of Berdowski et al. (1998).

For comparisons, the target of model simulations was not only Hungary but a few other countries in different regions of Europe – United Kingdom, The Netherlands, Spain, Austria, Romania and Poland (Figure 5). It is not surprising that cumulative lead deposition was much higher during the 30 years of the period 1955-1985 than that of 1985-2015. Regarding Hungary, the rate of total lead deposition was 320 mg m^{-2} during 1955-1985, while on the basis of model computations it is expected that it will not exceed 95 mg m^{-2} during the consecutive 30-year period (1985-2015). It can also be stated that in some selected countries (e.g. The Netherlands or Austria) the cumulative lead deposition rate was higher than in Hungary, while in the case of Romania and Spain lower cumulative deposition rates were estimated.

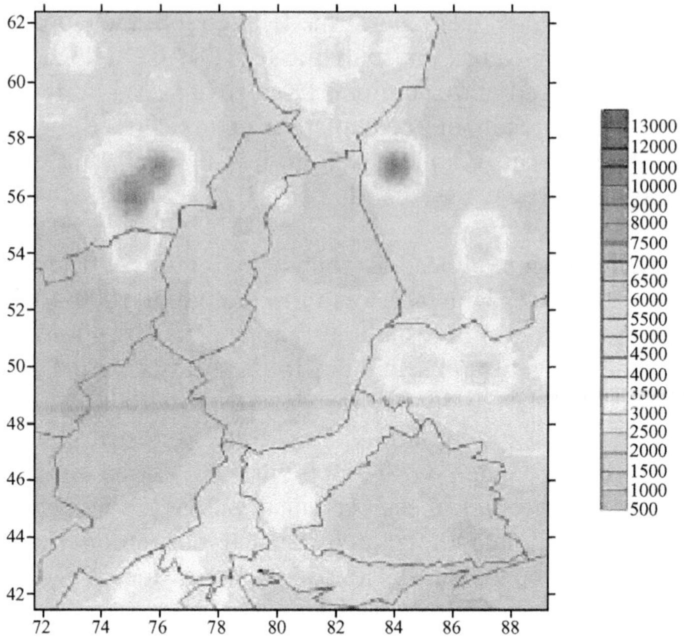

Figure 3: Atmospheric deposition of lead in East-Central Europe (g km^{-2} yr^{-1}).

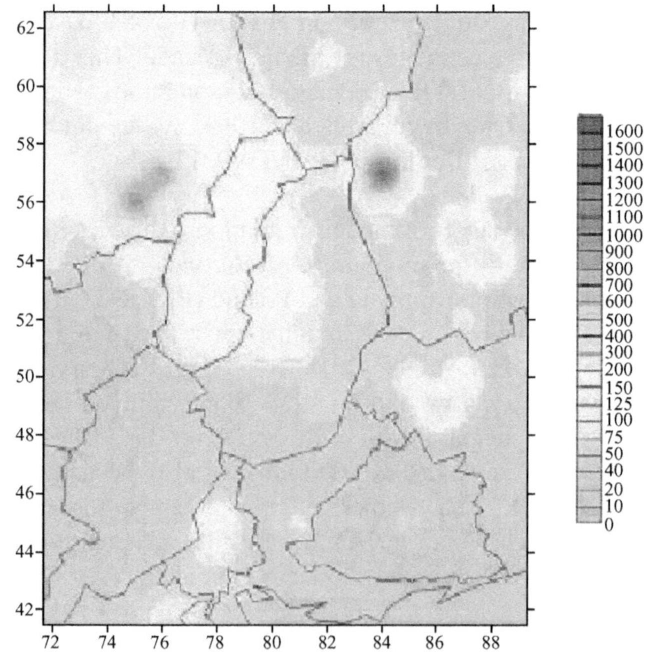

Figure 4: Atmospheric deposition of cadmium in East-Central Europe (g km^{-2} yr^{-1}).

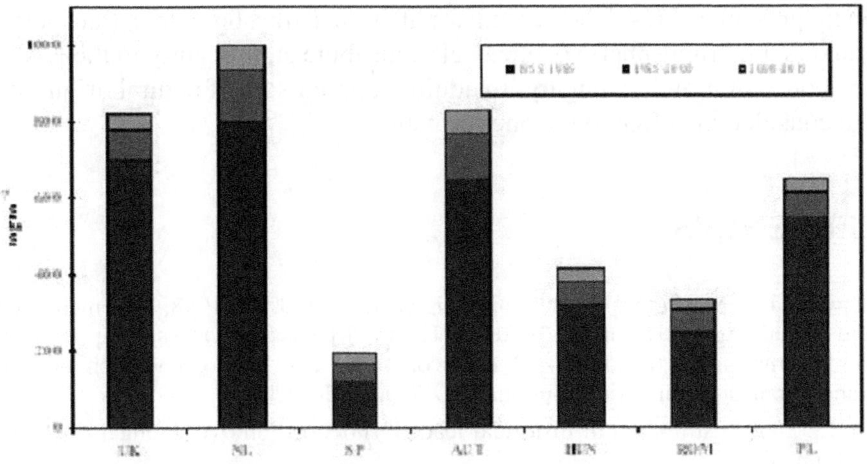

Figure 5: Cumulative lead depositions in selected European countries.

5. TROPOSPHERIC OZONE

In general, the concentration of atmospheric pollutants is lower under regional background conditions than in urban areas. Usually, this is not true for surface ozone. It has no direct natural and anthropogenic sources, tropospheric ozone is formed from NO_2 via complicated chemical processes in the presence of precursors and solar radiation. Volatile organic compounds (VOC's) speed up the rate of tropospheric ozone formation. It is typical that the ozone peaks occur within 50-100 km distance downwind from highly polluted urban areas. Most of the countries in the East-Central European region are coping with high ozone concentrations and threshold exceedences during summer period. Based on our regional background measurements carried out at K-puszta, Hungary, it was demonstrated that both 1 h and 8 h thresholds were exceeded: the numbers of annual exceedences – i.e., the number of the averaging periods with concentration values above the corresponding limit values in each year – were in the range of 2-37 for 1 h averages, and 79-138 for 8 h averages during the period of 1996-2003.

The problems of regulations of ozone level can be demonstrated by model computations (Havasi et al., 2001). Using the Danish Eulerian Model it was estimated, that even if the anthropogenic ozone precursors' emissions in Hungary were equal to zero, the AOT40 values would be at a level of 77-95% of their present values. The reason for it is that transboundary transport of precursors and tropospheric ozone itself contribute significantly to the level of ozone concentration in Hungary. In addition, precursors of natural origin have also a considerable effect on ozone formation.

REFERENCES

Berdowski, J. J. M., Pulles, M. P. J. and Visschedijk, A. J. H., 1998, Incremental cost and remaining emission in 2010 of Heavy Metals (HM) resulting from the implementation of the draft HM Protocol under UN ECE Convention on Long Range Transboundary Air Pollution. *TNO Report* - R 98/020.

Bozó, L., 2000, Estimation of historical lead (Pb) deposition over Hungary. *Időjárás* **104**, pp.161-172.

EMEP MSC-W, 2003, Transboundary acidification, eutrophication and ground level ozone in Europe. *MSC-W Status report* 1/2003, Part III.

Havasi, Á., Bozó, L., Zlatev, Z., 2001, Model simulation on the transboundary contribution to the atmospheric sulfur concentration and deposition in Hungary. *Időjárás* **105**, pp. 135-144.

Olendrzynski, K., Anderberg, S., Bartnicki, J., Pacyna, J. M. and Stigliani, W., 1995, Atmospheric Emissions of Cadmium, Lead and Zink in Europe During the Period 1955-1987. *IIASA Working paper* WP-95-35.

ESTIMATION OF THE EXCHANGE OF SULPHUR POLLUTION IN SOUTHEAST EUROPE

Hristo Chervenkov
National Institute of Meteorology and Hydrology - branch Plovdiv, Bulgarian Academy of Sciences, bul. "Ruski" 139, 4000 Plovdiv, Bulgaria.

Abstract: In the present paper the exchange of sulphur pollution between 12 countries in Southeast Europe for 1995 is estimated. As only sources from these countries are handled, the results can be treated as an estimate of their mutual pollution and their impact on the acidification of the region. The results of such calculations can be used in decision-making, negotiating and contamination strategies development.

Key words: Air pollution, dispersion modeling, sulphur dioxide, deposition budget matrix.

1. INTRODUCTION

The EMAP model is used to estimate the sulphur deposition over Southeast Europe for 1995 due to sources from 12 countries (Table 1). As only sources from these countries are handled, the results can be considered as an estimate of their impact on the acid pollution of the region as well as an estimate of the reciprocal pollution.

Albania	Bosnia and Herzegovina	Bulgaria	Croatia	Greece	Hungary
AL	BH	BG	HR	GR	HU
Moldavia	Romania	Turkey[1]	Slovenia	Serbia and Montenegro	The FYR Macedonia
MO	RO	TR	SL	YU	MK

(1) The part inside the model domain.

Table 1: List of countries and their notations.

I. Farago et al. (eds.), Advances in Air Pollution Modeling for Environmental Security, 41–50.

2. SHORT DESCRIPTION OF THE EMAP MODEL

EMAP (BC-EMEP, 1994 - 1997, Syrakov, 1995) is a simulation model that allows describing the dispersion of multiple pollutants. The processes of horizontal and vertical advection, horizontal and vertical diffusion, dry deposition, wet removal, gravitational settling (aerosol version) and the simplest chemical transformation (sulphur version) are accounted for in the model. Within EMAP, the semi-empirical diffusion-advection equation for scalar quantities is treated. The governing equations are solved in terrain-following coordinates. Non-equidistant grid spacing is settled in vertical directions.

The numerical solution is based on discretization applied on staggered grids using the splitting approach. Conservative properties are fully preserved within the discrete model equations. Advective terms are treated with the TRAP (Syrakov, 1996, Syrakov and Galperin, 1997b) scheme, which is a Bott-type one. Displaying the same simulation properties as the Bott scheme (explicit, conservative, positive definite, transportability, limited numerical dispersion), the TRAP scheme proves to be several times faster.

The advective boundary conditions are zero at income flows and "open boundary" - at outcome ones. Turbulent diffusion equations are digitized by means of the simplest schemes – explicit in horizontal, and implicit in vertical direction. The bottom boundary condition for the vertical diffusion equation is the dry deposition flux, the top boundary condition is optionally "open boundary" and "hard lid" type. The lateral boundary conditions for diffusion are "open boundary" type. In the surface layer (SL), a parameterization is applied permitting to have the first computational level at the top of SL. It provides a good estimate for the roughness level concentration and accounts also for the action of continuous sources on the earth surface (Syrakov and Yordanov, 1996, 1997).

A similarity theory based PBL model (Yordanov et al., 1983) is built in the model producing 3D velocity and turbulence fields on the base of minimum meteorological information – the wind and temperature at geostrophic level and the surface temperature.

The model is evaluated and validated during the ETEX-II intercalibration study - ranged 9th among 34 models (Syrakov and Prodanova, 1997). It is validated on the database of 1996 EMEP/MSC-E intercalibration of heavy metal models (Syrakov and Galperin, 1997a).

3. MODEL DOMAIN, PARAMETERIZATION AND INPUT DATA

The aim of this modeling is to estimate the sulphur pollution in the region of Southeast Europe, taking a territory of 38 x 28 EMEP 50x50 km^2 grid cells with Bulgaria in the center (see figures later on). Every cell is divided to four 25x25 km^2 cells. The chosen territory includes entirely all 12 countries of interest and partly other territories. In the created versions of EMAP, a 5-layer vertical structure is used. The first four layers have representative levels at 50, 200, 650 and 1450 m with layer boundaries 20-100, 100-375, 375-995, 995-1930 m. The 5th layer accounts parametrically for the free atmosphere. The volume of this layer is so big that the concentration tends to be zero there, although it can contain some mass.

Two kinds of input are necessary for EMAP performance: sources and meteorological data.

3.1 Emissions

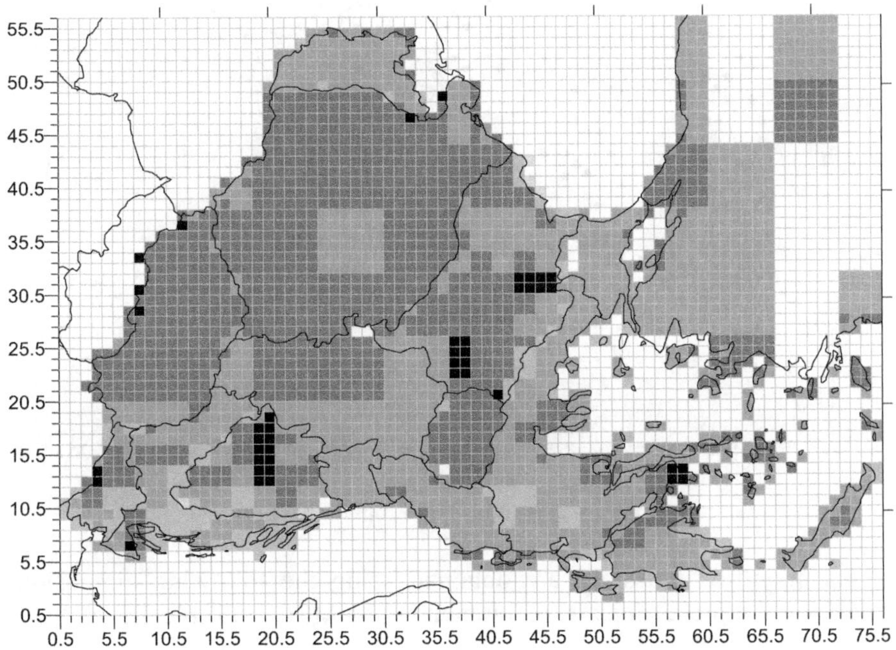

Figure 1: Distribution and strength (unit: μg (S)/s) of the area sources in the model domain 10% black: below 10^5, 20% black: $10^5 - 10^6$; 30% black: $10^6 - 10^7$; 40% black: $10^7 - 10^8$; black: over 10^8).

The sources are determined through an emission inventory based on the CORINAIR methodology. They correspond to the official 50x50 km^2 data reported by the corresponding governmental authorities to EMEPs MSC-West and can be downloaded from its web site. Additional redistribution of these data is made over the finer grid of 25-km space resolution. The emissions are provided in mass units per second. All sources are divided in two classes: high sources (like high and very strength industrial stacks etc.) called Large Point Sources (LPS) and area sources (AS) –the sum of all low and diffusive sources in the given grid cell. As all LPS are supplied with high stacks, the emission of these sources is prescribed to be released in layer 2, i.e., between 100 and 375 m. The distribution of the LPS and the AS in the model domain is shown in Figure 1 and Figure 2.

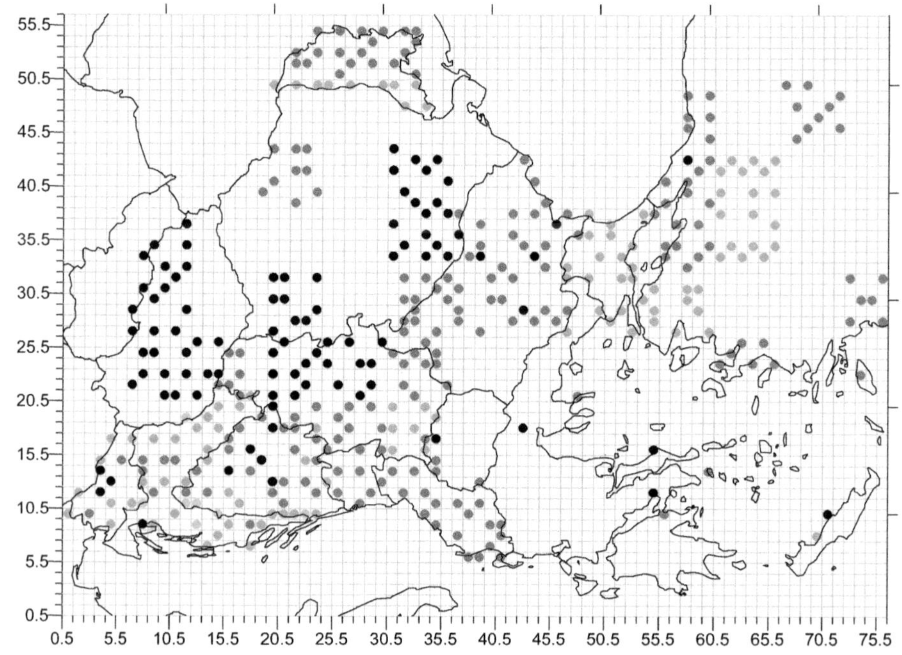

Figure 2: Same as Fig. 1, but for the LPS.

The monthly emissions are obtained using the following dimensionless annual variation coefficients, recommended by MSC-E (Table 2):

J	F	M	A	M	J	J	A	S	O	N	D
1.34	1.30	1.18	1.00	0.73	0.69	0.69	0.73	0.83	1.00	1.17	1.34

Table 2: Annual variation coefficients.

3.2 Meteorological data

An important advantage of the used model is that, due to the built in PBL model, it utilizes only numerical analysis and forecast data from the world weather centers, distributed via the Global Telecommunication System of the World Meteorological Organization.

For this task the meteorology input has a time resolution of 6 hours. It consists of the sequence of analyzed U_{850}, V_{850}, T_{850} and T_{surf} fields and 6-hour forecast for precipitation from the standard 50x50 km^2 output of the former "Europa-Model" of the German Met. Service (DWD). On this base, the PBL model calculates *U*-, *V*-, *W*- and K_z-profiles at each grid point. It provides also u_* and SL universal profiles necessary in SL parameterization. The roughness and the Coriolis parameter fields are pre-set additional input to the PBL model. Orography height, surface type (sea-land mask) and roughness height are to be provided for each grid location. Initial concentration field is optionally introduced (spin-up fields).

3.3 Sulphur parameterization and other model parameters

Two species of sulphur in the air are considered: gaseous sulphur dioxide SO_2 and particulate sulfate $SO_4^{=}$. Sources emit SO_2 only, in the air it is transformed to sulfate. The pollutant specific model parameters used are given in Table 3.

Pollutant	SO_2		$SO_4^{=}$	
Transformation rate constant α_{tr} [h^{-1}]	0.01 (winter) - 0.04 (summer)			
Wet removal constant γ [mm^{-1}],	0.3		0.2	
surface type	sea	land	sea	land
Dry deposition velocity V_d] [m/s	0.01	0.03	0.002	0.006

Table 3: Specific model parameters

The other model parameters are: horizontal diffusion coefficients $K_x = K_y = 5.10^4$ m^2/s and time step $Du = 0.25$ h.

4. CALCULATION RESULTS

Monthly runs with the above mentioned sources are performed. The output consists of the following fields: monthly dry (DD), wet (WD) and total (TD) depositions, monthly concentration mean in air (CA) and additionally, monthly sum of the precipitation (CP). Then, it is easy to calculate the annual total deposition (Figure 3), the mean annual surface concentration in air (Figure 4) and the mean concentration in the precipitation (Figure 5).

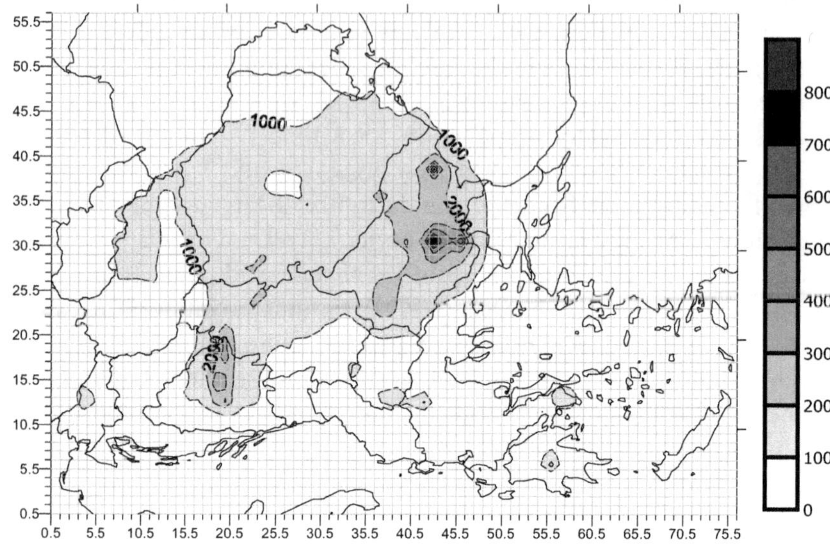

Figure 3: Total deposition (unit: mg (S)/m^2) of oxidized sulphur in 1995.

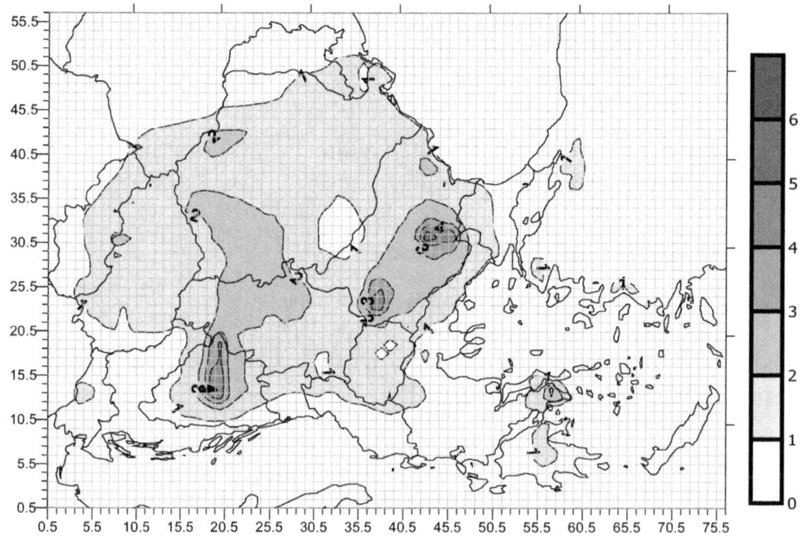

Figure 4: Mean annual concentration (unit: μg (S)/m^3) of oxidized sulphur in 1995.

The picture of the distribution of the total deposition shows that over the main part of the model domain it is below 1 g/m^2; over a part of Northern

Greece, Bulgaria, Romania, Serbia and Montenegro and Hungary it is between 1 and 2 g/m^2 and only over small areas in Bosnia and Herzegovina and in Bulgaria it is between 3 and 4 g/m^2. The most polluted region is that of Southeast Bulgaria, the place around the most powerful source in the "Maritsa Iztok" Thermal Power Plant. This TPP is a set of 3 neighboring coal-burning plants. They are so close to each other that they occupy a single 25 km cell and their emission rates are summed up giving an amount of 17.13 kg sulphur per second. In its vicinity the total deposition reaches 8.8 g/m^2.

The shape of isopleths of the mean annual surface concentration in air is similar. In the main part of the model domain it is below 1 μg/m^3; over a part of Northern Greece, Bulgaria, Romania, Serbia and Montenegro and Hungary it is between 1 and 2 μg/m^3 and only over small areas in Bosnia and Herzegovina, in Bulgaria (the region of Sofia) and in Greece (Athens) it is over 3 μg/m^3. Again, the most polluted place is the region around "Maritsa Iztok" TPP, where the concentration reaches 5.7 μg/m^3.

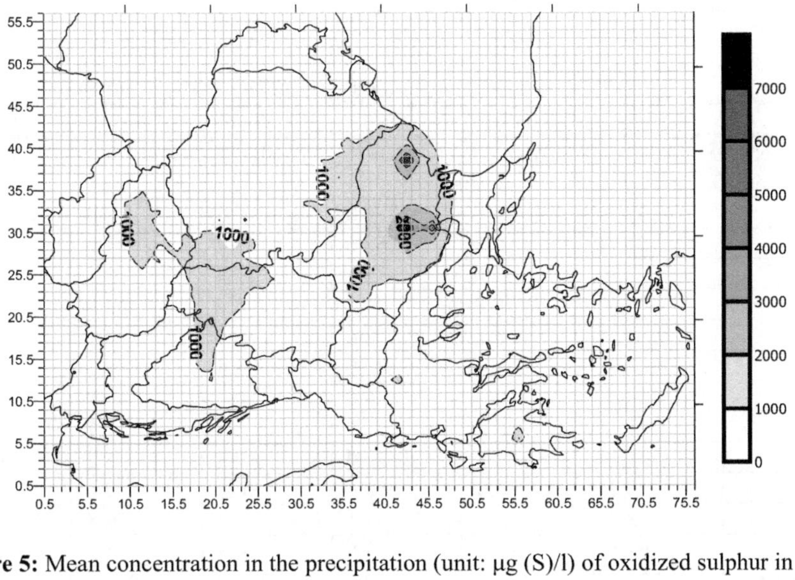

Figure 5: Mean concentration in the precipitation (unit: μg (S)/l) of oxidized sulphur in 1995.

The picture of the distribution of the concentration in the precipitation shows that in the main part of the model domain it is below 1 mg/l; over a part of Northern Greece, Bulgaria, Romania, Serbia and Montenegro, Bosnia and Herzegovina and Hungary it is between 1 and 2 mg/l. In spite of the irregular field of the precipitation, the highest concentration is again in the region of "Maritsa Iztok" TPP where it reaches 6.7 mg/l. A second maximum is observed over the Bulgarian Black Sea coast, in the region of "Varna" TPP.

In Table 4, the deposition budget matrix for oxidized sulphur is presented. It is obtained after multiplication of the total deposition field to an array containing the distribution of the territory of each country in the model domain. It shows the impact of each country to the sulphur pollution of the other countries.

emitter receiver	AL	BG	BH	GR	HR	HU
AL	79.3	20.9	9.9	42.6	0.7	3.3
BG	9.2	2147.0	39.2	53.9	2.2	37.9
BH	4.5	14.1	596.8	6.0	20.0	57.1
GR	22.0	281.8	10.9	503.8	0.7	6.9
HR	1.5	8.8	123.3	3.2	61.7	68.2
HU	2.0	25.9	104.7	3.6	22.6	792.5
MK	21.1	71.0	6.8	48.4	0.4	4.2
MO	0.5	40.3	8.2	1.5	0.6	14.5
RO	8.5	583.1	209.8	19.3	13.4	379.6
SL	0.2	1.7	9.3	0.8	14.7	17.9
TR	4.3	310.9	9.5	176.0	0.7	10.8
YU	30.8	160.5	327.2	30.5	11.7	131.2
total dep.	228.5	4727.0	1649.0	1539.0	199.3	2248.0
total emitt.	360.0	7485.0	2400.0	2780.0	315.0	3525.0
%	63	63	69	55	63	64
emitter receiver	MK	MO	RO	SL	TR	YU
AL	22.7	0.1	3.8	0.6	0.4	12.7
BG	54.7	5.2	268.6	3.4	17.0	120.2
BH	3.3	0.1	15.3	15.4	0.1	66.0
GR	35.0	1.3	24.4	0.9	13.0	15.3
HR	1.3	0.1	8.6	43.4	0.1	31.3
HU	2.9	0.4	80.0	38.4	0.3	94.6
MK	116.9	0.2	6.6	0.4	0.8	16.1
MO	1.1	33.2	117.1	1.0	1.1	11.7
RO	21.2	22.8	1832.0	19.3	6.2	424.8
SL	0.2	0.0	2.3	96.2	0.0	4.7
TR	8.2	3.3	52.1	1.2	579.8	16.4
YU	62.4	0.8	93.8	10.8	1.1	627.7
total dep.	373.5	122.3	3129.0	337.7	810.5	1620.0
total emitt.	541.0	295.0	4560.0	596.5	1603.0	2310.0
%	69	41	69	57	51	70

Table 4: Deposition budget matrix (unit 100 t). The shaded elements show the deposition quantity for each country due to its own sources.

Table 4 consists from two similar parts. The last three rows of each part show respectively the total deposited quantity due to the sources from the country in the column header, the total emitted sulphur for this country and the percentage of deposited quantities from the yearly emitted. The last value can be treated as the relative part of the sulphur that remained in the domain. It can be noticed that the main part of the sulphur pollution emitted by each country is deposited over the country itself. It can be seen also that the percentage of the total deposited quantity is between 41% and 70%, the rest goes out of the model domain. The minimum of this value is for Moldova - the possible explanation is that this territory is close to the east border of the model domain and the main tropospheric transport is west-east in these latitudes.

5. CONCLUSION

The paper shows that for long periods of time the part of sulphur pollution, released in one, but deposed over other countries and territories in Southeast Europe is significant. The obtained results are in good agreement with former calculations in NIMH - BAS. It can be seen from the status report of EMEP/MSC-W (Barrett and Berge, 1997) that according to their calculations the exchange of sulphur pollution between these countries is estimated in the correct order of magnitude, giving at the same time much more details in the time and space distribution of deposed quantities. The author's conclusion is that the model produces a reasonable picture of the concentrations and depositions in the 25-km grid for the sulphur components in the region of Southeast Europe.

The results of such calculations can be used in decision-making in negotiating and contamination strategies development.

Acknowledgements: The present work is supported by the NATO Collaborative Linkage Grant EST.CLG 979794, by the BULAIR project (EU 5FP Contract EVK2-CT-2002-80024) as well as by the Greek and Bulgarian ministries of education and science in the framework of an agreement for bilateral cooperation (Contract No. ES-1212-Gr/02 NScC).

REFERENCES

BC-EMEP 1994, 1995, 1996, 1997: Bulgarian contribution to EMEP, Annual reports for 1994, 1995, 1996, 1997, NIMH, EMEP/MSC-E, Sofia-Moscow.

Barret, K., Berge, E. (eds), 1996, Transboundary Air Pollution in Europe, EMEP/MSC-W status report 1997, Part One and Two: Numerical Adenddum, Norwegian Meteorological Institute, Oslo, Norway, pp. B3-B46.

Syrakov, D., 1995, On a PC-oriented Eulerian Multi-Level Model for Long-Term Calculations of the Regional Sulphur Deposition, in Gryning, S. E. and Schiermeier, F. A. (eds), Air

Pollution Modelling and its Application XI, NATO - Challenges of Modern Society, Vol. 21, Plenum Press, N.Y. and London, pp. 645-646.

Syrakov, D., 1996, On the TRAP advection scheme - Description, tests and applications, in Geernaert, G., Walloe-Hansen, A. and Zlatev, Z., Regional Modelling of Air Pollution in Europe. *Proceedings of the first REMAPE Workshop, Copenhagen, Denmark, September 1996*, National Environmental Research Institute, Denmark, pp. 141-152.

Syrakov, D., 1997, Influence of re-emission on pollution distribution: one-dimensional multi-level model, in Andreev V. (Eds), Bulgarian contribution to EMEP, Annual reports for 1997, NIMH, EMEP/MSC-E, Sofia-Moscow, pp. 21-25.

Syrakov, D., and Galperin, M., 1997a A model for airborne poli-dispersive particle transport and deposition, *Proc. of 22nd NATO/CCMS International Technical Meeting on Air Pollution Modelling and its Application, 2-6 June 1997, Clermont-Ferrand, France*, pp. 111-118.

Syrakov, D., and Galperin, M., 1997b On a new Bott-type advection scheme and its further improvement, in Hass, H. and Ackermann, I. J. (eds), *Proc. of the first GLOREAM Workshop, Aachen, Germany, September 1997*, Ford Forschungszentrum Aachen, pp. 103-109.

Syrakov, D., and Prodanova, M., 1998, Bulgarian Emergency Response Models - Validation against ETEX First Release, *Atmos. Environ.* **32**, No. 24, pp. 4367-4375.

Syrakov, D., and Yordanov, D., 1996, On the surface layer parameterization in an Eulerian multi-level model, *Proceedings of the 4th Workshop on Harmonisation within Atmospheric Dispersion Modelling for Regulatory Purposes, 6-9 May 1996, Oostende, Belgium*, v.1.

Syrakov, D., and Yordanov, D., 1997, Parameterization of SL Diffusion Processes Accounting for Surface Source Action, *Proc. of 22nd NATO/CCMS International Technical Meeting on Air Pollution Modelling and its Application, 2-6 June 1997, Clermont-Ferrand, France*, pp. 111-118.

Yordanov, D., Syrakov, D.,. Djolov, G., 1983 A barotropic planetary boundary layer, *Boundary Layer Meteorology* **25**, pp. 363-373.

IMPLEMENTING THE TRAJECTORY-GRID TRANSPORT ALGORITHM IN AN AIR QUALITY MODEL

David P. Chock, Margaret J. Whalen, Sandra L. Winkler, and Pu Sun
Ford Research and Advanced Engineering, P.O. Box 2053, MD-3083 SRL, Dearborn, MI 48121 USA

Abstract: Eulerian-based air quality models encounter a serious numerical problem in solving the advection equation. In addition, mass conservation is often violated when meteorological model output is used as input to air quality models. The Trajectory-Grid algorithm handles the advection and eddy-diffusion in the Lagrangian and Eulerian framework, respectively. It is very accurate and can be used to trivially solve the advection equation for mixing ratios to address (but not correct) the mass conservation issue. We incorporated the algorithm into the state-of-the-science Comprehensive Air Quality Model with Extensions (CAMx). Applications of the model reveal the inaccuracy of the commonly used Bott advection scheme, and the subsequent compensating errors of the model. The results clearly call for a more reliable description of eddy diffusivity and emissions inventory in order to truly improve the reliability and predictive capability of air quality models.

Key words: Trajectory-Grid algorithm, Bott scheme, advection equation, mass conservation, CAMx, eddy diffusivity, Eulerian, Lagrangian, MM5.

1. INTRODUCTION

Eulerian-based air quality models (AQMs) suffer a serious numerical problem in solving the advection equation, accounting for a large fraction of numerical inaccuracies in the models. This is because grid-discretization leads to artificial dispersion and diffusion of the concentration profiles. Correction attempts can lead to distortion of the profiles and introduce spurious nonlinearity, leading to unpredictable outcome. Another problem AQMs face is that the input flow field may not conserve mass. When meteorological-model output is used as input to an AQM, mass-conservation violation can be due to (1) mass-conservation violation of the meteorological

I. Farago et al. (eds.), Advances in Air Pollution Modeling for Environmental Security, 51–61.

model itself, (2) difference in spatial and temporal resolution, in terms of grid size and time step, between the AQM and the meteorological model, and/or (3) use of time interpolation in the AQM to approximate the instantaneous meteorological model output read in at large time intervals. To address this problem, models may recalculate the wind field, typically by modifying the vertical velocity to satisfy the advection equation, or introduce artificial local and time-dependent emission sources and sinks to cancel the mass non-conservation. Neither approach is very satisfactory.

Chock et al. (1996) presented the Trajectory-Grid (TG) approach for solving the physical transport equation. The approach is Lagrangian for the advective transport and Eulerian for the diffusive transport. The basic idea of the approach for the advective transport is very simple. It rewrites the advection equation for, say, concentration of a species, as one containing the full derivative of concentration with respect to time and a remaining term containing the velocity divergence. Thus, the equation is now describing the concentration profile of a species as this profile moves along the flow field. In other words, one can designate a set of points, or what we shall call pulses, on the concentration profile and trace the trajectories of these points along the wind field and modify their concentrations according to the velocity divergence along the trajectories. The method is sign-preserving and monotonic. It is also accurate because the species concentration is a solution to an ordinary differential equation and the trajectory can be made as accurate as necessary.

Since the flow field is the same for all species at a given location and time, the movement of a concentration pulse constitutes the advective transport of concentrations of all species represented by the pulse so that the advective transport of concentrations of different species need not be treated separately. This is an important timesaving feature of TG. However, interpolative errors may occur when one needs to estimate the cell-center concentrations of a species. But unlike the semi-Lagrangian methods, the errors do not propagate or grow because they are not fed back to the species concentrations at the pulse locations. Interpolation errors do occur in the diffusion step, but errors caused by the diffusion step are typically considerably smaller than those in Eulerian advection. Overall, the method yields a very accurate solution even for complex species concentration profiles.

An application of the approach to the European Chemical Transport Model was described by Chock et al. (1998). The approach has also been used to solve the condensation/evaporation equations in aerosol modeling accurately and with substantial time saving (Chock and Winkler, 2000; Gaydos et al., 2003). Here, we are describing our implementation of the TG approach in a state-of-the-science AQM called Comprehensive Air Quality Model with Extensions, or CAMx (Environ, 2004). The version we used is PMCAMx

3.01 (designated as CAMx hereafter), which contains an equilibrium aerosol module. In the course of this implementation, we also deal with the mass conservation issue.

2. THE ISSUE OF MASS CONSERVATION

Because we try to minimize the alterations of the meteorological model (MM5 in this case, see Grell et al., 1994.) output to be used in CAMx, we decided to retain the sigma coordinate instead of the physical coordinate used in CAMx. In addition, the vertical velocity, $\dot{\sigma}$, from MM5 can be used directly whenever both CAMx and MM5 refer to the same layer.

Many non-hydrostatic meteorological models, including MM5, do not conserve mass. In the case of MM5, there are two major reasons for failure to conserve mass. First, two terms corresponding to the heat fluxes from diabatic processes and subgrid-scale diffusion have been neglected in the pressure tendency equation (Dudhia, 1993). This will indirectly impact the accuracy of the estimated air density, though the terms are supposedly negligible compared to other terms in the equation. Second, and likely more important, the boundary condition, $\dot{\sigma} = 0$ at $\sigma = 1$ (bottom boundary) is generally not satisfied. This is because the σ levels are defined by reference (hydrostatic) pressures which are time independent, and for $\sigma = 1$ the reference pressure is prescribed according to the terrain height. But the actual surface pressure depends on location and time, which means that $\sigma = 1$ may be detached from (to be above or below) the surface, so that $\dot{\sigma}$ is no longer necessarily zero. This fact creates a lack of self-consistency in the model with mass non-conservation being one of the casualties.

As mentioned earlier, even if the meteorological model output is mass conserving, incorporating the output in an AQM need not guarantee mass conservation. An example of this was illustrated by Lee et al. (2004). To obviate such worries, CAMx determines the vertical velocity, *w*, by requiring that the continuity equation be satisfied, risking the possibility that the calculated vertical velocity may be unphysical. In addition, if the Bott scheme is used in the integration step to determine *w*, then the Bott scheme must also be used as the advection-equation solver to assure mass conservation. Therefore, the calculated vertical wind field is subject to the choice of the advection algorithm.

Another interesting thing is that even though the grid structure is assumed to be terrain following, no correction involving the horizontal gradient of the terrain height for the vertical velocity is made. In other words, *w* need not be zero at the surface but is assumed so. As a result, the vertical velocity is effectively determined not for a terrain-following coordinate but for a rectangular coordinate system, but consistency is maintained because the same coordinate system is used to solve the continuity equations in the model. This fact actually facilitates a direct comparison between $\dot{\sigma}$ of MM5 and w of CAMx. Thus, multiplying w of CAMx by $-\rho_0 g/p^*$ where ρ_0 is the dry-air reference density at the σ level of interest, p^* is the surface pressure minus the model-top pressure, and g is the gravitational acceleration, one can obtain $\dot{\sigma}$ corresponding to w of CAMx and compare it with $\dot{\sigma}$ of MM5.

Figure 1 shows an example of such a comparison for the top of the lowest layer at a specific time during a Southern California ozone episode. Though not identical, the two estimated $\dot{\sigma}$'s are strikingly similar. The $\dot{\sigma}$'s for other layers and other time periods also indicate that the vertical wind fields calculated by CAMx are generally quite similar to those of MM5 and thus, that MM5 does not violate mass conservation significantly. This observation is also true for the MM5 simulation of the NARSTO field study in the Northeast US in 1995.

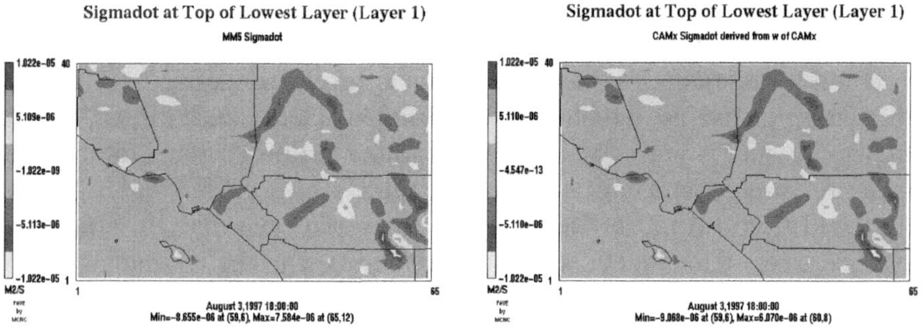

Figure 1: The $\dot{\sigma}$ of MM5 (left) and calculated $\dot{\sigma}$ of CAMx (right) at the top of the lowest layer at 1800 PST August 3, 1997.

The fact that MM5 need not violate mass conservation substantially in general is comforting. Furthermore, instead of requiring mass conservation precisely, one can take advantage of the fact that the mixing ratio, f, of a species, which is the ratio of the species concentration to that of air density, should satisfy the following advective equation regardless of whether the continuity equations for the density and for the species concentration truly satisfy mass conservation:

$$\frac{\partial f}{\partial t} + \mathbf{v} \cdot \nabla f = \frac{df}{dt} = 0,$$

where **v** is the 3D velocity. By requiring that this equation be satisfied, we avoid having to devise different artificial corrections to force conservation of mass. Furthermore, it also simplifies the TG applications because one simply constructs the pulse trajectory and multiplies the mixing ratios of the species carried by the pulse with the new local air density to obtain the updated local species concentrations.

3. IMPLEMENTATION OF TG IN PMCAMX

Because the TG advection is Lagrangian based, it is necessary to specify the identities and locations of the pulses or points, together with their associated species concentrations, and keep track of them as they move along their trajectories. To facilitate implementing this requirement, we converted the CAMx code from FORTRAN 77 to FORTRAN 90, allowing the use of dynamic memory allocation, structure, and pointer capabilities. Because TG allows as many location points as needed for concentration tracking, it potentially has "infinite" spatial resolution. Accordingly, by allowing more pulses as needed per grid cell, one can increase the resolution of the concentration profiles beyond that specified by the model grid.

For initial conditions, we assign four pulses per grid in the lowest two layers, located at mid-layer heights and evenly distributed in the horizontal planes. We also assign one pulse at each cell center in all other grid cells. Whenever the boundary cells become empty, they are refilled with the same number of pulses as originally assigned, each carrying boundary species concentrations.

Advection: TG is not constrained by the Courant number, but to construct an accurate trajectory in a potentially complex wind field, small time steps are needed. For each time step Δt, we allow the pulses to adjust their locations twice. So at every half time step, the pulses first add to their original concentrations the concentrations from local emissions for duration of $\Delta t/2$. The pulses then move according to a tri-linearly interpolated wind field, with the species concentrations at the trajectory end points updated by multiplying the emission-adjusted species mixing ratios by the local air densities at the end points, which is assumed constant within each model grid. This is a simple and accurate step to handle advection and emission.

Pulse management: Because of flow convergence and divergence, local accumulation and depletion of pulses will occur, necessitating the occasional addition or removal of pulses so that (1) all grids containing emission sources will not be missed, (2) cell concentrations can be constructed for diffusion calculation, and (3) execution time is not excessive. There are an infinite number of pathways to pulse management. For our applications, our criterion is to check every fifth time step that no cell has more than ten pulses. If it does, the pulses may be merged or removed according to their subgrid locations. After testing numerous options, we chose a very simple and fast procedure: only the six pulses closest to the cell center will be retained. Clearly, some information will be lost in this case, but the comparatively low resolution of the model grid implies that the subgrid concentration may be sufficiently smooth so that information lost is minimal.

At every time step, we add a pulse with spatially interpolated species concentrations to the cell center of each empty cell in the lowest layer. To further reduce the likelihood that the bottom-layer cells become empty, and to reduce potentially large interpolation errors, at every other time step, a pulse is also added to the cell center of each bottom-layer cell if the cell contains only one pulse and the pulse is also located at a radius greater than $\Delta x/4$ (assumed equal to $\Delta y/4$, both being ¼ of the horizontal grid sizes) from the cell center. Because the locations of the pulses are irregular, spatial interpolation to estimate the cell-center concentrations is not straightforward. Because the lowest layer is shallow in general, we simplify the interpolation by assuming that all pulses are on the central plane of the layer. In addition to the concentration of the pulse closest to the cell center, we use linear interpolations from nearby pulses to determine two more concentrations, one on each side of the cell center, and on the axis passing through the cell center and the nearest pulse.

We then determine the three coefficients of a quadratic polynomial that has the least difference (in the least-squares sense) between the area under the quadratic curve and the area under the piecewise linear concentration constructed from the three concentration points along the axis. The cell-center pulse's concentration is then estimated from the quadratic polynomial. We believe that this approach is superior to that based on linear or other interpolation schemes. When the conditions making the above procedure possible are not satisfied, other simple interpolation schemes like inverse-distance weighting are used. For all upper non-boundary layers, empty cells are identified every hour and filled with one cell-centered pulse bearing the cell concentrations of the previous time step for simplicity. Note that for the purpose of diffusion, cell concentrations must be available at every time step.

Diffusion: At each full time step Δt, diffusion for each species is accomplished by first taking the concentration average $\overline{C}^{old}$ from all the pulses within each cell, running the diffusion algorithm from CAMx, and adjusting the concentrations of the pulses according to the required concentration change, ΔC_D, dictated by the diffusion. But in the case where there is more than one pulse in the cell, simply adding the ΔC_D to the species concentration of each pulse may not be sufficient since diffusion also exists among the pulses within a grid cell. Therefore, for pulse i in a cell, the concentration after the diffusion step is $C_i^{new} = C_i^{old} + \Delta C_D + \Delta C_{SD,i}$, where C_i^{old} is the original concentration of pulse i, and $\Delta C_{SD,i}$ is the concentration change for pulse i due to subgrid diffusion, and is defined as $\Delta C_{SD,i}=D_{SG}(\overline{C}^{old}-C_i^{old})$, where $D_{SG} = A(\Delta t)(D_x/\Delta x^2+D_y/\Delta y^2)$. Here, D_{SG} is the non-dimensionalized subgrid diffusivity, assumed to be proportional to the horizontal eddy diffusivities (D_x and D_y) with A set equal to 5. We do not include vertical diffusivity in subgrid diffusion because only the lowest layers tend to have more pulses per grid cell by design but these layers generally have small layer thicknesses, while for upper layers, the grid cells tend to have no more than one pulse per cell. We further constrain D_{SG} to be no more than 0.1. Further investigation of D_{SG} is clearly needed.

Integration sequence: The handling of advection, emission, diffusion and pulse management have been described above. Other steps are identical to those used in CAMx. The integration sequence for each time step is as follows: (1) Add species concentrations from area and point emissions over a period of $\Delta t/2$ to the pulses. (2) Move the pulses for duration of $\Delta t/2$ to new locations. (3) Repeat step (1) based on the new pulse locations. (4) Repeat step (2) starting from the new pulse locations. (5) Turn on wet deposition for species within each pulse. (6) Update meteorological variables. (7) Add pulses as necessary to the side and top boundary cells and the lowest-layer cells. (For the lowest layer, pulses are also added at every other time step for cells containing only one pulse as described before.) (8) Determine concentration averages for all species in each cell. (9) Perform diffusion and subgrid diffusion. (10) Merge or remove pulses (every fifth time step). (11) Perform chemistry separately for each pulse. Separately, at every hour, cell-centered pulses are added to empty cells in the upper non-boundary layers.

4. MODEL RESULTS AND CONCLUSIONS

The cell concentrations for each species are obtained by averaging the concentrations of all pulses in each cell. By turning off emissions and all transport and chemistry steps except advection with an MM5 wind field and a known initial concentration profile, we notice that the CAMx concentration becomes substantially suppressed after several hours of advective transport compared to the TG-derived concentration. When emissions are turned on, we also notice that the CAMx-predicted primary-pollutant concentrations tend to drop off dramatically in the early evenings as emissions start to decline. The diffusive nature of the Bott scheme in a complex wind field is not generally appreciated and is certainly not obvious in well-designed wind fields used in many algorithm-comparison studies. Figure 2 shows the predicted CO hourly concentrations in Southern California at 1800 PST (1900 PDT) on August 3, 1997 using TG (left panels) and the Bott scheme (right panels, referred to as CAMx) separately. The model contains 65x40x10 cells, with a 5-km grid resolution. Emissions are turned on; chemistry and deposition are turned off. The top panels show the results of advection; the bottom panels show the results of both advection and diffusion. One can see that Bott seriously underestimates the high concentrations when emissions start to decline in the early evening. As a result, the predicted concentrations are considerably less than those of TG in the high-concentration regions. Even when diffusive smoothing is in effect, the underestimation of Bott remains significant.

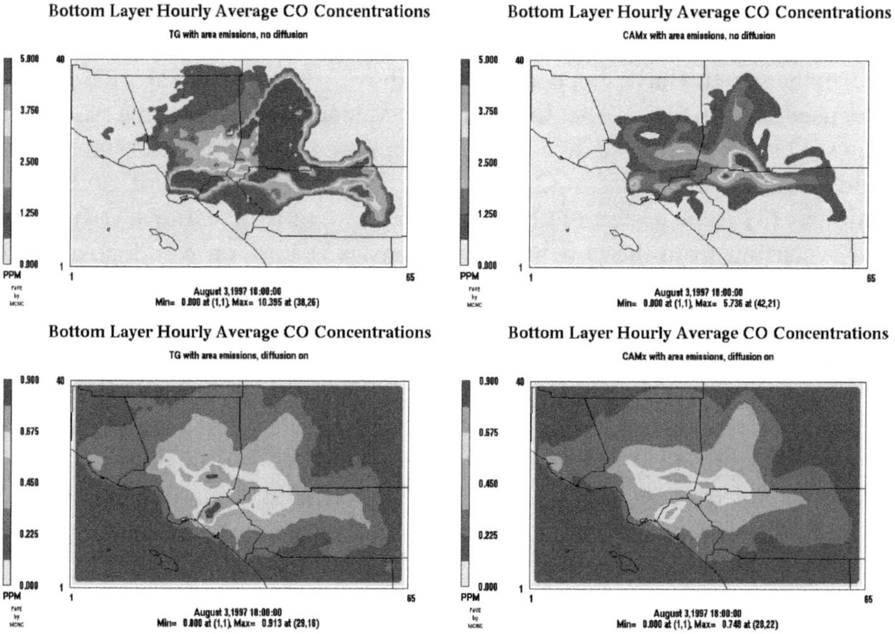

Figure 2: CAMx-predicted CO hourly concs using TG with $\dot{\sigma}$ (left) and Bott with CAMx *w* (right) at 1800 PST August 3, 1997. The top panels are with advection alone, the bottom are with advection and diffusion.

Figure 3 shows a comparison of the bottom-layer ozone hourly concentrations for the Northeast US predicted by CAMx using the TG (left panels) and Bott (right panels, referred to as CAMx) schemes for 1500 EST on July 10, 1995. The top panels are based on a 12-km grid, with a model domain of 152x131x12 cells. The bottom panels are based on a 36-km grid, with a model domain of 80x61x12 cells. The vertical structures of both grid resolutions are identical. Because ozone chemistry is nonlinear, it becomes difficult to determine if TG should necessarily predict higher ozone concentrations than Bott at a particular location. In the VOC-limited areas, it is possible that a predicted lower NO_x concentration can actually lead to a higher ozone concentration, but an artificially diffusive advective scheme can still suppress the ozone concentration significantly. Also, spurious diffusion can cause areas that otherwise have low concentrations of some species to have a substantial increase in the concentrations of those species. In the figure, even though the patterns look similar for both grid sizes, TG-based CAMx tends to predict higher ozone concentrations than the Bott-based CAMx.

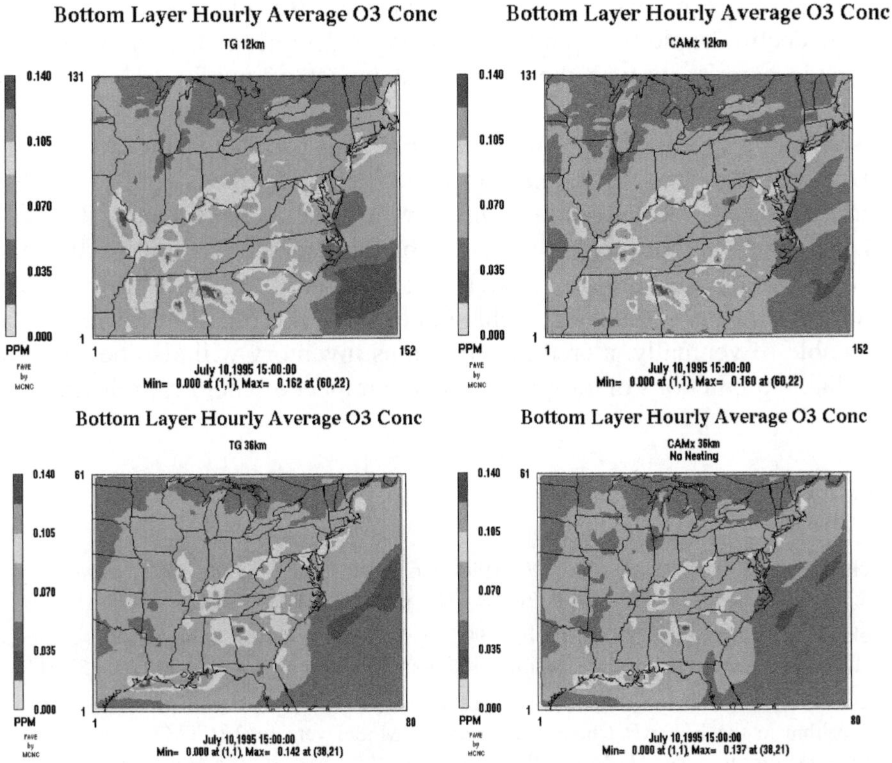

Figure 3: Comparison of ozone hourly concentrations in NE US at 1500 EST on July 10, 1995 predicted by TG (left panels) and Bott (CAMx, right panels) for 12-km grid (top panels) and 36-km grid (bottom panels).

For comparison purposes, in the case of the 36-km grid model domain for Northeast US, there are 58,560 cells, and a one-day CAMx simulation using the IEH chemistry solver (Sun et al., 1994; Chock et al., 1994) took 22.48 min. of execution time on a Sun Blade 1000 (900 Mhz) computer. The corresponding TG run had 87,840 pulses initially and 150,722 pulses at the end and took 58.00 min. of execution time.

We have implemented the Trajectory-Grid algorithm into a state-of-the-science AQM (CAMx) to substantially reduce the large inaccuracy generated by solving the advection equation. In the process, we also addressed the mass conservation problem, not by arbitrarily adjusting the wind field or introducing other corrections, but by scaling the species concentrations with air density, assuring that the species mixing ratios do not change as air parcels are transported. In comparing the concentration fields of relatively unreactive pollutants, like CO, with diffusion, chemistry and deposition turned off, we found that the Bott scheme adopted in CAMx spuriously diffuses the concentrations. This fact becomes evident when Bott-derived concentrations start to decline rapidly as emissions start to decline in the evening. This artifact does not occur with the TG-derived concentrations.

The numerical problems from the Eulerian-based scheme also greatly impact the prediction of ozone concentrations. Compensating errors have clearly occurred. As advection can now be accurately tackled with TG or perhaps other Lagrangian schemes, a better characterization of the eddy diffusivity in the meteorology models and eventually, AQMs is now critically important. In this connection, field studies with controlled tracer release are invaluable. Eventually, a credible emissions inventory will also be necessary for reliable predictions of air quality and the impact of chemical release.

REFERENCES

Chock, D. P., Winkler, S. L., Sun, P., 1994, Comparison of stiff chemistry solvers for air quality modeling. *Environ. Sci, & Technol.* **28**, pp. 1882-1892.

Chock, D. P., Sun, P., Winkler, S. L., 1996, Trajectory-Grid: An accurate sign-preserving advection-diffusion approach for air quality modeling. *Atmos. Environ.* **30**, pp. 857-868.

Chock, D. P., Winkler, S. L., Hass, H., Ackermann, I. J., 1998, Use of the Trajectory-Grid algorithm in the EURAD Chemistry-Transport Model version 2 (CTM2). In *Global and Regional Atmospheric Modeling, Proceedings of the First GLOREAM Workshop, Aachen, Germany, September 1997,* Eds. Hass, H. and Ackermann, I. J., pp. 185-192.

Chock, D. P., Winkler, S. L., 2000, A Trajectory-Grid Approach for Solving the Condensation and Evaporation Equations of Aerosols. *Atmos. Environ.* **34**, pp. 2957-2973.

Dudhia, J., 1993, A nonhydrostatic version of the Penn State-NCAR mesoscale model: validation tests and simulation of an Atlantic cyclone and cold front. *Mon. Wea. Rev.* **121**, pp. 1493-1513.

Environ, Novato, CA, 2004, http://www.camx.com/overview.html

Gaydos, T. M., Koo, B., Pandis, S. N. Chock, D. P., 2003, Development and application of an efficient moving sectional approach for the solution of the atmospheric aerosol condensation/evaporation equations. *Atmos. Environ.* **37**, pp. 3303-3316.

Grell, G. A., Dudhia, J., Stauffer, D. R., 1994, A description of the Fifth-Generation Penn State/NCAR Mesoscale Model (MM5), NCAR/TN-398 + STR, NCAR Technical Note, Boulder, CO, USA, December, 122 p.

Lee, S. M., Yoon, S. C., Byun, D. W., 2004, The effect of mass inconsistency of the meteorological field generated by a common meteorological model on air quality modeling. *Atmos. Environ.* **38**, pp. 2917-2926.

Sun, P., Chock, D. P., Winkler, S. L., 1994, An Implicit-Explicit Hybrid solver for a system of stiff kinetic equations. *J. Computational Phys.* **115**, pp. 515-523.

ESTIMATION OF AIR POLLUTION PARAMETERS USING ARTIFICIAL NEURAL NETWORKS

Hikmet Kerem Cigizoglu[(1)], Kadir Alp[(2)], Müge Kömürcü[(2)]
[(1)]Division of Hydraulics, Civil Engineering Faculty, Istanbul Technical University, Maslak, 34469, Istanbul, Turkey;[(2)]Environmental Engineering Department, Civil Engineering Faculty, Istanbul Technical University, Maslak, 34469, Istanbul, Turkey.

Abstract: The modeling of air pollution parameters is an issue investigated using different techniques. The pollution time series, however, are not continuous and contain gaps. Therefore, methods to infill the gaps providing satisfactory estimations are quite significant. In the presented study two ANN methods, feed forward back propagation, FFBP, and radial basis functions, RBF, were presented to estimate the SO2 values using the NO and CO values. It was seen that both ANN methods provided superior performances to conventional multi linear regression, MLR, method. The ANN performances were found satisfactory considering the selected performance criteria and the testing stage plots.

Key words: Air pollution parameters, artificial neural networks, estimation, radial basis functions.

1. INTRODUCTION

Air pollution is an extremely significant issue that should be focused on all around the globe as it affects human health, ecosystems and the environment, destroys aesthetical conditions, gives harm to matters, buildings, and in turn economy. It is defined as a condition, in which the substances that result from both natural and anthropogenic activities are present in the air at concentrations sufficiently high above their normal ambient levels producing considerable impacts on humans, animals, vegetation, or materials (Seinfeld and Pandis, 1998). Not all the studies on air pollution consider it as a

I. Farago et al. (eds.), Advances in Air Pollution Modeling for Environmental Security, 63–75.

consequence of both natural and anthropogenic activities, some studies consider define air pollution to be only of anthropogenic origin.

Various impacts on human health, including respiratory problems, hospitalization for heart or lung diseases, and premature death are caused by air pollution. As children spend more of their time outdoors and as their lungs are still developing, they are more susceptible to risk. Moreover, elderly people and people with heart or lung diseases are sensitive to several types of air pollution. (EPA, 2002)

Not only regional sources but also transport processes by weather patterns generate pollution. It is important to detect the most dominant process to take action. If transport processes are dominant, putting regional standards to industrial sources and limiting anthropogenic activities will not be adequate to prevent pollution. Regional sources of air pollution have several different sources categorized as natural and anthropogenic sources. Natural sources can be named as windblown dust, volcanoes, ocean sprays, evaporation and wildfire.

One of the most important natural pollutant sources is volcanic eruption. In volcanic eruptions, extremely significant amounts of gasses and particulate matter are released into the atmosphere. SO_xs are one of the major constituents of these gasses. The ultraviolet flux coming to earth is scattered and absorbed by these particles and gasses (Wayne, 2000).

Another natural source is ocean sprays. Sprays of salt from the oceans and seas are transferred to the atmosphere by evaporation and winds (Incecik, 1994). Evaporation is another significant source of natural air pollution. Numerous major gasses are transferred to the atmosphere via evaporation from water bodies with large surface areas such as oceans. In the development phase of the planktons, oceans are the sources of CO_2 emissions. Furthermore, forests release hydrocarbons by photochemical reactions (Incecik, 1994).

Although biomass burning can occur naturally, recently most of the biomass burning is done intentionally. Some of the purposes in intentional biomass burning are to control pests, to produce energy, to clear forest area, and to mobilize nutrients. Major species emitted during biomass burning are CO, CH_4, CH_3Br, CH_3Cl, NO_x, N_2O, Sulfur compounds, Volatile organic carbon species including non metal hydrocarbons, PAH and particulate matter. In dry seasons, biomass burning is intense enough to be seen from outer space. Biomass burning also increases ozone levels in the troposphere.

Aside from natural sources, there are also anthropogenic sources, which are divided into two categories: stationary, and mobile. Some stationary sources are factories, power plants, smelters, dry cleaners and degreasing operations; whereas some mobile sources are cars, buses, planes, trucks, and trains. (EPA, 2002; Incecik, 1994)

Like pollution sources, air pollutants are also categorized. Primary and secondary pollutants are the two categories of air pollutants. Primary pollutants are the pollutant chemicals that are directly emitted to the atmosphere from known sources, where as secondary pollutants are the pollutant species that are formed by the chemical reactions of the primary pollutants (Wayne, 2000). Primary pollutants can be listed as sulfur oxides (SO_x), nitrogen oxides (NO_x), carbon monoxide (CO), hydrocarbons (HC), and particulate matter (PM). One of the most important secondary pollutants is ozone (Incecik, 1994).

The artificial neural network (ANN) approach, which is a non-linear black box model, seems to be a useful alternative for modelling the complex hydro meteorological time series. The artificial neural networks (ANNs) became a speculated model in forecasting rainfall (Bodri and Cermak 2000; Sahai et al. 2000; Luk et al. 2000; Silverman and Dracup 2000), stream flow and suspended sediment forecasting (Cigizoglu 2003a; Cigizoglu 2003b; Cigizoglu, 2004; Cigizoglu and Kisi, 2005), as much as in air pollution researches (McKendry, 2002), in ozone concentration studies (Narasimhan et al. 2000; Gardner and Dorling 2001; Abdul-Wahab and Al-Alawi 2002), in temperature retrieval (Kuligowski and Barros 2001; Aires et al. 2002) and also in many other scientific fields.

In the majority of these studies feed forward error back propagation method (FFBP) was employed to train the neural networks. The performance of FFBP was found superior to conventional statistical and stochastic methods in continuous flow series forecasting (Cigizoglu 2003a; Cigizoglu 2003b). However the FFBP algorithm has some drawbacks. They are very sensitive to the selected initial weight values and may provide performances differing from each other significantly. Another problem faced during the application of FFBP is the local minima issue. In their work, Maier and Dandy (2000) summarized the methods used in the literature to overcome local minima problem as training a number of networks starting with different initial weights, the on-line training mode to help the network to escape local minima, inclusion of the addition of random noise, employment of second order (Newton algorithm, Levenberg-Marquardt algorithm) or global methods (stochastic gradient algorithms, simulated annealing). Other ANN methods such as conjugate gradient algorithms, radial basis function, cascade correlation algorithm and recurrent neural networks were described by ASCE Task Committee (2000). Levenberg-Marquardt algorithm was employed for FFBP applications in the presented study. The second ANN method employed in this work was the radial basis functions (RBF). In the presented the air pollution parameter estimation was carried out using both ANN methods.

In this study, the pollutant parameters measured in stations of Istanbul at chosen episode days with particulate matter in focal point will be estimated using two ANN methods (FFBP and RBF). The results were compared with multi-linear regression (MLR) method.

2. DESCRIPTION OF ANN METHODS

2.1 The feed forward back propagation (FFBP)

Figure 1 depicts the structure of FFBP neural networks. A FFBP distinguishes itself by the presence of one or more hidden layers, whose

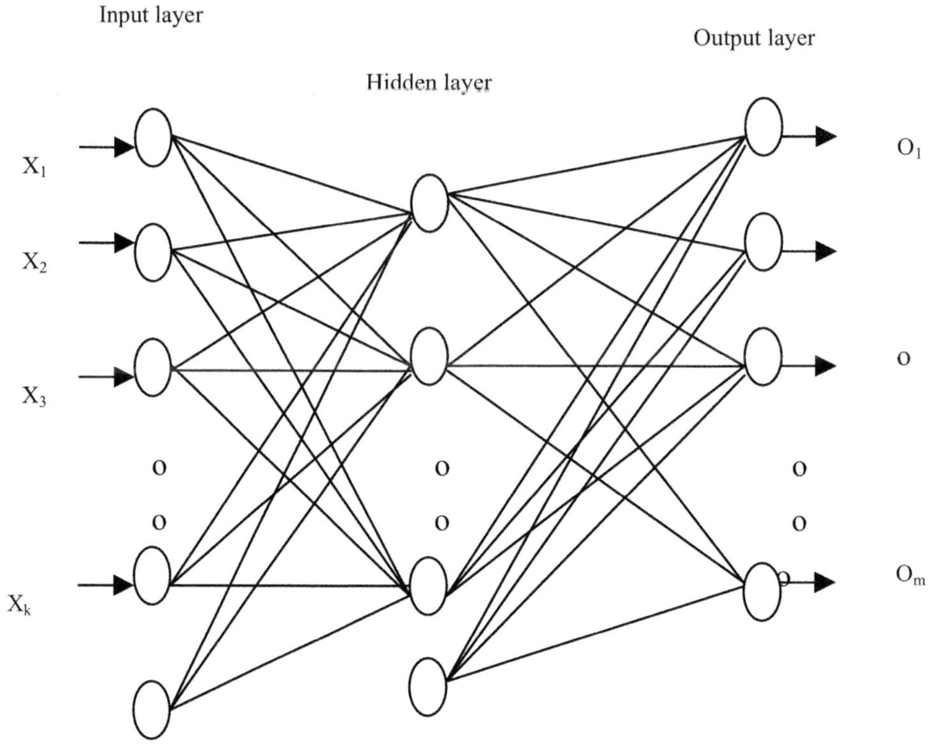

Figure 1: Structure of a feed forward neural network (FFBP).

computation nodes are correspondingly called hidden neurons of hidden units. The function of hidden neurons is to intervene between the external input and the network output in some useful manner. By adding one or more hidden layers, the network is enabled to extract higher order statistics. In a rather loose sense, the network acquires a global perspective despite its local connectivity due to the extra set of synaptic connections and the extra dimension of NN interconnections (Hagan and Menhaj, 1994).

The ability of hidden neurons to extract higher order statistics is particularly valuable when the size of the input layer is large. The source nodes in the input layer of the network supply respective elements of the activation pattern (input vector), which constitute the input signals applied to the neurons (computation nodes) in the second layer (i.e. the first hidden layer). The output signals of the second layer are used as inputs to the third layer, and so on for the rest of the network. Typically, the neurons in each layer of the network have as their inputs the output signals of the preceding layer only. The set of the output signals of the neurons in the output layer of the network constitutes the overall response of the network to the activation patterns applied by the source nodes in the input (first) layer. The FFBP was trained using Levenberg–Marquardt optimization technique. This optimization technique is more powerful than the conventional gradient descent techniques (Hagan and Menhaj, 1994).

FFBP algorithm exhibits two significant drawbacks. Firstly, the training simulation performance is dependent on the different random weight assignment in the beginning of each training simulation. The FFBP network can be trapped by different local error minima each time and the desired error value can not be attained. Another drawback of FFBP is the negative value generation for low value estimations. In the presented study different numbers of hidden layer nodes were tried and the configuration that gave the minimum mean square error (MSE) was determined.

2.2 The radial basis function-based neural networks (RBF)

Figure 2 depicts the structure of RBF neural networks. RBF networks were introduced into the neural network literature by Broomhead and Lowe (1988). The RBF network model is motivated by the locally tuned response observed in biological neurons. Neurons with a locally tuned response characteristic can be found in several parts of the nervous system, for example, cells in the visual cortex sensitive to bars oriented in a certain direction or other visual features within a small region of the visual field. These locally tuned neurons show response characteristics bounded to a small range of the input space. The theoretical basis of the RBF approach lies in the field of interpolation of multivariate functions.

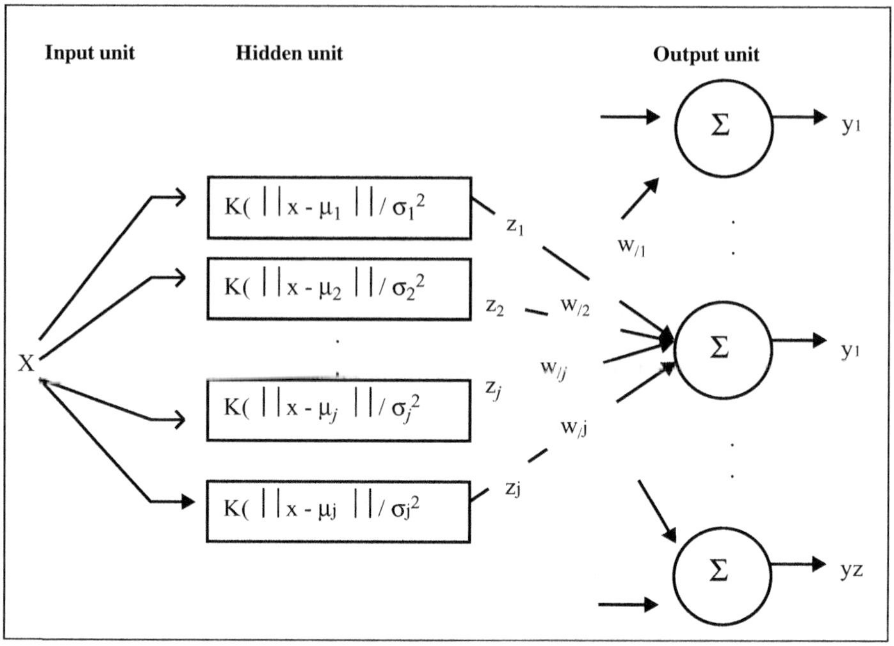

Figure 2: Structure of a Radial Basis Function Neural Network (RBF).

The solution of the exact interpolating RBF mapping passes through every data point. In the presence of noise, the exact solution of the interpolation problem is typically a function oscillating between the given data points. An additional problem with the exact interpolation procedure is that the number of basis functions is equal to the number of data points and so calculating the inverse of the NxN matrix ϕ becomes intractable in practice. The interpretation of the RBF method as an artificial neural network consists of three layers: a layer of input neurons feeding the feature vectors into the network; a hidden layer of RBF neurons, calculating the outcome of the basis functions; and a layer of output neurons, calculating a linear combination of the basis functions.

The RBF method provides a unique solution in a simulation with a constant spread values differing from FFBP. Similar to FFBP, RBF can also provide negative estimations for measurements having low values. Both FFBP and RBF algorithms have short training time. However, since multiple FFBP simulations are required until obtaining satisfactory performance criteria, this total duration is longer than the unique RBF application. Different numbers of hidden layer neurons and spread constants were tried in this study. The hidden layer neuron number that gave the minimum mean square error (MSE) was determined. The spread that gave the minimum MSE was also found simply by a trial-error approach adding some loops on program codes.

3. DATA

In the study the SO_2, NO and CO measurements in the Asian part of Istanbul, Ümraniye, for November 1999-April 2003 period were used (Figure 3). One measurement is taken during a day. The data is not continuous and contains gaps. The common measurement period for SO_2- NO pair covers 206 daily values whereas there are 169 values for SO_2- CO pair. Since the study aims the estimation of SO_2 values based on other

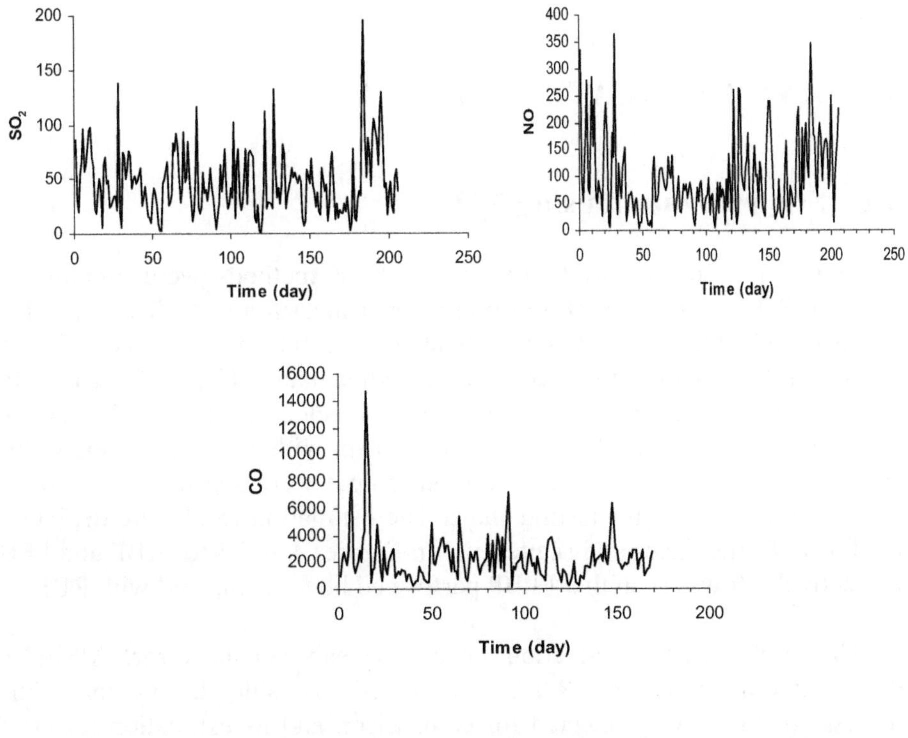

Figure 3: The observed air pollution records (November 1999 - April 2003).

pollution parameters the cross-correlations are computed in order to determine the time interval significant for the estimation work. The Table 1 reveals that only NO and CO values for times t, t-1 and t-2 have considerable correlations with SO_2 at time t. In fact the correlations at time t-2, r3, is quite low compared with r_1 and r_2.

Data measurement period	r_1	r_2	r_3
	SO_2-NO		
November 1999-April 2003 (206 days)	0.66	0.29	0.05
	SO_2-CO		
November 1999-April 2003 (169 days)	0.41	0.18	0.03

Table 1: First-, second- and third order cross-correlations.

4. ANN APPLICATION RESULTS

4.1 SO_2 estimation using NO

In this part of the study FFBP and RBF methods were employed to estimate SO_2 using the past and present NO measurements. The input layer consisted of three NO measurements at times t, t-1 and t-2 since only three cross-correlations two time series were higher than 0 (Table 1). The output layer had the unique SO_2 value for time t. A hidden layer with 3 nodes was found adequate for FFBP whereas the RBF provided best estimations for a spread equal to 0.70. The first 156 input-output were considered for training stage and the last 50 for testing stage. The estimation results are depicted in the form of time series and scatter plot in Figures.4 and 5 for RBF and FFBP, respectively. It was seen that FFBP performed better compared with RBF.

The performance evaluation criteria (mean square error, MSE, and determination coefficient, R^2) are presented in Table 2. The multi-linear regression, MLR, was selected for comparison and its estimation results for the testing period are presented in Figure 6 and Table 2. It is obvious that both ANN methods provided significantly superior performance compared with MLR.

Case	Evaluation criteria	R B F	F F B P	M L R
SO_2 estimation using NO	M S E	775.88	730.05	919.79
	R^2	0.477	0.512	0.385
SO_2 estimation using CO	M S E	1114.64	586.77	1678.67
	R^2	0.307	0.241	0.082

Table 2: ANN estimation results.

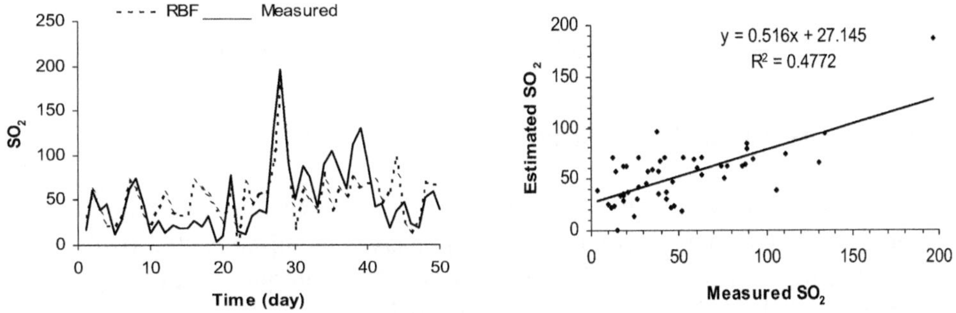

Figure 4: Estimation of SO_2 using NO with RBF.

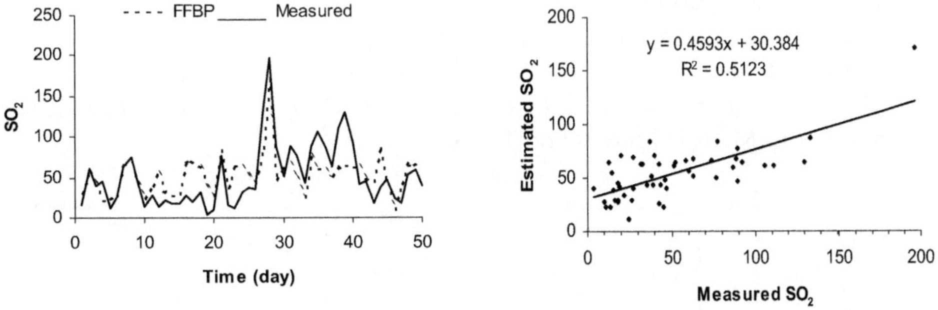

Figure 5: Estimation of SO_2 using NO with FFBP.

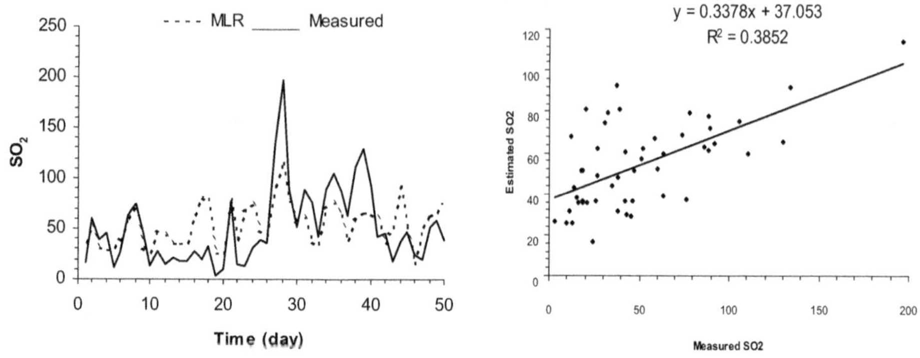

Figure 6: Estimation of SO_2 using NO with MLR.

4.2 SO_2 estimation using CO

The study was repeated again to estimate SO_2 but this time using the past and present CO measurements. The input layer again consisted of three CO measurements (times t, t-1 and t-2) and the unique output layer node represented the SO_2 value to be estimated for time t. A hidden layer with 4 nodes for FFBP and an RBF network with a spread equal to 0.08 provided best estimations. However, the order of training and testing periods were reversed this time. Accordingly the last 119 input-output patterns were considered for training and the first 50 for testing stage. The estimation results for the testing period are depicted in the form of time series and scatter plot in Figures 7 and 8 for RBF and FFBP, respectively. It was seen that FFBP performed better compared with RBF in terms of MSE but the R^2 value for RBF was higher than FFBP (Table 2). Both ANN methods were significantly superior to MLR (Figure 9 and Table 2).

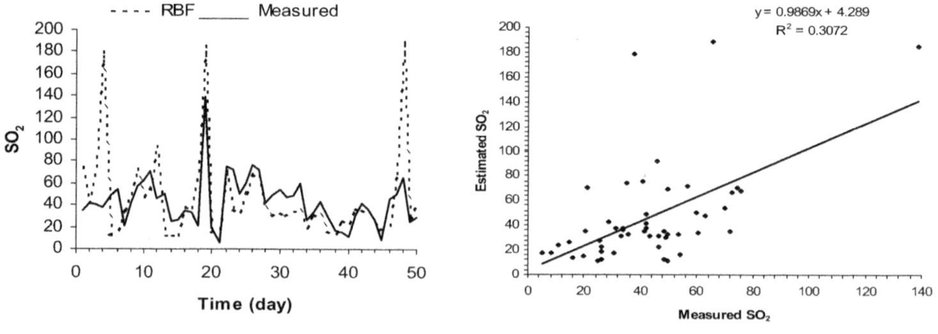

Figure 7: Estimation of SO_2 using CO with RBF.

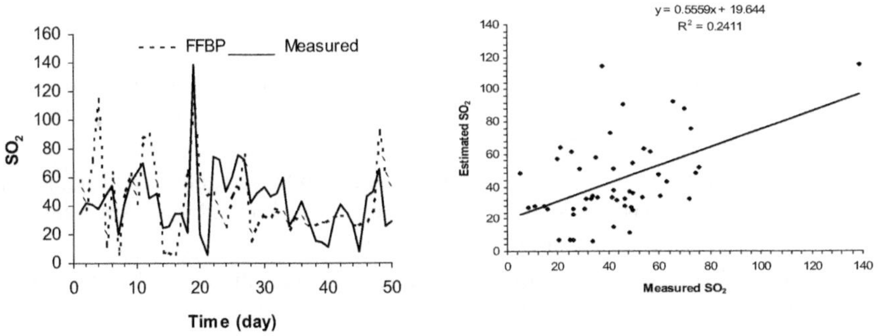

Figure 8: Estimation of SO_2 using CO with FFBP.

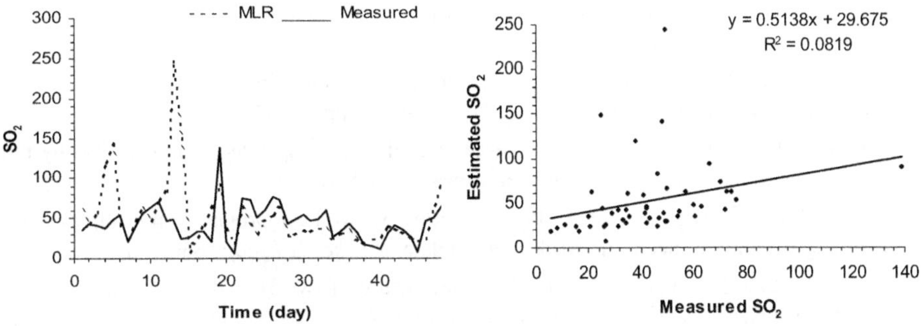

Figure 9: Estimation of SO_2 using CO with MLR.

5. CONCLUSIONS

In the presented study the air pollution parameter estimation based on two ANN methods, RBF and FFBP, was examined. Both methods provided significantly superior estimations compared with the conventional MLR method. The correlation statistics were employed successfully to determine the input layer node number. The performance evaluation criteria of FFBP for the testing period were slightly better than RBF. However, multiple FFBP simulations were required until obtaining satisfactory performance criteria and this total duration was longer than the unique RBF application. The air pollution series may have gaps during the observation period. The presented study has shown that both ANN methods could be successfully employed to infill the gaps in the air pollution record.

REFERENCES

Abdul – Wahab, S. A., and Al – Alawi, S. M., 2002, Assessment and prediction of tropospheric ozone concentration levels using artificial neural networks. *Environmental Modeling & Software* **17**(3), pp. 219-228.

Aires, F., Chedin, A., Scott, N. A., and Rossow, W. B., 2002, A regularized neural net approach for retrieval of atmospheric and surface temperatures with the IASI Instrument. *Journal of Applied Meteorology* **41**(2), pp. 144-159.

ASCE Task Committee, 2000, Artificial neural networks in hydrology I. *Journal of Hydrologic Engineering*, ASCE **5**(2), pp. 115-123.

Bodri, L., and Cermak, V., 2000, Prediction of extreme precipitation using a neural network: application to summer flood occurrence in Moravia. *Advances in Engineering Software* **31**, pp. 211 – 221.

Broomhead, D., and Lowe, D., 1988, Multivariable functional interpolation and adaptive networks. *Complex Syst.* **2**, pp. 321–355.

Cigizoglu, H. K., 2003a, Incorporation of ARMA models into flow forecasting by artificial neural networks. *Environmetrics* **14**(4), pp. 417-427.

Cigizoglu, H. K., 2003b, Estimation, forecasting and extrapolation of flow data by artificial neural networks. *Hydrological Sciences Journal* **48**(3), pp. 349-361.

Cigizoglu, H. K., 2004, Estimation and forecasting of daily suspended sediment data by multi layer perceptrons. *Advances in Water Resources* **27**, pp. 185-195.

Cigizoglu, H. K., and Kisi, O., 2005, Flow prediction by two back propagation techniques using k-fold partitioning of neural network training data. *Nordic Hydrology* **36**(1) (in press).

Gardner, M., and Dorling, S., 2001, Artificial neural network – derived trends in daily maximum surface ozone concentrations. *Journal of the Air & Waste Management Association* **51**(8), pp. 1202 – 1210.

Hagan, M. T., and Menhaj, M. B., 1994, Training feedforward techniques with the Marquardt algorithm. *IEEE Transactions on Neural Networks* **5**(6), pp. 989-993. Incecik, S., 1994, Air Pollution. Istanbul Technical University Press, Istanbul, Turkey, in Turkish.

Kuligowski, R. J., and Barros, A. P., 2001, Combined IR-microwave satellite retrieval of temperature and dewpoint profiles using artificial neural networks. *Journal of Applied Meteorology* **40**(11), pp. 2055-2067.

Luk K C, Ball, J. E., and Sharma, A., 2000, A study of optimal model lag and spatial inputs to artificial neural network for rainfall forecasting. *Journal of Hydrology* **227**(1 – 4), pp. 56-65.

Maier, H. R., and Dandy, G. C., 2000, Neural network for the prediction and forecasting of water resources variables: a review of modeling issues and applications. *Environmental Modeling and Software* **15**, pp. 101-124.

McKendy, I. G., 2002, Evaluation of artificial neural networks for fine particulate pollution (PM10 and PM2.5) forecasting. *Journal of the Air & Waste Management Association* **52**(9), pp. 1096 – 1101.

Narasimhan, R., J. Keller, Subramaniam, G., Raasch, E., Croley, B., Duncan, K., Potter, W. C., 2000, Ozone modeling using neural networks. *Journal of Applied Meteorology* **39**(3), pp. 291-296.

Sahai, A. K., Soman, M. K., and Satyan, V., 2000, All India summer monsoon rainfall prediction using an artificial neural network. *Climate Dynamics* **16**, pp. 291-302.

Silverman, D., and Dracup, J. A., 2000, Artificial neural networks and long-range precipitation prediction in California. *Journal of Applied Meteorology* **39**(1), pp. 57-66.

Seinfeld, J. H., Pandis, S. N., 1998, Atmospheric Chemistry and Physics. John Wiley & Sons Inc., United States.

U.S. Environmental Protection Agency, 2002, Latest Findings on National Air Quality: 2002 Status and Trends, Accessed: 15.02.04. Available: http://www.epa.gov/airtrends

Wayne, R. P., 2000, Chemistry of Atmospheres. Oxford University Press, New York, United States.

SOME ASPECTS OF INTERACTION BETWEEN OPERATOR SPLITTING PROCEDURES AND NUMERICAL METHODS

Petra Csomós
Eötvös Loránd University, H-1117 Budapest, Pázmány Péter sétány 1/C, Hungary

Abstract: For solving numerically the partial differential equations describing air pollution transport, numerical methods are usually applied together with operator splitting procedures. In the present paper the interaction between operator splitting procedures and numerical methods is investigated by analytical and numerical study of the order of the error oppressing the numerical solution.

Key words: Operator splitting procedures, order of the splitting error, numerical splitting.

1. INTRODUCTION

Air pollution processes are described by complex non-linear partial differential equations (Zlatev, 1995). Due to the complexity of these equations, there are no appropriate numerical methods which could be efficiently used to solve the problem numerically. In order to simplify the equations, *operator splitting procedures* have been introduced. The spatial differential operator appearing in the equations can be split into a sum of different sub-operators of simpler forms, with the corresponding equations being easier to solve. Instead of the original problem, a sequence of sub-models is solved, which gives rise to the so-called *splitting error*. The order of the splitting error can be estimated theoretically. In practice, splitting procedures are associated with different numerical methods for solving the sub-equations. In the present work the interaction between operator splitting procedures and numerical methods is investigated by analytical computations and through a numerical study of an air pollution model.

I. Farago et al. (eds.), Advances in Air Pollution Modeling for Environmental Security, 77–91.

Applying an operator splitting procedure and a numerical method together for solving an equation numerically represents a "new kind" of numerical method. We investigate the following two questions.

- How do the splitting procedures and the numerical methods interact?
- How can we characterize the splitting error in this case?

In this paper it will be shown that the behaviour of the splitting error is determined by the choice of the splitting procedure and the numerical method.

2. OPERATOR SPLITTING PROCEDURES

Operator splitting procedures are commonly used in modelling various physical phenomena, because the spatial differential operator appearing in them usually has very complex structure. Hence, there is not such a numerical method which can provide a numerical solution accurate enough, while taking reasonable integration time. In order to avoid this problem, the idea of *operator splitting procedure* has been initiated. The basic idea behind the procedure is dividing the spatial differential operator into a few sub-operators of simpler structure. For the sub-operators suitable numerical methods can be found. Then a sequence of equations corresponding to the sub-operators is solved. The connections between the sub-systems are the initial conditions. As an example, we introduce three possible splitting procedures (Lanser and Verwer, 1999, Csomós et al., 2002).

Let X denote a normed space of sufficiently smooth functions of type $R^N \to R^M$, and consider the following abstract Cauchy problem:

$$\left.\begin{aligned} \frac{\mathrm{d}u(t)}{\mathrm{d}t} &= \mathbf{M}u(t), \qquad t \in (0,T] \\ u(0) &= u_0, \end{aligned}\right\} \qquad (1)$$

where $u(t) \in X$ is the unknown function, $u_0 \in X$ is given, and $\mathbf{M}$ is an operator $X \to X$. Let us divide the time interval $[0,T]$ into m pieces of intervals with lenght τ, where τ is called *splitting timestep* ($T = m\tau$). For the sake of simplicity, we assume that the operator $\mathbf{M}$ can be written as the sum of two simpler sub-operators:

$$\mathbf{M} = \mathbf{A} + \mathbf{B}\,.$$

We note that the methods which will be discussed here can be extended in a natural way to the cases where more than two sub-operators are present.

The above decomposition of the operator **M** can be performed by applying different splitting procedures (e.g., Strang, 1968, Marchuk 1988, and Csomós et al., 2002). The simplest one is the so-called *sequential splitting*, defined by the following sequence of sub-problems:

$$\left.\begin{aligned} &\frac{\mathrm{d}u_1^{(k)}(t)}{\mathrm{d}t} = \mathbf{A}u_1^{(k)}(t) \qquad t \in ((k-1)\tau, k\tau] \\ &u_1^{(k)}((k-1)\tau) = u_2^{(k-1)}((k-1)\tau) \end{aligned}\right\} \\ \left.\begin{aligned} &\frac{\mathrm{d}u_2^{(k)}(t)}{\mathrm{d}t} = \mathbf{B}u_2^{(k)}(t) \qquad t \in ((k-1)\tau, k\tau] \\ &u_2^{(k)}((k-1)\tau) = u_1^{(k)}(k\tau) \end{aligned}\right\} \tag{2}$$

with $k = 1,\ldots,m$, and $u_1^{(0)} = u_0$. Applying this scheme, first the system with sub-operator **A** is solved using the original initial condition, and then, applying the obtained solution at time $t = \tau$ as an initial condition, the system with operator **B** is solved. This procedure is then repeated cyclicly.

The second introduced type of operator splitting procedure is called *weighted splitting*, which can be obtained by using two sequential splittings, once in the order **A** - **B**, once in **B** - **A**. At time $t = \tau$ the numerical solution is computed as a weighted average of the solutions obtained by the two sequential splitting steps:

$$u_{spl}(\tau) = \theta \cdot u_{spl,\mathbf{AB}}(\tau) + (1-\theta) \cdot u_{spl,\mathbf{BA}}(\tau),$$

where $\theta \in [0,1]$ is the weight parameter, and $u_{spl,\mathbf{AB}}(\tau)$ and $u_{spl,\mathbf{BA}}(\tau)$ are the solutions of the two sequential splittings, respectively. The case $\theta = 1/2$ is called *symmetrical weighted splitting*. We note that the symmetrically weighted splitting is of higher order if the condition

$$[[\mathbf{A},\mathbf{B}],\mathbf{A}-\mathbf{B}] = \mathbf{0} \tag{3}$$

holds, namely, the commutator of the two sub-operators and the difference of the sub-operators commute.

Another splitting technique is the *Marchuk-Strang splitting*, where at one splitting timestep three problems are solved. We note that the Marchuk-Strang splitting can also be weighted, and weighted splittings can be efficiently used on parallel computers.

3. DIFFERENT KINDS OF SPLITTING ERROR

In order to define the different kinds of splitting errors investigated in the present paper, the original problem and its analytical and numerical solutions will be introduced.

3.1 Definitions of splitting errors

Before defining the three kinds of splitting error, which we will investigate, we introduce some notations, where τ is the splitting timestep and $h = \tau/n$ is the step size of the numerical method.

Notation 1

1. *u(T) is the solution of the original continuous problem (1) at time* $t = T$*;*
2. $u_{spl}(m\tau)$ *is the solution of the continuous split problem (e.g., (2)) at time* $t = T$*;*
3. *y(mnh) is the solution of the discretized form of the original problem at time* $t = T$*, i.e,. the numerical solution of (1);*
4. $y_{spl}(m\tau)$ *is the solution of the discretized form of the split problem at time* $t = T$.

Definition 1

1. *Theoretical splitting error:* $S_{theor}(\tau) := \left\| u(T) - u_{spl}(m\tau) \right\|$
2. *Numerical splitting error:* $S_{num}(\tau) := \left\| u(T) - y_{spl}(m\tau) \right\|$
3. *Practical splitting error:* $S_{prac}(\tau) := \left\| y(mnh) - y_{spl}(m\tau) \right\|$

Up to now, there are conditions for vanishing only for the theoretical splitting error. In practice, we are interested in the numerical splitting error because it tells us how much the numerical solution differs from the analytical solution of the original problem. In real cases neither S_{theor} nor S_{num} can be determined, because the analytical solutions *u(T)* and $u_{spl}(m\tau)$ are not known. That is why the so-called practical splitting error has been introduced. The value of this error can always be computed from the numerical solutions *y(mnh)* and $y_{spl}(m\tau)$.

Before using the practical splitting error instead of the numerical splitting error, it should be examined how much they differ. In order to answer this question, we have done analytical computations and numerical experiments as well.

3.2 Discretization of the continuous problem

As an example, we introduce the discretization of equation (1) in case of using the *explicit Euler* method for solving it numerically, and the sequential splitting procedure. The discrete form of the original problem yields:

$$\frac{y^{i+1} - y^i}{h} = (\mathbf{A} + \mathbf{B})y^i \qquad i = 0,\ldots,mn \qquad (4)$$

where $y^0 = u_0$ and $h = \tau/n$ is the step size of the numerical method. Using the explicit Euler method, and the sequential splitting, the kth split problem can be written as:

$$\begin{aligned} \frac{y_1^{(k),j+1} - y_1^{(k),j}}{h} &= \mathbf{A}y_1^{(k),j} \qquad j = 0,\ldots,n \\ \frac{y_2^{(k),j+1} - y_2^{(k),j}}{h} &= \mathbf{B}y_2^{(k),j} \qquad j = 0,\ldots,n\,, \end{aligned} \qquad (5)$$

where $k = 1,\ldots,m$ ($T = m\tau$) and $y_1^{(1),0} = u_0$. The connection between the two equations in (5) is

$$y_1^{(k),j} = y_2^{(k),n} \quad \text{and} \quad \mathrm{y}_2^{(\mathrm{k}),0} = y_1^{(k),n}$$

for each splitting step (so, for all values of *k*). Equations (4) and (5) can be formulated also in the cases of other splitting procedures and other numerical methods. The "new method" for solving the model numerically can be easily understood from *Figure 1*.

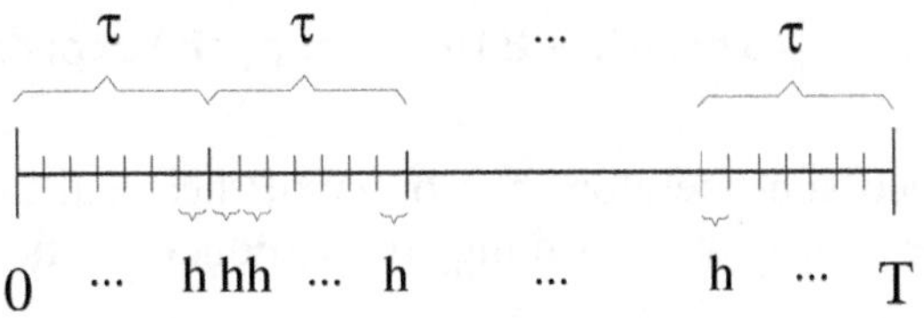

Figure 1: Timesteps of the "new method": τ is the splitting timestep, and h is the step size of the numerical method.

4. ANALYTICAL COMPUTATIONS

In this section the order of the above introduced three kinds of splitting error will be examined.

Definition 2

Let $S(m\tau)$ be one of the three kinds of splitting error defined in Definition 1 at time $t = m\tau$. In the case when $m = 1$, $S(\tau)$ is called local splitting error. The order $p-1$ of the local splitting error is defined as

$$\sup_{q\in N}\left(\lim_{\tau\to 0}\frac{S(\tau)}{\tau^q} = c < +\infty\right) =: p. \qquad (6)$$

4.1 Local theoretical splitting error

Definition 3

The order of the local theoretical splitting error is called theoretical order of the splitting.

As an example, the theoretical order of the sequential splitting will be determined. The analytical solution of the original problem (1) and the split problem applying sequential splitting at time $t = \tau$ reads

$$\begin{aligned} u(\tau) &= \exp(\tau(\mathbf{A}+\mathbf{B}))u_0 \\ u_{spl}(\tau) &= \exp(\tau\mathbf{B})\exp(\tau\mathbf{A})u_0. \end{aligned} \qquad (7)$$

The local theoretical splitting error is the difference of the two solutions in (7):

$$S_{theor}(\tau) = \| \exp(\tau(\mathbf{A}+\mathbf{B}))u_0 - \exp(\tau\mathbf{B})\exp(\tau\mathbf{A})u_0 \|.$$

In case of bounded sub-operators, the order of the local theoretical splitting error can be determined by expanding the exponential function into Taylor series up to $O(\tau^3)$.

$$S_{theor}(\tau) \le \left\| \left(I + \tau(\mathbf{A}+\mathbf{B}) + \frac{\tau^2}{2}(\mathbf{A}+\mathbf{B})^2 + O(\tau^3) \right) - \right.$$

$$-\left(I+\tau\mathbf{B}+\frac{\tau^2}{2}\mathbf{B}^2+\mathrm{O}(\tau^3)\right)\left(I+\tau\mathbf{A}+\frac{\tau^2}{2}\mathbf{A}^2+\mathrm{O}(\tau^3)\right)\Bigg\|\cdot\|u_0\|=$$

$$=\frac{\tau^2}{2}\cdot\|\mathbf{AB}-\mathbf{BA}\|\cdot\|u_0\|+\mathrm{O}(\tau^3).$$

One can see that in the general case when the two sub-operators do not commute, the local theoretical splitting error is of first order. We note that it can be shown, see for example Havasi et al. (2001), that if the sub-operators L-commute, the theoretical splitting error vanishes.

Table 1 contains the theoretical order in case of different splitting procedures (Faragó and Havasi, 2004).

Splitting	**Theoretic order**	*p*
Sequential	first	2
Marchuk-Strang	second	3
Weighted in general	first	2
Symmetrically weighted	second	3
Symmetrically weighted with the condition (3)	third	4

Table 1: Theoretical orders in case of different splitting procedures.

4.2 Local numerical and practical splitting errors

The computations of the orders of numerical and practical splitting errors will be introduced through one example. The results of the other cases will be mentioned without proving them.

Let us consider an abstract Cauchy problem (1), with operator $\mathbf{M}=\mathbf{A}+\mathbf{B}$, and $0\le t\le\tau$, and apply the weighted splitting scheme. The analytical solution of the original problem (1) and the split problem applying weighted splitting, can be determined as

$$u(\tau)=\exp(\tau(\mathbf{A}+\mathbf{B}))u_0$$

$$u_{spl}(\tau)=\theta\cdot\exp(\tau\mathbf{B})\exp(\tau\mathbf{A})u_0+$$
$$+(1-\theta)\cdot\exp(\tau\mathbf{A})\exp(\tau\mathbf{B})u_0.$$

Applying the explicit Euler method, the numerical solutions can be written in the following form:

$$h \qquad y(n\ ) = \left(I + h(\mathbf{A} + \mathbf{B})\right)^n u_0$$

$$y_{spl}(\tau) = \theta \cdot (\mathbf{I} + h\mathbf{B})^n (\mathbf{I} + h\mathbf{A})^n u_0 +$$
$$+ (1-\theta) \cdot (\mathbf{I} + h\mathbf{A})^n (\mathbf{I} + h\mathbf{B})^n u_0 .$$

Taking into account *Definition 1* and that $h = \tau / n$, the expressions of the splitting errors read

$$S_{num}(\tau) = \Big\| \exp(\tau(\mathbf{A} + \mathbf{B}))u_0 -$$
$$- \left(\theta \cdot (\mathbf{I} + h\mathbf{B})^n (\mathbf{I} + h\mathbf{A})^n u_0 + (1-\theta) \cdot (\mathbf{I} + h\mathbf{A})^n (\mathbf{I} + h\mathbf{B})^n u_0\right)\Big\|$$

$$S_{prac}(\tau) = \Big\| (I + h(\mathbf{A} + \mathbf{B}))u_0 -$$
$$- \left(\theta \cdot (\mathbf{I} + h\mathbf{B})^n (\mathbf{I} + h\mathbf{A})^n u_0 + (1-\theta) \cdot (\mathbf{I} + h\mathbf{A})^n (\mathbf{I} + h\mathbf{B})^n u_0\right)\Big\|$$

Since we are interested in the order of $S_{num}(\tau)$ and $S_{prac}(\tau)$, we can expand the exponential function into Taylor-series up to $\mathrm{O}(\tau^3)$, and we can use the binomial theorem to expand the other terms. After these steps we obtain the following expressions:

$$S_{num}(\tau) = \frac{\tau^2}{2} \cdot \left\| \frac{1}{n}(\mathbf{A}^2 + \mathbf{B}^2)u_0 + (1-2\theta)(\mathbf{BA} - \mathbf{AB})u_0 \right\| +$$
$$+ \mathrm{O}(\tau^3)$$

$$\tag{8}$$

$$S_{prac}(\tau) = \frac{\tau^2}{2} \cdot \left\| \frac{1}{n}(\mathbf{BA} + \mathbf{AB})u_0 + (1-2\theta)(\mathbf{BA} - \mathbf{AB})u_0 \right\| +$$
$$+ \mathrm{O}(\tau^3).$$

Rewriting formulae (8) using $h = \tau / n$, it follows that

$$S_{num}(\tau) \leq \frac{h\tau}{2} \cdot \left\| (\mathbf{A}^2 + \mathbf{B}^2) u_0 \right\| + \\ + \frac{\tau^2}{2} \cdot \left\| (1-2\theta)(\mathbf{BA} - \mathbf{AB}) u_0 \right\| + \mathrm{O}(\tau^3)$$

$$S_{prac}(\tau) \leq \frac{h\tau}{2} \cdot \left\| (\mathbf{BA} + \mathbf{AB}) u_0 \right\| + \\ + \frac{\tau^2}{2} \cdot \left\| (1-2\theta)(\mathbf{BA} - \mathbf{AB}) u_0 \right\| + \mathrm{O}(\tau^3). \tag{9}$$

From (9) it can be seen that the expression of the local splitting errors can be separated into two terms, which correspond to the numerical and the splitting error, respectively. The numerical error can be set to $\mathrm{O}(\tau^3)$ by choosing the numerical step size small enough, i.e., $h = \mathrm{O}(\tau^2)$. In this case the weighted splitting is in general of first order, and the symmetrically weighted splitting ($\theta = 1/2$) is of second order.

The above analytical computations can also be done for the other splitting procedures and numerical methods. The results of our computations using the explicit and the implicit Euler method, are listed in Table 2.

Splitting	Suitable magnitude of h	Order of $S_{num}(\tau)$ and $S_{prac}(\tau)$
Sequential	$\mathrm{O}(\tau)$ – always	2
Marchuk-Strang	$\mathrm{O}(\tau^2)$	3
Weighted in general	$\mathrm{O}(\tau)$ – always	2
Symmetrically weighted	$\mathrm{O}(\tau^2)$	3

Table 2: Results of our analytical computations regarding the suitable magnitude of the numerical step size having the theoretical values for the orders of the local umerical and practical splitting error. The explicit and the implicit Euler method were used.

Corollary

Let $h = \tau/n$. Choosing $h = \mathrm{O}(\tau^{p-1})$, we get the theoretical value of the orders of $S_{num}(\tau)$ and $S_{prac}(\tau)$.

The condition $h = \mathrm{O}(\tau)$ always holds in real cases, therefore, regarding the first order splittings, there is no new restriction for the value of the numerical step size. However, in the case of the second order splittings, the numerical step size has to be chosen $h = O(\tau^2)$ to get the theoretical values of the order of the numerical and the practical splitting errors. The explanation of the observed behaviour is the following. Both the explicit and the implicit Euler methods are first order numerical methods. According to the expectations, a second order splitting procedure is applied in vain if a first order numerical method is used together with it. If h is chosen small enough (i.e., $h = O(\tau^{p-1})$), the numerical error becomes negligible compared to the splitting error. In the latter case, the order of either the numerical and the practical splitting error equals the theoretical order.

5. NUMERICAL EXPERIMENTS FOR AN AIR POLLUTION TRANSPORT MODEL

Let $c = c(\mathbf{x}, t)$ denote the concentration of an air pollutant, c is a function of the location ($\mathbf{x} \in R^N$) and the time ($t \in R$). If we neglect the chemical reactions, the time evolution of c can be described mathematically by the partial differential equation (Zlatev, 1995)

$$\left.\begin{aligned} &\frac{\partial c}{\partial t} = -\nabla(\mathbf{u}c) + \nabla(K\nabla c) + E - \sigma c \qquad && t \in (0,T] \\ &c(\mathbf{x},0) = c_0(\mathbf{x}) && \end{aligned}\right\} \tag{10}$$

where $\mathbf{u} = \mathbf{u}(\mathbf{x}, t)$ is a function of type $R^{N+1} \to R^3$ describing the wind velocity, $K = K(\mathbf{x}, t)$ is the diffusion coefficient function, $E = E(\mathbf{x}, t)$ is the function of emission, and $\sigma = \sigma(\mathbf{x}, t)$ desribes the deposition. The initial function $c_0(x)$ is given. Using these notations, the terms in equation (10) have the following physical meaning. The first term on the right-hand side describes the transportation due to the velocity field, which is called *advection*. The second term expresses the *turbulent diffusion*, the third term the *emission*, and the last term describes the *deposition* of the pollutant. In our example we choose $N = 2$ ($\mathbf{x} = (x, y) \in [0,1] \times [0,1]$), $K = K_0(x, y)$ and $\sigma = \sigma_0(x, y)$ constant in time. The advection is chosen to be the *Molenkampf-Crowley advection* described as follows:

$$-\nabla(\mathbf{u}c) = y\frac{\partial c}{\partial x} - x\frac{\partial c}{\partial y}.$$

The emission field is constant in time, and it has a spatial distribution as shown in Figure 2. So, at the centre of the domain $E = E_0$, and at the other points it is equal to zero (this can serve as a model of emission from a chimney).

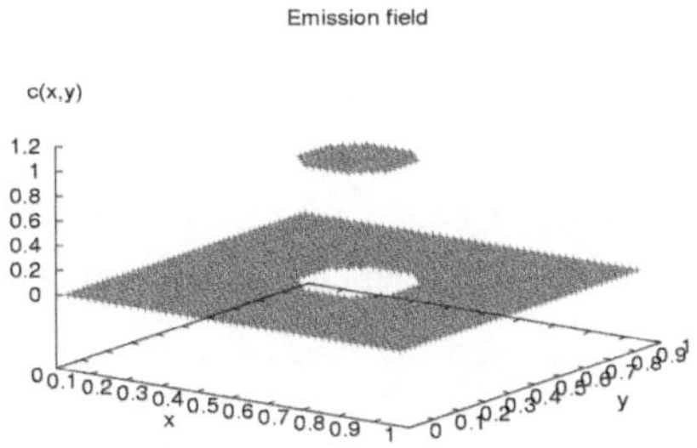

Figure 2: The emission field used in the air pollution model.

In order to solve equation (10) by a certain splitting procedure, the operator on the left-hand side is split into the sub-operators describing the physical sub-processes: advection, diffusion, emission, and deposition. We have applied sequential, Marchuk-Strang, weighted and symmetrically weighted splittings again. For time discretization the explicit Euler method (with step size $h = 10^{-4}$), and for spatial discretization the finite difference method (with resolution $\Delta x = \Delta y = 0.02$) has been used.

5.1 Solution of the air pollution model

First we solved equation (10) using time interval $t \in [0,1]$ with step size $h = 10^{-4}$, and with the following values of parameters:

$$u_{1_0} = u_{2_0} = 1, \quad K_0 = 0.1, \quad E_0 = 1, \quad \sigma_0 = 0.1.$$

In Figure 3 the numerical solutions are shown without using splitting procedure. The left panel of Figure 3 shows the spatial distribution of the concentration of the air pollutant in three dimensions at time $t = 1$. In the right panel the same distribution can be seen, but only in two dimensions.

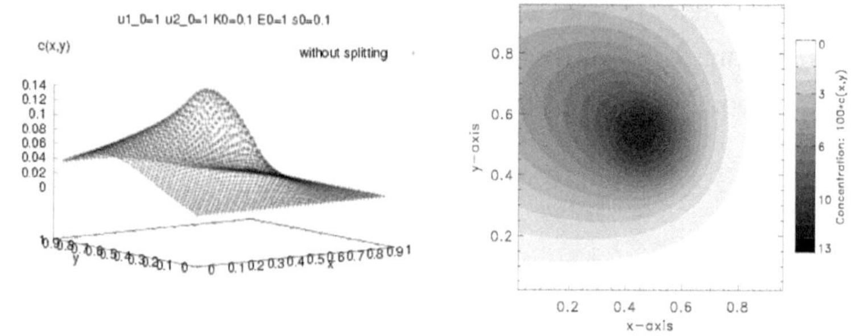

Figure 3: Spatial distribution of concentration without applying splitting procedure.

5.2 Practical splitting error

The difference between the solutions obtained by applying and without applying splitting procedure is shown in Figure 4. In this example the sequential splitting has been used, however, the figures in the cases of the other splittings are similar. The values of the difference are negative because the split solution has lower values of concentration. One can see that the structure of the splitting error field is similar to the concentration field.

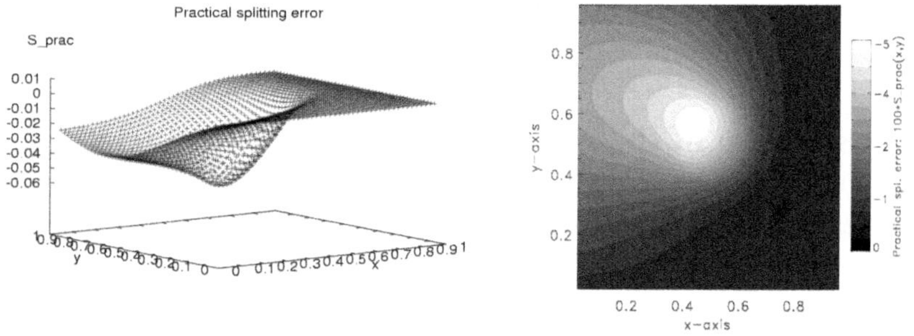

Figure 4: Spatial distribution of the practical splitting error.

The value of the practical splitting error at each τ can be defined as follows.

Definition 4
The local practical splitting error at each value of τ (denoted by $s(\tau)$) is computed by

$$s(\tau) := \sqrt{\Delta x \cdot \Delta y} \cdot \sqrt{\sum_{i,j}^{m} \left(S_{prac}(\tau)_{(i,j)}\right)^2},$$

where Δx and Δy are the grid steps, m is the number of grid points, and $S_{prac}(\tau)_{(i,j)}$ denotes the value of the practical splitting error for a certain value of τ at the grid point (i,j).

5.3 Order of the practical splitting error

The order of the local practical splitting error has been computed analytically in the previous section in a few examples. Numerically, the order of the local practical splitting error can be determined as follows. From *Definition 2* of the order, it follows that the method has an order $p-1$ if p is the supremum of those numbers for which

$$\frac{s(\tau)}{\tau^p} \approx c, \qquad (11)$$

if τ is small enough. Taking the logarithm of both sides in (11), we obtain:

$$\log s(\tau) \approx p \cdot \log \tau + \log c,$$

which defines a linear function on a logarithmic scale. From the fitted lines we can calculate the slope of the line, which can be regarded as the order of the practical splitting error obtained by numerical experiments.

Definition 5
The slope of the fitted line on a logarithmic scale is called the numerical order of the practical splitting error.

Having computed the value of $s(\tau)$ for several values of τ, we could determine numerically the order of the practical splitting error using this method. The values of the numerical orders are listed in Table 3.

Splitting	Numerical order	p
Sequential	1.22208	2
Marchuk-Strang	2.80306	3
Weighted in general	1.82665	2
Symmetrically weighted	2.55551	3

Table 3: Numerical and analytical orders in case of an air pollution model for the investigated splitting procedures.

From Table 3 it can be seen that the numerical order is close to the theoretical order. The reason why the differences arise is, on the one hand, that in the solution of partial differential equations the spatial discretization also causes a certain amount of error. On the other hand, solving this model describing the air pollution transport, the round-off error becomes large due to the large amount of arithmetical calculations of the CPU.

6. CONCLUSIONS

Applying a splitting procedure and a numerical method together means a "new numerical method" for solving partial differential equations numerically. The analytical and numerical properties of this new method are determined by the choice of both the splitting procedure and the numerical method.

Through analytical computations we have shown that the numerical order of the splittig error equals the analytical order if the timestep h of the numerical method is chosen small enough: $h = O(\tau^{p-1})$, where τ denotes the splitting timestep, and $p-1$ is the analytical order of the applied splitting procedure. The numerical study of an air pollution model has also shown that the numerical order is very close to the analytical order.

Further investigations are needed to know the relation between the numerical order and the magnitude of the splitting error. While the topic of the paper is the interaction of splitting error and the error caused by the numerical method, in practical situations other sources of errors also exiest (for example those caused by uncertainties in the input data and the mechanisms used for the descriptions of the physical processes). However, it is very difficult to establish the interference between these errors and the errors due to the numerical solution (splitting and numerical error). Therefore more investigations in this direction might lead to useful results.

Acknowledgement: The work is partially supported by NATO Collaborative Linkage Grant 980505 / NATO ARW project "Impact of future climate changes on pollution levels in Europe".

REFERENCES

Csomós, P., Faragó, I., Havasi, Á., 2002, Weighted sequential splittings and their analysis, *Comp. Math. Appl.*, to appear.

Faragó, I., Havasi, Á., 2004, On the convergence and local splitting error of different splitting schemes. *Progress in Computational Fluid Dynamics*, to appear.

Havasi, Á., Bartholy, J., Faragó, I., 2001, Splitting method and its application in air pollution modelling. *Időjárás* **105**, pp. 39-58.

Lanser, D., Verwer, J. G., 1999, Analysis of operator splitting for advection-diffusion-reaction problems in air pollution modelling, *J. Comput. Appl. Math.* **111**, No. 1-2, pp. 201-216.

Marchuk, G. I., 1988, Methods of Splitting, Nauka, Moscow.

Strang, G., 1968, On the construction and comparison of different splitting schemes, *SIAM J. Numer. Anal.* **5**, No. 3, pp. 506-517.

Zlatev, Z., 1995, Computer treatment of large scale air pollution models. Kluwer Academic Publishers.

MATHEMATICAL ASPECTS OF DATA ASSIMILATION FOR ATMOSPHERIC CHEMISTRY MODELS

Gabriel Dimitriu[(1)], Rodica Cuciureanu[(2)]
University of Medicine and Pharmacy "Gr. T. Popa", Faculty of Pharmacy,[(1)]Department of Mathematics and Informatics,[(2)]Department of Environmental and Food Chemistry

Abstract: This study analyzes some mathematical aspects of data assimilation for atmospheric chemistry models based on variational approaches. Of interest here are the so called adjoint methods which are particularly efficient to mathematically solve the variational problem. Details concerning operator splitting, adjoint and gradient computations are also discussed.

Key words: Atmospheric chemistry model, data assimilation, variational approach, adjoint method, operator splitting.

1. INTRODUCTION

The purpose of data assimilation is to incorporate measured observations into a dynamical system model in order to produce accurate estimates of all the current (and future) state variables of the system. In the atmospheric chemistry field, the knowledge of initial conditions (and sometimes boundary conditions) of chemical concentrations is a very challenging task. Having a good model at our disposal, the first and most important problem to solve for numerical predictions is the approximation of initial conditions.

This study analyzes some mathematical aspects of data assimilation for atmospheric chemistry models based on variational methods. In the variational data assimilation, information provided by the observations is used to find an optimal set of model parameters through a minimization process.

The paper is organized as follows. In Section 2 we formally introduce a functional framework for the linear case of the variational data assimilation.

I. Farago et al. (eds.), Advances in Air Pollution Modeling for Environmental Security, 93–104.

Section 3 contains the description of a 3D atmospheric transport-chemistry model and we define the 4D-Var assimilation problem associated with it. In the context of the variational approach, a general algorithm to compute the gradient of the cost functional is also presented. The main part of the paper is contained in Section 4, which presents in details the adjoint computations for a 3-stage Rosenbrock method. The last section contains some computational issues and conclusions.

2. DATA ASSIMILATION. ABSTRACT SETTING

The approach of adjoint equations has been efficiently used in numerical solution of control problems. Although there is a considerable literature dealing with adjoint formalism (Talagrand and Courtier, 1987; Marchuk et al., 1996; Giering and Kaminski, 1998; Wang et al., 2001; Silva et al., 2002), we present in this section a functional framework to the method as well as a brief discussion of its mathematical meaning.

Let U and V be vectorial spaces with inner products $<\cdot,\cdot>_U$ and $<\cdot,\cdot>_V$, respectively, being $U=\{x:[t_0,t_1\}\to V\}$. The linear version of the variational data assimilation problem is formally given by the following problem (P): Find the solution $Y \in U$ of

$$(P)\qquad \begin{cases} \dfrac{\partial y}{\partial t} = \dfrac{\partial F}{\partial x} y, \text{ that minimizes the functional} \\ J{:}U \to IR, \text{ given by } J(y) = \int_0^{t_1} T(y(t))\, dt \end{cases}$$

with $F:U\to U$ being a differentiable operator, $T:U\to IR$ given by

$T(x)=\frac{1}{2}\langle w(t)(x(t)-\tilde{x}(t)),x(t)-\tilde{x}(t)\rangle_U$, with $w\in U$ and $\tilde{x}\in U$ given.

We introduce the operators $A:U\to U$ and its adjoint $A^*:U\to U$ defined by $A(y)=\dfrac{\partial y}{\partial t}-\dfrac{\partial F}{\partial x}y$ and $A^*(y)=-\dfrac{\partial y}{\partial t}-\left(\dfrac{\partial F}{\partial x}\right)^* y$, respectively. In general, the adjoint depends on the definition of spaces. It always exists and it is unique, assuming spaces of finite dimension. It is actually true for all continuous operators in Hilbert spaces. Hence, coding the adjoint does not raise questions about its existence, only questions of technical implementation. Let $x\in U$ and $y\in U$ be such that $A(x)=0$ and $A^*(y)=\nabla_x T$. Now, by taking the inner product between $A(x)$ and y, and

$A^*(y)$ and x, and subtracting the latter from the former, we have successively:

$$< A(x), y >_U - < A^*(y), x >_U = - < \nabla_x T, x >_U$$

$$< \frac{\partial x}{\partial t} - \frac{\partial F}{\partial x} x, y >_U - < - \frac{\partial y}{\partial t} - \left(\frac{\partial F}{\partial x} \right)^* y, x >_U = - < \nabla_x T, x >_U$$

$$< \frac{\partial x}{\partial t}, y >_U + < \frac{\partial y}{\partial t}, x >_U - < \frac{\partial F}{\partial x} x, y >_U + < y, \frac{\partial F}{\partial x} x >_U = - < \nabla_x T, x >_U$$

$$\frac{\partial}{\partial t} < x, y >_U = - < \nabla_x T, x >_U .$$

Denoting $x_k = x(t_k)$ and $y_k = y(t_k)$, $k = 0,1$, and integrating the last expression between t_0 and t_1, one obtains

$$< x_1, y_1 >_V - < x_0, y_0 >_V = - \int_0^{t_1} < \nabla_x T, x >_U \, dt = - < \nabla_x J, x >_U .$$

Choosing now y such that $y(t_1) = y_1 = 0$, the expression above becomes: $< y_0, x_0 >_V = < \nabla_x J, x >_U$, for all $x \in U$.

What we want to emphasize here is the fact that in a vectorial space U, endowed with an inner product, it is possible to represent the differential of a linear functional f defined in U in the following form: $f'(x) \cdot v = < \nabla f(x), v >_U$, for all $v \in U$, what has been used in the expression of the differential of the functional J above. Therefore, as well as x_0 is the projection of x onto the vectorial space V, y_0 is the projection of $\nabla_x J$ over V, which can be denoted by $\nabla_{x_0} J$, whence $y_0 = \nabla_{x_0} J$.

Therefore, the control of the functional J with respect to the data in U (initial conditions) is obtained from the adjoint problem $(P)^*$ of the problem (P). The adjoint problem $(P)^*$ is defined as follows: Find the solution of

$$(P)^* \qquad \begin{cases} -\frac{\partial y}{\partial t} - \left(\frac{\partial F}{\partial x} \right)^* y = \nabla_x T, \text{ that verifies the condition} \\ y(t_1) = 0. \end{cases}$$

We refer to Silva et al. (2002) for details concerning the interpretation of the adjoint formalism and applications. Also, for a complete description of the various assumptions used by the data assimilation techniques, we refer to Wang et al. (2001). A rigorous mathematical framework of the adjoint parameter estimation, identifiability issues and regularization techniques are presented in Navon (1998), and Dimitriu (1999).

3. A 3D ATMOSPHERIC TRANSPORT-CHEMISTRY MODEL

We consider a 3D atmospheric transport-chemistry model given by the system of differential equations (Daescu et al., 2000):

$$\frac{\partial}{\partial t} c_i = -\nabla \cdot (\mathrm{u} c_i) + \nabla \cdot (\mathrm{K} \cdot \nabla c_i) + f_i(c) + E_i, \qquad i = 1, \ldots, S. \tag{1}$$

The initial condition is $c(t_0) = c_0$ and appropriate boundary values are prescribed. The solution $c(t,x,y,z) \in I^{S}$ of problem (1) represents the concentration vector of the chemical species in the model, u is the wind field and K is the eddy diffusivity tensor; the chemical reactions are governed by the nonlinear stiff terms $f_i(c) = P_i(c) - D_i(c)c_i$, with $P_i(c)$, $D_i(c)$ the chemical production and destruction terms; E_i represents the source term, and depositions are modeled as a boundary condition at the surface of the earth: $-n_h \cdot (\mathrm{K} \cdot \nabla c_i) = Q_i - v_i c_i$ with n_h the inward vector normal to the surface, Q_i and v_i the surface emission rate and deposition velocity of species i, respectively. We will refer to problem (1) as the forward model and $c(t)$, $t \in [t_0, T]$ will represent the "forward trajectory". Typical choices for the set of control variables in data assimilation are the boundary values, initial concentrations, emissions and deposition rates.

We consider here the 4D variational data assimilation problem associated to (1) with the set of control variables given by the initial state of the model, c_0. Under suitable assumptions, forward model (1) has a unique solution, viewed as a function of the initial conditions, $c = c(x, y, z, c_0)$.

If space discretization is applied to problem (1) on a grid (N_x, N_y, N_z), the resulting ODE system of dimension $N = S \times N_x \times N_y \times N_z$ is:

$$\frac{dc}{dt} = F_A(c) + F_D(c) + F_R(c), \qquad c(t_0) = c_0, \tag{2}$$

where F_A represents the advection and horizontal diffusion, F_D is the vertical diffusion, and the reaction terms are given by F_R. We assume that a "background estimate" c^b of c_0 with the error covariance matrix B, has been generated from the output of a previous analysis, using some assumptions of consistency in time of the model state, like stationarity (hypothesis of persistence) or the evolution predicted by a forecast model. The measurements c_k^0, $k = 1,\ldots,m$ of the concentrations at moments t_k are scattered over the interval $[t_0, T]$. The errors in measurements and model representativeness are given by the covariance matrices R_k, $k = 1,\ldots,m$. The covariance matrices B and R_k are symmetric and positive definite such that B^{-1}, R_k^{-1} are well defined. In practice, B and R_k are often taken diagonal, which corresponds to the assumption that there is no spatial and chemical correlation in the background errors, and measurement and model errors are uncorrelated in space and time.

The 4D-Var data assimilation determines an initial state c_0 that minimizes the distance between the model predictions and observations expressed by the cost function:

$$F(c_0) = \frac{1}{2}(c_0 - c^b)^T \mathrm{B}^{-1}(c_0 - c^b) + \frac{1}{2}\sum_{k=1}^{m}(c_k - c_k^0)^T \mathrm{R}_k^{-1}(c_k - c_k^0) \quad (3)$$

Most of the powerful optimization techniques require the evaluation of the gradient $\nabla_{c_0} F$ of the cost function. In a comprehensive atmospheric chemistry model the dimension of the vector c_0 can easily be of order 10^6, which makes the optimization a very expensive computational process.

In the variational approach, the gradient of the functional F is computed by using the "adjoint method". The general theory of adjoint equations and the derivation of the adjoint model are given in Marchuk et al. (1996), Giering and Kaminski (1998), and Wang et al. (2001). Below we sketch the basic ideas following Daescu et al. (2000). The gradient of the cost function is

$$\nabla_{c_0} F(c_0) = \mathrm{B}^{-1}(c_0 - c^b) + \sum_{k=1}^{m}\left(\frac{\partial c_k}{\partial c_0}\right)^T \mathrm{R}_k^{-1}(c_k - c_k^0). \quad (4)$$

Using now the chain rule $(\partial c_k / \partial c_0)^T = (\partial c_{k-1} / \partial c_0)^T (\partial c_k / \partial c_{k-1})^T$ in its transpose form, we can construct the algorithm to compute the necessary gradient:

Step 1. Initialize $gradient = 0$

Step 2. for $k = m, 1, -1$ do

$$gradient = (\partial c_k / \partial c_{k-1})^T [\mathrm{R}_k^{-1}(c_k - c_k^0) + gradient]$$

Step 3. $gradient = \mathrm{B}^{-1}(c_0 - c^b) + gradient$

The main advantage of the adjoint method is that explicit computation of the Jacobian matrices $\partial c_k / \partial c_{k-1}$ is avoided and the matrix-vector products can be computed directly at Step 2 (Giering and Taminski, 1998).

Problem (2) is usually solved by using the operator splitting (Bartholy et al., 2001; Dimov et al., 2001; Botchev et al., 2004). This technique has the advantage that processes such as advection, vertical diffusion and chemical reactions can be treated with different numerical schemes. In a second order accurate Strang splitting approach (Strang, 1968) with the time step $h = t_{n+1} - t_n$ the solution c_{n+1} is obtained from c_n as follows:

$$c_{n+1} = \overline{F}_A\left(t_{n+\frac{1}{2}}, \frac{h}{2}\right)\overline{F}_D\left(t_{n+\frac{1}{2}}, \frac{h}{2}\right)\overline{F}_R(t_n, h)\overline{F}_D\left(t_n, \frac{h}{2}\right)\overline{F}_A\left(t_n, \frac{h}{2}\right)c_n \tag{5}$$

where the operators $\overline{F}$ are defined by the numerical method used to solve the corresponding processes. If $\overline{J}$ denotes the Jacobian matrix associated to $\overline{F}$, the adjoint algorithm to compute the gradient (4) of the cost function requires products of the form $\overline{J}^T u$, with u an arbitrary seed vector. For large systems, constructing the adjoint code by hand can be a frustrating process and raises questions of technical implementation. In this respect, automatic tools have been developed (Damian-Iordache and Sandu, 1995; Giering and Kaminski, 1998).

4. ADJOINT COMPUTATIONS FOR A 3-STAGE ROSENBROCK METHOD

Following the basic ideas in (Daescu et al., 2000) applied to a 2-stage Rosenbrock method, this section develops the adjoint formulas for a general 3-stage Rosenbrock method targeting at an efficient implementation, suitable for automatization. We consider now the problem:

$$\frac{dc}{dt} = f(c), \qquad c(t_0) = c_0 \tag{6}$$

with $c(t)$, $c_0 \in IR^n$ and $f : IR^n \to IR^n$, $f = (f_1, f_2, \ldots, f_n)^T$. One step from t_0 to t_1 with $h = t_1 - t_0$ of a 3-stage Rosenbrock method reads:

$$\begin{aligned}
&(\frac{1}{\gamma_{11}h}I - J_0)k_1 = f(c_0) \\
&(\frac{1}{\gamma_{22}h}I - J_0)k_2 = f(c_0 + \alpha_{21}k_1) + \frac{\beta_{21}}{h}k_1 \\
&(\frac{1}{\gamma_{33}h}I - J_0)k_3 = f(c_0 + \alpha_{31}k_1 + \alpha_{32}k_2) + \frac{\beta_{31}}{h}k_1 + \frac{\beta_{32}}{h}k_2 \\
&c_1 = c_0 + m_1k_1 + m_2k_2 + m_3k_3 \\
&\hat{c}_1 = c_0 + \hat{m}_1k_1 + \hat{m}_2k_2 + \hat{m}_3k_3
\end{aligned} \tag{7}$$

where J_0 is the Jacobian matrix of f evaluated at c_0, $J_0 = (\partial f_i / \partial c_j)_{ij} |_{c=c_0}$ and the coefficients γ_{11}, γ_{22}, γ_{33}, α_{21}, α_{31}, α_{32}, β_{21}, β_{31}, β_{32}, m_1, m_2, m_3 are chosen to obtain a desired order of consistency and numerical stability (see Table 1, Lang and Verwer, 2001).

γ = 7.886751345948129e-01	
α_{21} = 1.267949192431123e+00	β_{21} = -1.607695154586736e+00
α_{31} = 1.267949192431123e+00	β_{31} = -3.464101615137755e+00
α_{32} = 0.000000000000000e+00	β_{32} = -1.732050807568877e+00
m_1 = 2.000000000000000e+00	$\hat{m}_1$ = 2.113248654051871e+00
m_2 = 5.773502691896258e-01	$\hat{m}_2$ = 1.000000000000000e+00
m_3 = 4.226497308103742e-01	$\hat{m}_3$ = 4.226497308103742e-01

Table 1: Set of coefficients for the 3-stage ROS3P method.

We consider the case when $\gamma_{11} = \gamma_{22} = \gamma_{33} = \gamma$, since of special interest are the methods that require only one LU decomposition of $(1/\gamma_{ii})I - J_0$ per step. The difference $c_1 - \hat{c}_1$ can be used as a local error estimator. For the adjoint computations we have from (7):

$$\left(\frac{\partial k_1}{\partial c_0}\right)^T = \left(J_0^T + (\frac{\partial J_0}{\partial c_0} \times k_1)^T\right)\left((\frac{1}{\gamma h}I - J_0)^T\right)^{-1}$$

$$\left(\frac{\partial k_2}{\partial c_0}\right)^T = \left((I + \alpha_{21}\frac{\partial k_1}{\partial c_0})^T J_1^T + \frac{\beta_{21}}{h}(\frac{\partial k_1}{\partial c_0})^T + \left(\frac{\partial J_0}{\partial c_0} \times k_2\right)^T\right) \cdot$$

$$\left((\frac{1}{\gamma h}I - J_0)^T\right)^{-1} \qquad (8)$$

$$\left(\frac{\partial k_3}{\partial c_0}\right)^T = \left((I + \alpha_{31}\frac{\partial k_1}{\partial c_0})^T J_2^T + \frac{\beta_{31}}{h}(\frac{\partial k_1}{\partial c_0})^T + \frac{\beta_{32}}{h}(\frac{\partial k_2}{\partial c_0})^T + \left(\frac{\partial J_0}{\partial c_0} \times k_3\right)^T\right) \cdot$$

$$\left((\frac{1}{\gamma h}I - J_0)^T\right)^{-1}$$

where J_1 is the Jacobian evaluated at $c_0 + \alpha_2 k_1$, J_2 is the Jacobian evaluated at $c_0 + \alpha_{31}k_1 + \alpha_{32}k_2$ and the terms $(\partial J_0/\partial c_0) \times k_i$, $i = 1,2,3$ are $n \times n$ matrices whose j th column is $(\partial J_0/\partial c_0^j)k_i$, $i = 1,2,3$. We took into account in the evaluation of the last equality in (8) that $\alpha_{32} = 0$ (see Table 1). We want to point out here the fact that these matrices are not symmetric. From (7) and using (8), for an arbitrary seed vector $u \in IR^n$, we have:

$$\left(\frac{\partial c_1}{\partial c_0}\right)^T u = u + m_1\left(J_0^T + (\frac{\partial J_0}{\partial c_0} \times k_1)^T\right)\left((\frac{1}{\gamma h}I - J_0)^T\right)^{-1} u$$

$$+ m_2\left((I + \alpha_{21}\frac{\partial k_1}{\partial c_0})^T J_1^T + \frac{\beta_{21}}{h}(\frac{\partial k_1}{\partial c_0})^T + \left(\frac{\partial J_0}{\partial c_0} \times k_2\right)^T\right)\left((\frac{1}{\gamma h}I - J_0)^T\right)^{-1} u$$

$$+ m_3\left((I + \alpha_{31}\frac{\partial k_1}{\partial c_0})^T J_2^T + \frac{\beta_{31}}{h}(\frac{\partial k_1}{\partial c_0})^T + \frac{\beta_{32}}{h}(\frac{\partial k_2}{\partial c_0})^T + \left(\frac{\partial J_0}{\partial c_0} \times k_3\right)^T\right) \cdot$$

$$\left((\frac{1}{\gamma h}I - J_0)^T\right)^{-1} u.$$

We are going to show that the computations can in fact be arranged in a much more efficient way. In order to avoid frequent recomputations and to exploit the particular properties of the method, the order of the operations in the formula above becomes important. Below we describe the following efficient algorithm:

Step 1. Solve for v the linear system $(\frac{1}{\gamma h}I - J_0)^T v = u$. Then,

$$\left(\frac{\partial c_1}{\partial c_0}\right)^T u = u + m_1\left(J_0^T + (\frac{\partial J_0}{\partial c_0}\times k_1)^T\right)v + m_2 J_1^T v + m_2\left(\frac{\partial J_0}{\partial c_0}\times k_2\right)^T v$$
$$+ m_2(\frac{\partial k_1}{\partial c_0})^T(\alpha_2 J_1^T + \frac{\beta_{21}}{h}I)v + m_3(\frac{\partial k_1}{\partial c_0})^T(\alpha_3 J_2^T + \frac{\beta_{31}}{h}I)v \qquad (9)$$
$$+ m_3 J_2^T v + m_3\frac{\beta_3}{h}(\frac{\partial k_2}{\partial c_0})^T v + m_3\left(\frac{\partial J_0}{\partial c_0}\times k_3\right)^T v$$

Step 2. Compute $\omega_1 = J_1^T(m_2 v)$; $\omega_{11} = \alpha_{21}\omega_1 + (m_2\beta_{21}/h)v$;
$\omega_2 = J_2^T(m_3 v)$; $\omega_{22} = \alpha_{31}\omega_2 + (m_3\beta_{31}/h)v$;
$\omega_3 = J_1^T((m_3\beta_{32}/h)v)$; $\omega_{33} = \alpha_{21}\omega_3 + (m_3\beta_{21}/h)v$;
$\omega = \omega_1 + \omega_2 + \omega_3$.

Step 3. Solve for θ_1, θ_2, and θ_3 the linear systems: $(\frac{1}{\gamma h}I - J_0)^T\theta_1 = \omega_{11}$, $(\frac{1}{\gamma h}I - J_0)^T\theta_2 = \omega_{22}$, and $(\frac{1}{\gamma h}I - J_0)^T\theta_3 = \omega_{33}$ respectively . Then, compute $\theta = \theta_1 + \theta_2 + \theta_3$. After replacing in (9), one obtains:

$$\left(\frac{\partial c_1}{\partial c_0}\right)^T u = u + m_1\left(J_0^T + (\frac{\partial J_0}{\partial c_0}\times k_1)^T\right)v + \omega_1 + \left(J_0^T + (\frac{\partial J_0}{\partial c_0}\times k_1)^T\right)\theta_1$$
$$+ m_2\left(\frac{\partial J_0}{\partial c_0}\times k_2\right)^T v + \omega_2 + \left(J_0^T + (\frac{\partial J_0}{\partial c_0}\times k_1)^T\right)\theta_2 \qquad (10)\,\mathrm{a}$$
$$+ m_3\left(\frac{\partial J_0}{\partial c_0}\times k_3\right)^T v + \omega_3 + \left(J_0^T + (\frac{\partial J_0}{\partial c_0}\times k_1)^T\right)\theta_3$$

nd after arranging the terms we obtain:

Step 4. Compute

$$\left(\frac{\partial c_1}{\partial c_0}\right)^T u = u + \omega + J_0^T(m_1 v + \theta) + (\frac{\partial J_0}{\partial c_0} \times k_1)^T (m_1 v + \theta)$$
$$+ m_2(\frac{\partial J_0}{\partial c_0} \times k_2)^T v + m_3(\frac{\partial J_0}{\partial c_0} \times k_3)^T v. \qquad (11)$$

5. COMPUTATIONAL ISSUES AND CONCLUSION

From relation (11) we can deduce that is enough to call the routine to compute the product $J_0^T s$ (s a seed vector) once, with the seed vector $m_1 v + \theta$. The same remark is made for the products $[(\partial J_0 / \partial c_0) \times k_1]^T m_1 v$ and $[(\partial J_0 / \partial c_0) \times k_1]^T \theta$.

We now concentrate on the terms $[(\partial J_0 / \partial c_0) \times k]^T v$ whose evaluation dominate the computational cost of the algorithm given by *Steps* 1-4. Here $k, v \in IR^n$ are arbitrary constant vectors. For the i th component we have:

$$\left((\frac{\partial J_0}{\partial c_0} \times k)^T v\right)_i = \left(\frac{\partial J_0}{\partial c_0^i} k\right)^T v = k^T\left(\frac{\partial (J_0^T v)}{\partial c_0^i}\right) = \left(\frac{\partial (J_0^T v)}{\partial c_0^i}\right)^T k \qquad (12)$$

Consider now the function $g: IR^n \to IR^n$, $g(c_0) = J_0^T v$. Using (7) results in:

$$[(\partial J_0 / \partial c_0) \times k]^T v = [\partial g(c_0) / \partial c_0] k .$$

Notice that the Jacobian matrix of g is symmetric (for details, see Daescu et al., 2000). The symmetry of the Jacobian matrix of the function g plays a significant role in the implementation of the adjoint code.

The integration of the forward model (1) using implicit methods, together with the performance analysis was carried out in Sandu et al. (1997), proving that when the sparsity of the system is efficiently exploited, Rosenbrock methods outperform traditional explicit methods like QSSA and CHEMEQ. Implementation of the Rosenbrock methods can be done in the symbolic kinetic preprocessor KPP environment (Damian-Iordache and Sandu, 1995), which generates the sparse matrix factorization LU required in (2), (3) and the routine to forward-backward solve the linear systems without indirect addressing.

Step 2 claims the evaluation of the product $J_1^T v$, which is automatically computed by KPP using sparse multiplications. This task introduces some extra work (J_1 is evaluated at $c_0 + \alpha_2 k_1$ and J_2 is evaluated at $c_0 + \alpha_{31}k_1 + \alpha_{32}k_2$), but its cost is appreciated as relatively low. The efficiency of the adjoint code is then dominated by the implementation of *Step* 4, given by the formula (11). The computation of the terms of the form $[(\partial J_0 / \partial c_0) \times k]^T v$ in formula (11) appears to require the following order: forward automatic mode to compute $J_0 k$, reverse mode to compute $[(\partial J_0 / \partial c_0) \times k]^T v$. Relation (12) can be used to rewrite (11) as:

$$\left(\frac{\partial c_1}{\partial c_0}\right)^T u = u + \omega + g_1(c_0) + \left(\frac{\partial g_1(c_0)}{\partial c_0}\right) k_1 + m_2 \left(\frac{\partial g_2(c_0)}{\partial c_0}\right) k_2 + m_3 \left(\frac{\partial g_2(c_0)}{\partial c_0}\right) k_3 \tag{13}$$

with $g_1, g_2 : IR^n \to IR^n$, $g_1(c_0) = J_0^T (m_1 v + \theta)$, $g_2(c_0) = J_0^T v$.

The functions g_1, g_2 can be generated via KPP, taking full advantage of the sparsity of the Jacobian matrix J_0. In (13) we have then to compute the $Jacobian \cdot vector$ products for the functions g_1, g_2 which can be done by *forward* automatic differentiation (Rostaing, 1993). By default, during the computation of $Jacobian \cdot vector$, forward automatic differentiation computes the value of the function. Automatic differentiation for g_1 provides then the value $g_1(c_0)$ and there is no need to compute it separately. We also notice that the computations related to g_1 and g_2 are independent, allowing parallel implementation (Ostromsky and Zlatev, 2003).

To conclude, the algorithm presented in Section 4 has the great benefit that the adjoint of the chemistry integration can be generated completely automatically, taking full advantage of the sparsity of the system. Moreover, since symbolic computations can be used, rounding errors are avoided and the accuracy of the results goes up to the machine precision.

Finally, we want to mention that the recent trend in data assimilation is the adaptive location of the observations by using the adjoint sensitivity method, as well as to combine the advantages of 4D-Var and the Kalman filter techniques. However, one should not forget the importance of other issues like physical consistency of the assimilation, including bias correction, which can be of great importance for the quality of the assimilation system taken as a whole.

Acknowledgement: The work is partially supported by NATO Collaborative Linkage Grant 980505 / NATO ARW project "Impact of future climate changes on pollution levels in Europe".

REFERENCES

Bartholy, J., Faragó, I., Havasi, Á., 2001, Splitting Method and its Application in Air Pollution Modeling, *Időjárás* **105**, pp. 39-58.

Botchev, M., Faragó, I., and Havasi, Á., 2004, Testing Weighted Splitting Schemes on a One-Column Transport-Chemistry Model, I. Lirkov et al. (Eds.): LSSC 2003, LNCS 2907, pp. 295-302.

Dimitriu, G., 1999, *Parameter Identification for Some Classes of Nonlinear Systems*, PhD thesis, (in Romanian).

Dimov, I., Faragó, I., Havasi, Á., and Zlatev, Z., 2001, L-Commutativity of the Operators in Splitting Methods for Air Pollution Models, *Annales Univ. Sci. Sec. Math.* **44**, pp. 127-148.

Daescu, D., Carmichael, G. R., and Sandu, A., 2000, Adjoint Implementation of Rosenbrock Methods Applied to Variational Data Assimilation Problems, *J. Comput. Phys.* **165**, No. 2, pp. 496-510.

Damian-Iordache, V., and Sandu, A., 1995, KPP - A Symbolic Preprocessor for Chemistry Kinetics – User's guide. Tech. Rep., Univ. of Iowa, Department of Mathematics.

Giering, R., and Kaminski, T., 1998, *Recipes for Adjoint Code Construction*, ACM Trans. Math. Software.

Lang, J., and Verwer, J., 2001, ROS3P – An Accurate Third-Order Rosenbrock Solver Designed for Parabolic Problems, *BIT* **41**, No. 4, pp. 731-738.

Marchuk, G. I., Agoshkov, I. V., and Shutyaev, P. V., 1996, Adjoint Equations and Perturbation Algorithms in Nonlinear Problems, CRC Press.

Navon, I. M., 1998, Practical and Theoretical Aspects of Adjoint Parameter Estimation and Identifiability in Meteorology and Oceanography, *Dynamics of Atmospheres and Oceans.* Special Issue in honor of Richard Pfeffer **27**, No. 1-4, pp. 55-79.

Ostromsky, T., and Zlatev, Z., 2003, Flexible Two-Level Parallel Implementations of a Large Air Pollution Model, I. Dimov et al. (Eds.): NMA 2002, LNCS 2542, pp. 545-554.

Rostaing, N., Dalmas, S., and Galligo, A., 1993, Automatic Differentiation in Odysee. *Tellus* **45**, pp. 558-568.

Sandu, A., Verwer, J. G., Blom, J. G., Spee, E. J., and Carmichael, G. R., 1997, Benchmarking Stiff ODE Solvers for Atmospheric Chemistry Problems II: Rosenbrock Solvers, *Atmos. Environ.* **31**, pp. 3459-3472.

Silva, P. S. D., Navon, I. M., and Landau, L., 2002, Adjoint Formalism and the Configuration Retrieval of an Evolutionary System Represented by Burgers' Equations, Pre-print.

Strang, G., 1968, On the Construction and Comparison of Difference Schemes, *SIAM Journal on Numerical Analysis* **5**, pp. 506-517.

Talagrand, O., and Courtier, P., 1987, Variational Assimilation of Meteorological Observations with the Adjoint of the Vorticity Equations. Part I. Theory. *Q. J. R. Meteorol. Soc.* **113**, pp. 1311-1328.

Wang, K. Y., Lary, D. J., Shallcross, D. E., Hall, S. M., and Pyle, J. A., 2001, A Review on the Use of the Adjoint Method in Four-Dimensional Atmospheric-Chemistry Data Assimilation. *Q.J.R. Meteorol. Soc.* **127** (576) Part B, pp. 2181-2204.

FIGHTING THE GREAT CHALLENGES IN LARGE-SCALE ENVIRONMENTAL MODELLING

Ivan Dimov [(1)], Gerald Geernaert [(2)] and Zahari Zlatev [(3)]

[(1)] *Insitute for Parallel Processing, Bulgarian Academy of Sciences, Sofia, Bulgaria;* [(2)] *Insitute of Geophysics and Planetary Physics, Los Alamos National Laboratory, MS C-305, Los Alamos, NM 87545,USA;* [(3)] *National Environmental Research Institute, Frederiksborgvej 399, P. O. Box 358, DK-4000 Roskilde, Denmark*

Abstract: High pollution levels may cause different damages. Therefore, critical levels have been established by different authorities (such as WHO, EU and environmental agencies in many countries). The critical levels should not be exceeded in the efforts to avoid damaging effects from harmful pollutants. Mathematical models can potentially be very useful tools when efficient measures have to be taken in order to bring the pollution levels under the prescribed critical levels and/or to keep them there. However, there are still many unresolved problems connected with the application of mathematical models in the attempts to design reliable and robust control strategies for reduction of the pollution to the critical levels and/or for keeping the pollution under the prescribed critical levels. Three of the unresolved problems, which are very challenging, will be discussed in this paper.

Key words: High pollution levels, mathematical models, fine resolution grids, biogenic emissions, climatic changes.

1. THREE CHALLENGING PROBLEMS

The air pollution levels should be kept under prescribed critical levels in order to avoid different damaging effects. This a great challenge imposing requirements for (a) more details about the air pollution levels and, what is even more important, (b) more reliable information about trends and

I. Farago et al. (eds.), Advances in Air Pollution Modeling for Environmental Security, 105–114.

relationships related to the high air pollution levels. Mathematical models are indispensable tools in the efforts to resolve successfully these two tasks. However, there are still many unresolved problems in air pollution modelling. Three such problems will be discussed in this paper:

- running high-resolution mathematical models,
- studying the influence of biogenic emissions on high ozone levels,
- investigating the effect of climate changes on the pollution levels.

2. RUNNING HIGH-RESOLUTION MODELS

It is necessary to run the mathematical models on high-resolution grids in order to obtain more details about the air pollution levels in a large space domain (the whole of Europe). This leads to huge computational tasks, which are very challenging even when modern high-speed computers are available. This is illustrated in Table 1, where computing times, which were obtained by running the 10-layer UNI-DEM on a coarse 50 km x 50 km horizontal grid and on a fine 10 km x 10 km horizontal grid, are given. UNI-DEM stands for "Unified Danish Eulerian Model" (see Dimov and Zlatev, 2004). This model is an improved version of the Danish Eulerian Model (DEM), which is discussed in detail in Zlatev (1995). The runs were performed on 32 processors of a Sun Fire 15k computer at DCSC (the Danish Centre for Scientific Computing, see WEB-site DCSC, 2004).

Grid	Number of equations	Number of time-steps	Computing time
96x96x10	3 225 600	35 520	2.48
480x480x10	80 640 000	213 120	122.83

Table 1: Results obtained by running UNI-DEM for 1997 on 32 processors of a Sun Fire 15k computer. The times are given in CPU hours. Some details about the performance of UNI-DEM on parallel architectures can be found in Alexandrov et al. (2004).

It is clearly seen that UNI-DEM can be run on fast PCs when the coarse grid is used, while this is practically not possible when the fine grid is used (some crude computations indicate that the big job, UNI-DEM discretized on a 480x480x10 grid, will run about half a year on a fast PC).

The results given in Figure 1 and Figure 2 explain why fine resolution is highly desirable. The upper right-hand-side plots in Figure 1 and Figure 2 show the distribution of the $\mathbf{NO_2}$ concentrations and $\mathbf{SO_2}$ concentrations in Europe for 1997. The results are obtained by the fine resolution option of UNI-DEM. Very similar plots were obtained (these plots are not given here) when the coarse grid is used. The conclusion is that if one is only interested in the distribution of the pollution levels on a very large space domain (in this case, the domain containing the whole of Europe), then computations on a coarse grid could give sufficiently good results.

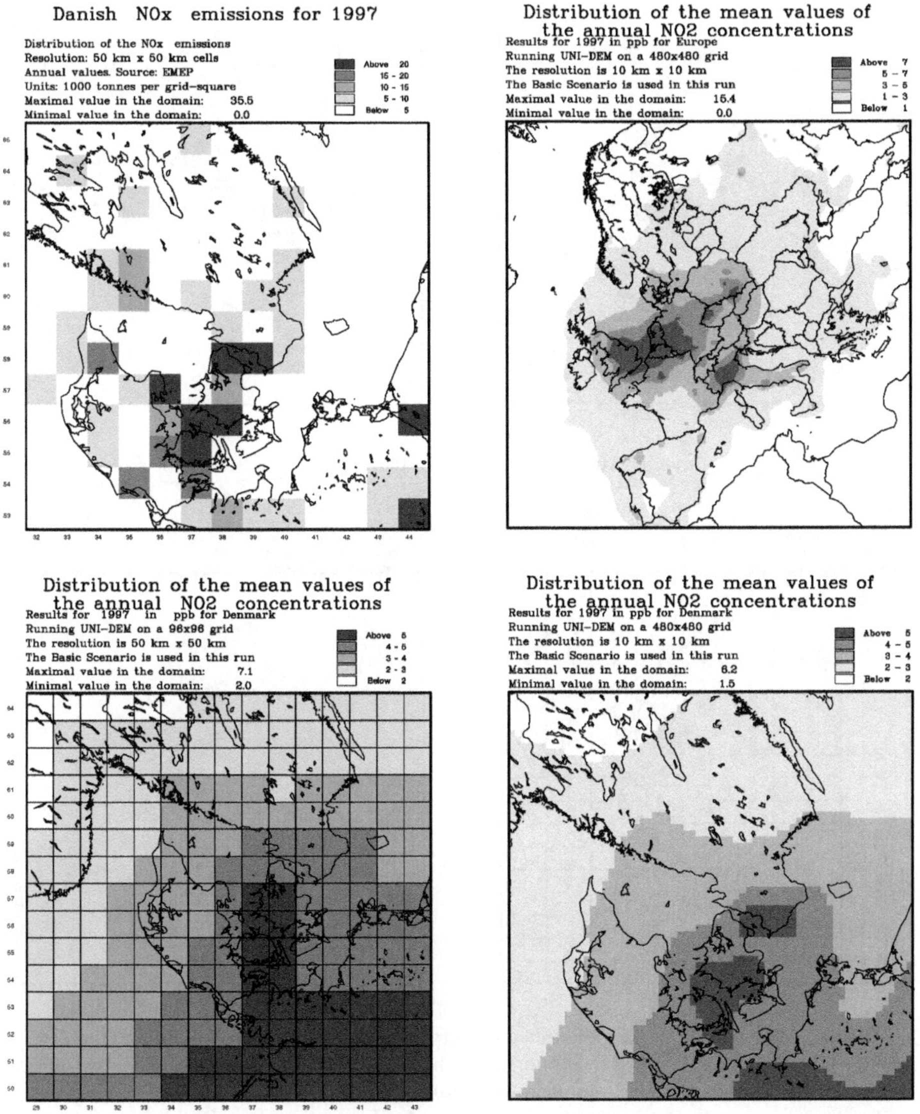

Figure 1: Results for nitrogen dioxide concentrations in Denmark, which are obtained by running UNI-DEM on two grid resolutions, are shown in the lower two plots. The distribution of the nitrogen dioxide concentrations in the whole of Europe is given, for the fine resolution grid, on the upper right-hand-side plot. Emissions of nitrogen oxides in Denmark are given on the upper left-hand-side plot.

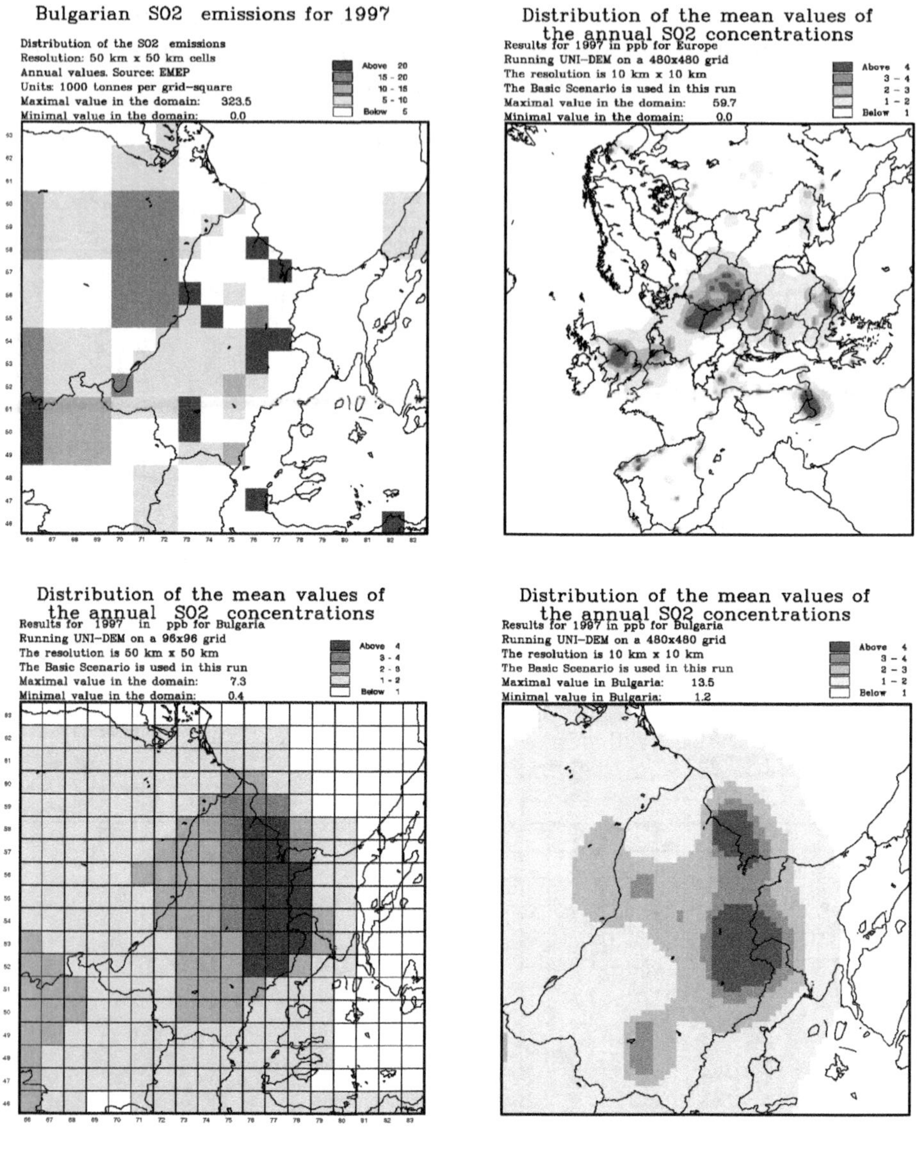

Figure 2: Results for sulphur dioxide concentrations in Bulgaria, which are obtained by running UNI-DEM on two grid resolutions, are shown in the lower two plots. The distribution of the sulphur dioxide concentrations in the whole of Europe is given, for the fine resolution grid, on the upper right-hand-side plot. Emissions of sulphur dioxide in Bulgaria are given on the upper left-hand-side plot.

The situation changes when some details on a small area are needed. The two plots on the upper left-hand-side of Figure 1 and Figure 2 show the locations of the major sources of (a) $\mathbf{NO_x}$ emissions in Denmark (Figure 1) and (b) $\mathbf{SO_2}$ emissions in Bulgaria (Figure 2). The model should reproduce the pattern (showing high concentrations in the same locations). It is seen on the lower left-hand-side plots on Figure 1 and Figure 2 that the pattern is not correctly reproduced when the coarse grid is used (the pattern produced by the code is smoothed in this case). On the other hand, the lower right-hand-side plots in Figure 1 and Figure 2 show that the pattern is well reproduced when the fine resolution grid is used. The SO_2 emissions in Bulgaria are very high. Therefore high gradients should be expected. Comparing the lower plots in Figure 2, it is clear that higher gradients are obtained when the fine resolution is used (the highest level of the SO_2 concentrations in Bulgaria for 1997 when the fine grid is used is twice higher than that obtained by using the coarser grid).

3. INFLUENCE OF BIOGENIC EMISSIONS ON HIGH OZONE LEVELS

It is well known that the production of inventories for biogenic emissions is connected with many uncertainties (see Simpson et al., 1995). However, there are two very important facts: (a) the anthropogenic emissions in Europe have been decreased during the last two decades and they will further be decreased in the next several years (see Table 2) and (b) the major part of the biogenic emissions is emitted during summer time (see Table 3). This implies that (i) the relative part of the biogenic emissions is increasing, (ii) in year 2010 the amount of the biogenic VOC emissions will be comparable with the amount of the anthropogenic emissions and (iii) the influence of the biogenic emissions on summer time concentrations of related compounds (as ozone) will be increasing. Therefore, the use of scenarios involving only anthropogenic emissions is probably not sufficient when control strategies are to be designed and used.

The sensitivity of the quantities related to high ozone concentrations to the variations of both anthropogenic and biogenic emissions has been studied in Geernaert and Zlatev (2004) by using 24 scenarios. The importance of the biogenic emissions is demonstrated in Figure 3. The quantity actually studied was AOT40 (accumulated over threshold of 40 ppb ozone). The AOT40 values were accumulated for the period May-June-July. If these values exceed 3000 ppb.hours, then this leads to corn loses (see Zlatev et al., 2001). The areas, where the critical value of 3000 ppb.hours is exceeded, are shown on Figure 3 for four scenarios. Four major conclusions can be drawn from the results shown in the four plots of Figure 3.

Year	NOx emis. (1000 tonnes/year)	VOC emis. (1000 tonnes/ year)
1990	28040	26751
1995	23369	21766
1998	21186	19854
2010	14979	13799

Table 2: Trends for anthropogenic emissions in Europe.

Month	Forests	In percent	Crops	In percent
January	265	2.3 %	0.606	0.38 %
February	260	2.3 %	0.716	0.45 %
March	327	2.8 %	1.150	0.72 %
April	605	5.2 %	4.450	2.80 %
May	1411	12.3 %	20.800	13.10 %
June	2130	18.5 %	38.200	24.00 %
July	2341	20.3 %	43.900	27.60 %
August	1989	17.3 %	32.100	20.20 %
September	1015	8.8 %	11.300	7.10 %
October	658	5.7 %	3.800	2.40 %
November	295	2.6 %	1.130	0.71 %
December	217	1.9 %	0.672	0.42 %
1995	**11513**		**159.00**	

Table 3: Biogenic VOC emissions in Europe for 1995.

- Assume that the biogenic emissions are as those that are shown in Table 3. It is said that *normal biogenic emissions* are used when this is the case. The transition from the Basic Scenario (anthropogenic emissions and meteorology for 1995) to Scenario 2010 (predicted emissions for 2010 and meteorology for 1995) leads to a considerable reduction of the areas where the critical limit of 3000 ppb.hours for the AOT40 values is exceeded (compare the two left-hand-side plots in Figure 3).
- Assume that the biogenic emissions are about five times higher than those that are shown in Table 3. It is said that *high biogenic emissions* are used when this is the case. The transition from the Basic Scenario (anthropogenic emissions and meteorology for 1995) to Scenario 2010 (predicted anthropogenic emissions for 2010 and meteorology for 1995) leads, also in the case where *high biogenic emissions* are used, to a considerable reduction of the areas where the critical limit of 3000

ppb.hours is exceeded (compare the two right-hand-side plots in Figure 3).

- The increase of the biogenic emissions (by using the *high biogenic emissions* instead of the *normal biogenic emissions*) leads to a considerable increase of the areas where the critical limit of 3000 ppb.hours for the AOT40 values is exceeded (compare the upper two plots of Figure 3).

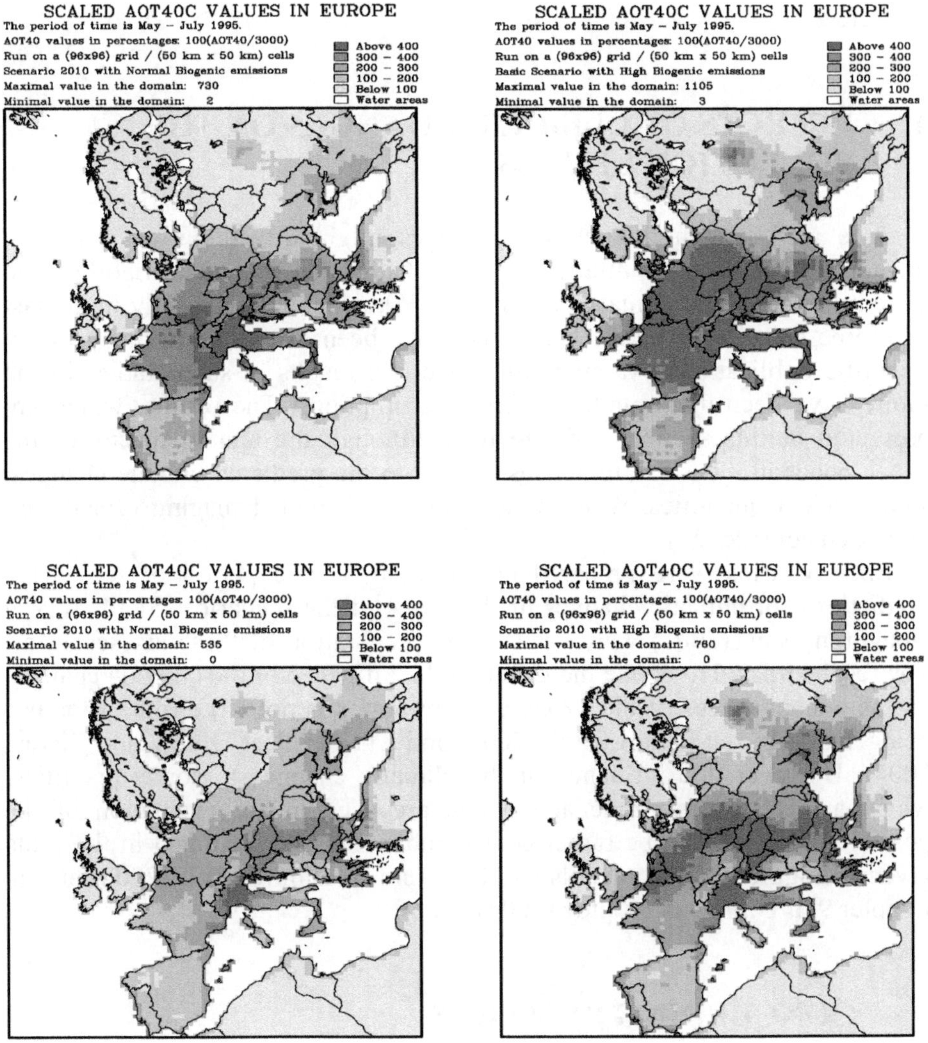

Figure 3: Distribution of AOT40 values in Europe obtained by different scenarios for anthropogenic and biogenic emissions.

- The comparison of the results in the upper left-hand-side plot of Figure 3 with the lower right-hand-side plot indicate that the reduction of the anthropogenic emissions (i.e. the use of Scenario 2010) combined by using *high biogenic emissions* is giving approximately the same results as the Basic Scenario when *normal biogenic emissions* are used. In other words, if the claims of some scientists that the biogenic emissions used at present are greatly underestimated (by a factor up to ten), then the reduction of the anthropogenic emissions only might not have the desired effect.

4. EFFECTS OF CLIMATE CHANGES ON THE POLLUTION LEVELS

It is well known that climate changes might affect our environment, including here the air pollution levels. Changes in climate variability and extreme weather and climate events in the 20th century, especially in the last two-three decades of the 20th century, have been discussed in many recent scientific publications. Attempts to project the results of such studies in the future have been made under different assumptions. The results, which are presented in this section, indicate that, although the annual means of the ozone concentrations are rather insensitive to the predicted climate changes, some related quantities, which might cause different damaging effects, are increased considerably.

Six scenarios were prepared and run on a time period of ten years. Predictions for climatic changes made in one of the scenarios (SRES A2) in Houghton et al. (2001) were used in the preparation of these six scenarios. The results related to ozone indicate that the influence of the climatic changes on the annual values of the ozone concentrations is not large (the difference being 1-3 ppb; see Dimov et al., 2002, and Chapter 9 of Dimov and Zlatev, 2004). However, the influence of the climatic changes to some quantities, which are related to ozone and which are used in the evaluation of the possible damages, can be rather considerable. This is illustrated in the plots given in Figure 4. More details can be found in Dimov et al. (2002) and in Chapter 9 of Dimov and Zlatev (2004).

5. CONCLUDING REMARKS

The results, which were presented in the previous sections of this paper show clearly that

- fine resolution computations will be very useful in many comprehensive studies,

- scenarios with biogenic emissions are to be included in many studies related to the attempts to keep the concentrations of damaging pollutants under prescribed critical levels

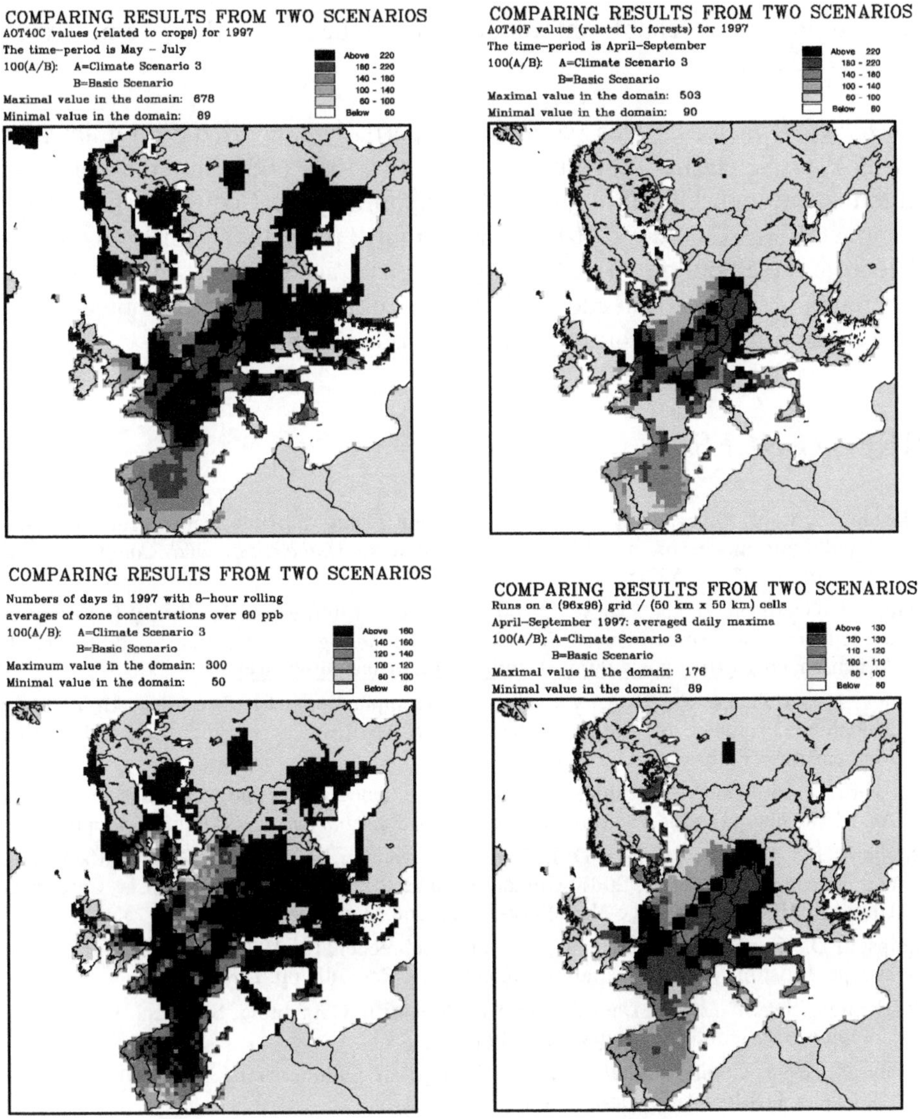

Figure 4: Comparison of results obtained by the Basic Scenario and the Climatic Scenario 3 for four quantities that are related to high ozone concentrations.

and

- the climate changes may have a significant influence on some quantities related to high ozone levels.

Much more experiments, especially experiments based on the use of the fine resolution options of UNI-DEM, are necessary in order to obtain more detailed and more reliable information about the cases where some critical levels are exceeded.

Acknowledgement: This research was partially supported by a grant from the NATO Scientific Programme (CRG 960505) and from the European Commission under the Programme for supporting the Centres of Excellence (contract ICA1-CT-2000-70016). Two grants, from the Danish Centre for Scientific Computing (DCSC) gave us access to many Danish supercomputers, both Sun and IBM parallel computers. These computers were used in the runs needed for the studies discussed in this paper.

REFERENCES

Alexandrov, V. N., Owczarz, W., Thomsen, P. G. and Zlatev, Z., 2004, Parallel runs of a large air pollution model on a grid of Sun computers, *Mathematics and Computers with Appliccations* **65**, pp. 557-577.

Dimov, I., Geernaert, G. and Zlatev, Z., 2002, Influence of future climate changes in Europe on exceeded ozone critical levels. In: Nordic Meteorological Meeting 23, 2002 (H. E. Jørgensen, ed.). **http://www.dams.dk/nmm2002/proceedings.htm**.

Dimov, I. and Zlatev, Z., 2004, Numerical and Computational Challenges in Air Pollution Modelling, Frontiers in Applied Mathematics, SIAM (Society for Industrial and Applied Mathematics), Philadelphia, to appear.

Geernart, G. and Zlatev, Z., 2004, Studying the influence of the biogenic emissions on the AOT40 levels in Europe, *International Journal of Environment and Pollution*, to appear.

Houghton J. T., Ding, Y., Griggs, D. J, Noguer, M., van der Linden, P. J., Dai, X., Maskell, K. and Johnson, C. A., eds., 2001, Climate Change 2001: The Scientific Basis, Cambridge University Press, Cambridge-New York-Melbourne-Madrid-Cape Town.

Simpson, D., Guenter, A., Hewitt, C. N. and Steinbrecher, R., 1995, Biogenic emissions in Europe: I. Estimates and uncertainties, *J. Geophys. Res.* **100**, pp. 22875-22890.

WEB-Site of DCSC (2002), Danish Centre for Scientific Computing, Sun High Performance Computing Systems, **http://www.hpc.dtu.dk**.

Zlatev, Z., 1995, Computational treatment of large air pollution models, Kluwer Academic Publishers, Dordrecht-Boston-London.

Zlatev, Z., Dimov, I. Ostromsky, Tz. Geernaert, G., Tzvetanov, I. and Bastrup-Birk, A., 2001, Calculating losses of crops in Denmark caused by high ozone levels, *Environmental Modeling and Assessment* **6**, pp. 35-55.

CHALLENGES IN USING SPLITTING TECHNIQUES FOR LARGE-SCALE ENVIRONMENTAL MODELING

Ivan Dimov[(1)], Tzvetan Ostromsky[(1)], and Zahari Zlatev[(2)]
[(1)]Institute for Parallel Processing, Bulgarian Academy of Sciences, Acad. G Bonchev str., bl. 25-A, 1113 Sofia, Bulgaria; [(2)]National Institute for Environmental Research, Frederiksborgvej 399, Roskilde, Denmark

Abstract: Splitting techniques are, to our knowledge, used in all operationally run large scale air pollution models with many scenarios. The modellers believe that really huge computational tasks can be made tractable on the available computers by dividing them into a sequence of smaller and simple sub-tasks. Probably, the first simple splitting procedure for partial differential equations was proposed by Bagrinovski and Godunov in 1957. Since then many different splitting schemes were proposed and studied. A significant progress in splitting analysis was done during the last 3-4 years when the Laplace transformations technique was replaced by more powerful techniques. Several splitting schemes for large scale air pollution models are analysed and tested in a set of numerical experiments with respect to the splitting error. The computational properties of splitting schemes are analysed from algorithmic point of view. Parallel properties of these splitting schemes are also discussed.

Key words: Air quality, weather, numerical forecast, atmospheric chemistry, atmospheric transport.

1. INTRODUCTION

The application of splitting procedures in the treatment of large scientific and engineering problems is an excellent tool (and, very often, the only tool) by which huge computational tasks can be made tractable on the available computers. This is achieved by dividing the original problem into a sequence of simpler tasks.

I. Farago et al. (eds.), Advances in Air Pollution Modeling for Environmental Security, 115–131.

A simple splitting procedure for partial differential equations (PDE) was proposed, as an example, by Bagrinovskii and Godunov (1957). This was probably the first attempt to introduce the splitting technique. Different splitting procedures have been developed and/or studied in many scientific papers, for example: Dyakonov (1962), Yanenko (1962), Marchuk (1968), Strang (1968), Penenko and Obraztsov (1976), Tikhonov and Samarski (1977), Marchuk (1980,1986), Lanser and Verwer (1999), Faragó and Havasi (2002), Csomós et al. (2003). A detailed theoretical study and analysis of some splitting procedures can be found in Marchuk (1980), Marchuk (1988), Faragó and Havasi (2002), Faragó et al. (2002). Some convergence results were recently presented in Faragó and Havasi (2003).

Splitting techniques are successfully used in large-scale air pollution modeling (see McRae (1982), Marchuk (1986), Zlatev (1995), Dimov et al. (1999,2001), Hunsdorfer and Verwer (2003)). Results related to the use of some splitting procedures in a large air pollution model (DEM in particular) as well as the advantages and disadvantages of these procedures are discussed in this paper.

2. GENERAL DESCRIPTION OF SPLITTING PROCEDURES

Consider the differential equation

$$\frac{\partial c(x,t)}{\partial t} = A(c(x,t)) \qquad (1)$$

where $t \in [0, T]$, $x \in D$ and A is a differential operator.
We shall assume that operator $A(c)$ can be represented as a sum of m operators, which are in some sense simpler than the original operator $A(c)$:

$$A(c) = \sum_{i=1}^{m} A^{[i]}(c)$$

It should be noted here that the division of the operator A(c) to a sum of simpler operators is in general not unique.

This means that the original problem can be split into the following m sub-problems:

$$\frac{\partial c^{[i]}(x,t)}{\partial t} = A^{[i]}(c^{[i]}(x,t)), \quad i=1,2,\ldots m. \qquad (2)$$

Moreover, the time-interval $[0,T]$ is divided to $N=T/\tau$ "small" sub-intervals, where τ is called the splitting time-step. The particular splitting procedure depends on the way in which the operator $A(c)$ is represented as a sum of simpler operators as well as on the order in which the smaller problems are treated.

Assume that the representation of the original operator A(c) as a sum of simpler operators $A^{[l]}(c)$ is fixed. Various types of splitting procedures can be constructed by ordering the simpler operators in different ways. Four of them, e.g. sequential splitting, symmetric splitting, weighted sequential splitting and weighted symmetric splitting are considered and analyzed in this paper.

2.1 Sequential splitting procedure

Consider, for example, the Danish Eulerian Model (DEM), described in detail in Zlatev (1995). It is formally represented by the following PDE system:

$$\frac{\partial \mathbf{c}(\mathbf{x},t)}{\partial t} = -\frac{\partial (u\mathbf{c})}{\partial x} - \frac{\partial (v\mathbf{c})}{\partial y} - \frac{\partial (w\mathbf{c})}{\partial z} + \\ + \frac{\partial}{\partial x}\left(K_x \frac{\partial \mathbf{c}}{\partial x}\right) + \frac{\partial}{\partial y}\left(K_y \frac{\partial \mathbf{c}}{\partial y}\right) + \frac{\partial}{\partial z}\left(K_z \frac{\partial \mathbf{c}}{\partial z}\right) + \\ + E + Q(\mathbf{c}) - \mathbf{c}\mathbf{k}^T \tag{3}$$

where $\mathbf{x} = (x,y,z)$ is the position vector in the 3-D space domain, $\mathbf{c} = (c_1, c_2 \ldots, c_q)^T$ are the concentrations of different pollutants, $E = (E_1, E_2 \ldots, E_q)^T$ are the emissions of different pollutants, $Q = (Q_1, Q_2 \ldots, Q_q)^T$ is the chemical reactions function and $\boldsymbol{k} = (k_{11}+k_{21}, k_{12}+k_{22}, \ldots, k_{1q}+k_{2q})^T$ are the dry/wet deposition coefficients.

(3) was split in 5 separate submodels in the older versions of DEM, when the pseudospectral algorithm was applied in the numerical treatment of the horizontal transport (advection) and the horizontal diffusion (see for more details Zlatev (1995)).

In the new splitting procedure used in DEM finite elements are used in the discretization of both the horizontal advection and the horizontal diffusion. Therefore, there is no need to split these two processes and they are combined in one (advection-diffusion) submodel. Furthermore, the deposition process is added to the chemical part. In this way, by implementing the new splitting procedure, the number of the sub-models is reduced from five to three. These three sub-models are given below:

$$\frac{\partial \mathbf{c}^{[1]}(\mathbf{x},t)}{\partial t} = -\frac{\partial (w\mathbf{c}^{[1]})}{\partial z} + \frac{\partial}{\partial z}\left(K_z \frac{\partial \mathbf{c}^{[1]}}{\partial z}\right) \tag{4}$$

$$\frac{\partial \mathbf{c}^{[2]}(\mathbf{x},t)}{\partial t} = -\frac{\partial (u\mathbf{c}^{[2]})}{\partial x} - \frac{\partial (v\mathbf{c}^{[2]})}{\partial y} + \frac{\partial}{\partial x}\left(K_x \frac{\partial \mathbf{c}^{[2]}}{\partial x}\right) + \frac{\partial}{\partial y}\left(K_y \frac{\partial \mathbf{c}^{[2]}}{\partial y}\right) \tag{5}$$

$$\frac{\partial \mathbf{c}^{[3]}(\mathbf{x},t)}{\partial t} = E + Q(\mathbf{c}^{[3]}) - \mathbf{c}^{[3]}\mathbf{k}^{T}. \tag{6}$$

These three systems are solved successively at every time-step ($t = t_n$, $n=1,2, ...,N$), with $t_0=0$ and $t_n - t_{n-1} = \tau$. Assume that some approximation of the concentration vector $\mathbf{c}$ is known for $t = t_n$. Then it is used as an initial value for solving (4) at time-step $n+1$. The solution $\mathbf{c}^{[1]}$ of (4) at $t = t_{n+1}$ is used as initial value when solving (5) at time-step $n+1$ (i.e., when $\mathbf{c}^{[2]}$ is computed at $t = t_{n+1}$). Similarly, the solution $\mathbf{c}^{[2]}$ of (5) at $t = t_{n+1}$ is used as initial value when solving (6) at time-step $n+1$ (i.e., when $\mathbf{c}^{[3]}$ is computed at $t = t_{n+1}$). Finaly, the solution $\mathbf{c}^{[3]}$ of the last sub-model at $t = t_{n+1} = t_n+\tau$ is accepted as an approximation of the solution $\mathbf{c}$ of the original system at time-step $n+1$. This means that the splitting time-step $n+1$ is completed and everything is ready to carry out the same kind of computations at splitting time-step $n+2$. The simple splitting procedure, described above and used by default in the current version of DEM, is called sequential splitting. It is very often useful to describe how the splitting procedure is used by (i) giving the order, in which the operators $A^{[i]}$ are applied, and (ii) indicating the splitting time-step τ, which is actually used. For the particular case discussed in this sub-section (i.e., the sequential splitting procedure) such a description is given by the following sequence:

$$(A^{[1]})_\tau\,(A^{[2]})_\tau \;......\; (A^{[m-1]})_\tau\,(A^{[m]})_\tau\,. \tag{7}$$

2.2 Symmetric splitting procedure

Assume that the first $m-1$ sub-problems in (2) are solved by using a splitting time-step $\tau/2$. Assume that the last of the simpler sub-problems, the m-th sub-problem, is after that treated by using a splitting time-step τ. Finally, assume that the computations at the splitting time-step under consideration are completed by performing (again, but in reverse order) the computations with the first $m-1$ simpler problems by using a splitting time-step $\tau/2$. It is easy to express these computations by showing just the sequence of operators, as in (7) for the sequential splitting. This is done in (8) for the symmetric splitting (note that only the index numbers of the operators and the splitting time-steps are given; in other words all $c^{[i]}$, $i=1,2,..., m$, are omitted in the formula given below):

$$(A^{[1]})_{\tau/2}\,(A^{[2]})_{\tau/2} \;......\; (A^{[m-1]})_{\tau/2}\,(A^{[m]})_\tau\,(A^{[m-1]})_{\tau/2} \;......\; (A^{[2]})_{\tau/2}\,(A^{[1]})_{\tau/2} \tag{8}$$

2.3 Weighted sequential splitting procedure

Consider again (2). Assume that the computations at the splitting time-step under consideration are carried out by applying the following two procedures ($c^{[i]}$, i=1,2,..., m, are again, as in the previous section, omitted):

$$(A^{[1]})_\tau (A^{[2]})_\tau \ldots\ldots (A^{[m-1]})_\tau (A^{[m]})_\tau \tag{9}$$

and

$$(A^{[m]})_\tau (A^{[m-1]})_\tau \ldots\ldots (A^{[2]})_\tau (A^{[1]})_\tau \,. \tag{10}$$

It is seen from (9) and (10) that only the order in which the simpler sub-problems are treated is different. In the first of these two formulae the computations are performed by taking sequentially the operators (the index numbers of the operators being increasing). The splitting procedure in the second formula (10) is also sequential, but now the index numbers of the operators are decreasing.

Denote by $c_{forward}$ and $c_{backward}$ the results obtained when (9) and (10) are applied at a given splitting time-step. Then an approximation:

$$c = \theta\, c_{forward} + (1-\theta)\, c_{backward} \tag{11}$$

of the solution of the original problem (1) at the splitting time-step under consideration is calculated by using some weighting parameter θ and one can proceed by the calculations for the next splitting time-step.

The splitting procedures based on the performance of the calculations at each splitting time-step by using the formulae (9), (10) and (11) are called weighted sequential splitting procedures.

The definition of weighted sequential splitting procedures given in this sub-section can be slightly generalized. One can apply ordering with upper indices {r1, r2,..., r_{m-1}, r_m} instead of ordering with upper indices {m, m-1,..., 2, 1}, where indices {r_1, r_2,... , r_{m-1}, r_m} are any permutation of {1, 2,..., m-1, m}. Then the procedure defined by (15) and (16) is a special case of the generalized weighted splitting procedures. Probably, it is most natural to use in practice that special case.

2.4 Weighted symmetric splitting procedure

Assume that the calculations at each splitting time step are carried out by using the following two formulae:

$$(A^{[1]})_{\tau/2} (A^{[2]})_{\tau/2} \ldots\ldots (A^{[m-1]})_{\tau/2} (A^{[m]})_\tau (A^{[m-1]})_{\tau/2} \ldots\ldots (A^{[2]})_{\tau/2} (A^{[1]})_{\tau/2} \tag{12}$$

and

$$(A^{[m]})_{\tau/2} (A^{[m-1]})_{\tau/2} \ldots\ldots (A^{[2]})_{\tau/2} (A^{[1]})_\tau (A^{[2]})_{\tau/2} \ldots\ldots (A^{[m-1]})_{\tau/2} (A^{[m]})_{\tau/2} \,. \tag{13}$$

It is seen that the same principles as in the previous section are used. However, symmetric splitting procedures are used both in the forward mode and in the backward mode; in fact, the formula (12) is precisely the same as (8).

Denote again (as in the previous subsection) by $c_{forward}$ and $c_{backward}$ the results obtained when (12) and (13) are applied at a given splitting time-step. Then, as in the previous section, the approximation:

$$c = \theta\, c_{forward} + (1-\theta)\, c_{backward} \tag{14}$$

of the solution of the original problem (1) at the splitting time-step under consideration is calculated by using some weighting parameter θ and one can proceed by the calculations for the next splitting time-step.

3. ADVANTAGES AND DISADVANTAGES OF THE SPLITTING PROCEDURES

The replacement of the original problem (1) by a series of simpler problems (2) has several advantages:

- It is in general much easier to find optimal (or, at least, good) numerical methods for each of the simpler systems in (2) than for the big system arising after the direct discretization of operator $A(c(x,t))$ in the right-hand side of (1).

- It is easy to change numerical algorithms and physical mechanisms used in the different sub-problems arising as a results of the implementation of the splitting procedure chosen. Thus, improvements of the code can easily be achieved.

- The application of splitting procedure facilitates the preparation of efficient codes for different parallel computers. In many cases, the splitting procedure is leading directly to parallel tasks. Consider, for example, the first two sub-models defined by (4) and (5) in the default DEM splitting procedure. It is easily seen that each of these sub-models consists of q parallel tasks (where q is the number of chemical species studied by the model under consideration).

4. COMPARISON OF THE SPLITTING PROCEDURES

In this section we are trying to compare the above splitting procedures and to give some recommendations about the choice of a suitable splitting procedure according to two important criteria: **the computational cost** and **the accuracy of the results.**

4.1 Comparing the computational costs

- The sequential splitting procedures have the cheapest computational cost (among the four types of splitting procedures) when the number of the simple operators m is fixed.
- If a symmetric splitting procedure is to be selected, then it is most efficient to put the most expensive operator in the middle of (8). In this way the total computational cost of the symmetric splitting procedure is reduced, but this cost will always be higher than the cost of the corresponding sequential procedure.
- The computational cost of a weighted sequential splitting procedure is higher, by a factor approximately equal to two, than that of the corresponding sequential procedure. If two processors are available when the weighted sequential procedure is used and only one processor is applied with the sequential splitting procedure, then the computer time needed will be nearly the same for the two splitting procedures. This is due to the fact that the two tasks (9) and (10) can be carried out in parallel. However, if the code based on a sequential splitting procedure is parallelized and if the fact that two processors are available is utilized when the sequential splitting procedure is run, then this is no more true. It will be shown in Section 5 that an efficient parallel code can be developed in the case when the sequential splitting procedure is used. Of course, a similar conclusion can be drawn if 2p processors are available when the weighted sequential splitting procedure is used, while the number of processors available when the sequential splitting procedure is used is only p. If the parallel codes for the two splitting procedures are scalable, then approximately the same computing time will be used for the two splitting procedures.

Similar conclusions (as those for the relationship between weighted sequential splitting procedures and sequential splitting procedures) are valid for the relationship between weighted symmetric splitting procedures and symmetric splitting procedures. Some numerical results in support of these conclusions are given in Table 3 in the next section (Section 5).

4.2 Comparing the accuracy properties

- All sequential splitting procedures are of first order of accuracy, i.e., of order $O(\tau)$. This means that the errors due to these splitting procedures are proportional to the splitting time-step τ. Thus, if the time-step τ is sufficiently small and if the error due to the use of the splitting procedure chosen is dominant, then one should expect the errors caused by the splitting procedure to be reduced by a factor of two when the splitting time-step is reduced from τ to $\tau/2$.

- Symmetric splitting procedures of second order of accuracy, $O(\tau 2)$, can be constructed. This means that if the splitting time-step τ is sufficiently small and if the error due to the use of the splitting procedure chosen is dominant, then one should expect the errors caused by the splitting procedure to be reduced by a factor of four when the splitting time-step is reduced from τ to $\tau/2$.
- Weighted sequential and symmetric splitting procedures of order higher than two can be constructed.

The discussion in this sub-section shows that if the accuracy is the most important issue, then symmetric splitting procedures and even weighted (either sequential or symmetric) splitting procedures are to be selected.

Combining the conclusions in this sub-section with those in the previous sub-section makes it clear that the selection of a splitting procedure is a trade off between two requirements (the requirement for reducing the computing time and the requirement for achieving sufficiently accurate results). Unfortunately, these two requirements are working in opposite directions. This is why in general it is very difficult to find a choice which satisfies sufficiently well both requirements.

If some assumptions are satisfied, then the errors due to the splitting procedures can vanish. Such results were proved by Lanser and Verwer (1999) by using the concept of "L-commutativity of two operators". This property, which is rather important from a practical point of view, will be studied in further works.

5. REAL DATA EXPERIMENTS WITH DEM – COMPARISON WITH OBSERVATIONS

One-year experiments with real data were carried out by using DEM and the results were compared with the observations, obtained from a number of measurement stations scattered throughout Europe. The results for two major pollutants, NO_2 and SO_2, are presented in Table 1 and Table 2, respectively.

The scalability and the performance of DEM on a Sunfire 6800 parallel machine for the above one-year experiments are shown in Table 3. The chemistry-deposition stage is perfectly parallelized as it is independently executed for each grid point. Moreover, it shows superlinear speed-up because of the cache utilization of the processors.

NO_2 concentrations in air for 1998: DEM results in comparison with measured values													
Splitting method:	Jan	Feb	Mar	Apr	May	June	July	Aug	Sept	Oct	Nov	Dec	**Year**
	NO_2 [mol/l] (averaged over 35 places with measurement stations)												
Sequential	**3.20**	4.96	1.83	**1.76**	2.05	2.33	1.86	1.99	2.42	2.38	3.97	4.10	2.740
Symmetric	2.35	**3.83**	1.38	1.39	**1.66**	**1.95**	**1.56**	**1.81**	**2.11**	**1.85**	**3.04**	**3.10**	**2.169**
Weighted	3.21	4.97	**1.84**	**1.76**	2.05	2.34	1.86	1.99	2.42	2.39	3.97	4.11	2.742
Measured	***2.79***	***3.24***	***2.18***	***1.98***	***1.81***	***1.58***	***1.57***	***1.70***	***1.84***	***1.96***	***3.10***	***3.33***	***2.258***
	DISCREPANCY [%]												
Sequential	13	35	-19	-12	12	32	15	15	24	18	22	19	17
Symmetric	-19	15	-58	-42	-9	19	-1	6	13	-6	-2	-8	-4
Weighted.	13	35	-19	-12	12	32	16	15	24	18	22	19	17
	CORRELATION FACTOR												
Sequential	0.67	0.61	0.69	0.59	0.64	0.56	0.50	0.59	0.65	0.57	0.65	0.70	0.692
Symmetric	0.62	0.59	0.75	0.49	0.66	0.47	0.49	0.57	0.56	0.53	0.70	0.71	0.679
Weighted	0.66	0.61	0.69	0.59	0.64	0.56	0.50	0.59	0.65	0.57	0.65	0.69	0.690

Table 1: The concentrations of NO_2 in the air for 1998, calculated by DEM by using three different splitting schemes. The results are compared with the values, obtained from 35 permanent measurement stations in Europe at the corresponding grid points. The (96 x 96) grid is used in the discretization of the spatial domain.

SO_2 concentrations in air for 1998: DEM results in comparison with measured values													
Splitting method:	Jan	Feb	Mar	Apr	May	June	July	Aug	Sept	Oct	Nov	Dec	**year**
	SO_2 [mol/l] (averaged over 35 places with measurement stations)												
Sequential	2.70	2.02	1.49	1.06	1.13	0.72	0.63	**0.73**	**0.85**	0.68	1.59	2.89	1.375
Symmetric	**2.49**	**1.91**	**1.41**	**1.01**	**1.12**	0.72	**0.62**	0.75	0.86	**0.67**	**1.58**	**2.64**	**1.317**
Weighted.	2.72	2.04	1.51	1.07	1.14	0.72	0.64	0.74	0.88	0.69	1.60	2.91	1.387
Measured	***1.36***	***1.41***	***1.11***	***0.80***	***0.73***	***0.51***	***0.49***	***0.56***	***0.60***	***0.58***	***1.32***	***1.41***	***0.906***
	DISCREPANCY [%]												
Sequential	50	30	26	25	35	29	22	23	29	16	17	51	34
Symmetric	46	26	22	21	35	30	20	25	30	15	17	47	31
Weighted	50	31	26	25	36	30	23	24	29	17	18	52	34
	CORRELATION FACTOR												
Sequential	0.83	0.61	0.64	0.60	0.71	0.72	0.82	0.63	0.82	0.75	0.29	0.39	0.712
Symmetric	0.83	0.69	0.64	0.64	0.72	0.70	0.83	0.71	0.81	0.81	0.45	0.52	0.775
Weighted	0.83	0.61	0.64	0.60	0.71	0.72	0.83	0.63	0.82	0.75	0.28	0.39	0.712

Table 2: The concentrations of SO_2 in the air for 1998, calculated by DEM by using three different splitting schemes. The results are compared with the values, obtained from 35 permanent measurement stations in Europe at the corresponding grid points. The (96 x 96) grid is used in the discretization of the spatial domain.

	Time [sec.] ***(Speed-up)*** of 2-D version of DEM on Sunfire 6800				
Splitting method:	1 proc.	2 proc.	4 proc.	8 proc.	16 proc.
1. Sequential	37893	18570 ***(2.04)***	8406 ***(4.51)***	4588 ***(8.26)***	2362 ***(16.0)***
2. Symmetric	45864	23116 ***(1.98)***	11780 ***(3.89)***	5684 ***(8.07)***	2942 ***(15.6)***
3. Weighted seq.	68746	33465 ***(2.05)***	16227 ***(4.24)***	7700 ***(8.93)***	4127 ***(16.7)***

Table 3: Time in seconds and speed-up (in brackets) of the three splitting schemes for a one-year experiment with the 2-D version of DEM on a (96 x 96 x 1) grid.

The superlinear speed-up of the chemistry-deposition stage has stronger influence on the total speed-up in the sequential and weighted sequential schemes, where this stage takes more than 70% of the total time. In the symmetric scheme it is put in the middle and is surrounded by 2 advection-diffusion stages with a half time-step, so its relative weight decreases as well as the total speed-up.

6. EXPERIMENTS ON TEST PROBLEMS WITH KNOWN SOLUTION

Three important cases are studied by these experiments:

- The behaviour of the results in the case where only splitting errors appear in the computations.
- The behaviour of the results in the case where the splitting errors are appearing together with errors due to the spatial discretization and the errors due to the time integration.
- The behaviour of the results in the case where there are no splitting errors (i.e., all pairs of the involved operators are L-commuting).

6.1 Information about the presentation of the results

Let $\mathbf{c}_{i,j,n} \approx \mathbf{c}(x_i, y_j, t_n)$; $i \in \{1,2,...,N_x\}, j \in \{1,2,...,N_y\}, n \in \{1,2,...,N\}$. The error ERR_n at time-step n is evaluated by:

$$ERR_n = \frac{DIFFERENCE_n}{MAXVALUE_n}, \quad \text{where}$$

$$DIFFERENCE_n = \max_{i \in \{1,2,...,N_x\}, j \in \{1,2,...,N_y\}} (|\, \mathbf{c}_{i,j,n} - \mathbf{c}(x_i, y_j, t_n) \,|),$$

$$MAXVALUE_n = \max_{i \in \{1,2,...,N_x\}, j \in \{1,2,...,N_y\}} (|\, \mathbf{c}(x_i, y_j, t_n) \,|).$$

The global error $ERRMAX$ (which is the maximal error over the whole time-integration interval) is evaluated by $ERRMAX = \max_{n \in \{1,2,...,N\}} (ERR_n)$.

The two-dimensional case is treated above. It is clear that values of $ERRMAX$ for three-dimensional problems can be calculated in a quite similar manner. The values of $ERRMAX$, rounded to the third digit, will be presented in Tables 4-7.

6.2 Only splitting errors appear in the computations

Consider the following problem:

$$\frac{\partial \mathbf{c}}{\partial t} = y\frac{\partial \mathbf{c}}{\partial x} - x\frac{\partial \mathbf{c}}{\partial y} + 1 + x - y, \quad t \in (0,1], \quad \mathbf{c}(x,y,0) = x + y, \tag{15}$$

with exact solution

$$\mathbf{c}(x,y,t) = x+y+t \ . \tag{16}$$

Assume that a sequential splitting procedure is applied in the solution of (15) by using the following two simpler operators:

$$\frac{\partial \mathbf{c}^{[1]}}{\partial t} = y\frac{\partial \mathbf{c}^{[1]}}{\partial x} - x\frac{\partial \mathbf{c}^{[1]}}{\partial y}, \tag{17}$$

$$\frac{\partial \mathbf{c}^{[2]}}{\partial t} = 1 + x - y. \tag{18}$$

Linear finite elements are used in the semi-discretization of the first sub-problem (17). The resulting system of ODE's is treated by a third-order predictor-corrector method. The second sub-problem (18) is solved exactly. Since the exact solution (16) of (15) is a linear function in x, y and t, this means that the numerical methods are not causing errors. Thus, the only error is caused by the splitting procedure.

Two important conclusions can be drawn from the results given in Tables 4 and 5:

- As mentioned above, there are no errors caused by the numerical algorithms. Therefore, all errors (excepting the rounding errors, which are negligible, because the computations were carried out in double pre cision) are caused by the use of the splitting procedure. This implies that refining the grids should have no effects on the accuracy of the approximate solution. This statement is fully confirmed by the results shown in Table 4.
- Since the sequential splitting procedure used is of first order, one should expect the errors to depend linearly on the splitting time-step. The results shown in Table 5 indicate that this is true (reducing the splitting time-step by a factor of ten is leading to a reduction of the errors of the numerical solution by a factor approximately equal to ten).

Grid	Without splitting	With splitting
8 x 8	3.34E-14	1.70E-4
16 x 16	3.79E-14	1.80E-4
32 x 32	3.99E-14	1.80E-4
96 x 96	7.21E-14	1.97E-4
288 x 288	4.57E-14	1.99E-4
480 x 480	4.28E-14	1.99E-4

Table 4: The values of *ERRMAX* obtained when (15) is solved without splitting and with splitting (the number of time-steps is 1500 for all grids, but the values of *ERRMAX* are rather insensitive to variations of the number of the time-steps when no splitting is used). The values of *ERRMAX* are close to the machine precision when no splitting is used. The values of *ERRMAX* are of order $O(10^{-4})$ for all grids when the splitting procedure is applied.

Steps	Step	Without splitting	With splitting
10	0.1	4.35E-14	2.42E-2
100	0.01	4.36E-15	2.39E-3
1000	0.001	5.76E-14	2.39E-4
10000	0.0001	7.24E-13	2.39E-5

Table 5: The values of $ERRMAX$ obtained when (15) is solved by using different numbers of time-steps without splitting and with splitting. The 16 x 16 grid is used in the discretization of the spatial derivatives. The experiment shows that the values of $ERRMAX$ do not depend on the number of time-steps when no splitting is used. In the case where splitting is applied, the values of $ERRMAX$ are reduced when the number of steps is increased, because, as mentioned above, $\Delta t = \tau$.

6.3 Splitting errors versus other numerical errors

The combined effect of splitting errors and other errors, caused by the numerical algorithms applied in the discretization of the spatial derivatives and by the numerical methods applied in the treatment of the systems of ODE's arising after the semi-discretization, is illustrated in this sub-section.

Consider the following three-dimensional problem:

$$\frac{\partial c}{\partial t} = y\frac{\partial c}{\partial x} - x\frac{\partial c}{\partial y} + \frac{\partial c}{\partial z} + y(\cos z - \sin z) + x(\sin z + \cos z), \quad t\in[0,1], \tag{19}$$

with an initial condition

$$c(x,y,z,0) = e^{z}(x^2+y^2). \tag{20}$$

The analytical solution of this problem is

$$c(x,y,z,t) = e^{z+t}(x^2+y^2) + t[\, y(\cos z - \sin z) + x(\sin z + \cos z)\,]. \tag{21}$$

A sequential splitting procedure is used in the solution of this problem. This procedure is based on the sequential treatment of the following sub-problems:

$$\frac{\partial c^{[1]}}{\partial t} = y\frac{\partial c^{[1]}}{\partial x} - x\frac{\partial c^{[1]}}{\partial y}, \tag{22}$$

$$\frac{\partial c^{[2]}}{\partial t} = y(\cos z - \sin z) + x(\sin z + \cos z), \tag{23}$$

$$\frac{\partial c^{[3]}}{\partial t} = \frac{\partial c^{[3]}}{\partial z}. \tag{24}$$

Results obtained with this splitting procedure for $\tau = \Delta t$ for different resolution grids and different values of Δt are given in Table 6.

The results in Table 6 indicate that the error due to the spatial discretization is very quickly becoming dominant (except the results for the finest grid). Further reductions of the time-step Δt do not lead to reduction of the total error, because the errors caused by the spatial discretization become dominant.

Time-step $\tau=\Delta t$	**Grid size**			
	16x16x16	**32x32x32**	**64x64x64**	**128x128x128**
8.00E-3	1.58E-2	1.92E-2	unstable	unstable
4.00E-3	7.99E-3	9.03E-3	unstable	unstable
2.00E-3	4.03E-3	4.55E-3	5.07E-3	unstable
1.00E-3	2.22E-3	2.33E-3	2.53E-3	unstable
5.00E-4	1.32E-3	1.22E-3	1.28E-3	1.35E-3
2.5E-4	1.OOE-3	6.73E-4	6.58E-4	6.74E-4 •
1.25E-4	1.02E-3	4.02E-4	3.47E-4	3.43E-4
6.25E-5	1.03E-3	2.96E-4	1.91E-4	1.77E-4
3.12E-5	1.03E-3	2.95E-4	1.14E-4	9.32E-5
1.56E-5	1.03E-3	2.94E-4	8.11E-5	5.15E-5
7.81E-6	1.03E-3	2.94E-4	8.09E-5	3.07E-5
3.91E-6	1.03E-3	2.94E-4	8.08E-5	2.15E-5

Table 6: Values of *ERRMAX* obtained when the problem defined by (19) and (20) is solved with the splitting procedure (22) – (24) by using different numbers of time-steps and different spatial grids.

The splitting time-step and the discretization time-step used in the integration of the system of ODE's arising after the semi-discretization were, as stated above, equal when the results shown in Table 6 were calculated (i.e., $\tau = \Delta t$). This does not allow us to distinguish between errors due to the use of different splitting steps τ and errors caused by use of different discretization steps Δt in the solution of the system of ODE's. That is why some experiments, in which τ is not equal to Δt, are also needed and were carried out. More precisely, the number of splitting steps was kept constant and the number of time-steps per splitting step was varied. Some results for *ERRMAX* in dependence with τ and Δt, obtained in such experiments, are given in Table 7 for the $(16\times16\times16)$ grid.

Δt	τ					
	5.0E−3	**4.0E−3**	**2.0E−3**	**1.0E−3**	**5.0E−4**	**2.5E−4**
8.00E-3	1.59E-2	-	-	-	-	-
4.00E-3	1.59E-2	7.99E-3	-	-	-	-
2.00E-3	1.59E-2	7.99E-3	4.03E-3	-	-	-
1.00E-3	1.59E-2	7.99E-3	4.03E-3	2.22E-3	-	-
5.00E-4	1.59E-2	7.99E-3	4.03E-3	2.22E-3	1.32E-3	-
2.50E-4	1.59E-2	7.99E-3	4.03E-3	2.22E-3	1.33E-3	1.00E-3
1.25E-4	1.59E-2	7.99E-3	4.03E-3	2.22E-3	1.33E-3	1.01E-3
6.25E-5	1.59E-2	7.99E-3	4.03E-3	2.22E-3	1.32E-3	1.01E-3
3.12E-5	1.59E-2	7.99E-3	4.03E-3	2.22E-3	1.32E-3	1.00E-3
1.56E-5	1.59E-2	7.99E-3	4.03E-3	2.22E-3	1.32E-3	1.01E-3
7.81E-6	1.59E-2	7.99E-3	4.03E-3	2.22E-3	1.32E-3	1.01E-3
3.91E-6	1.59E-2	7.99E-3	4.03E-3	2.22E-3	1.32E-3	1.01E-3

Table 7: Values of *ERRMAX* obtained when the problem defined by (19) and (20) is solved by using the splitting procedure (22) – (24) with different numbers of discretization time-steps Δt per splitting time-step τ.

It is seen from the results presented in Table 7 that if splitting time-step τ is fixed, then it does not matter how many discretization time-steps of size Δt will be carried out per splitting time-step. In other words, the results shown in Table 6 indicate that the errors due to the splitting (with time-step τ) are dominant in this example.

7. CONCLUSIONS AND PLANS FOR FUTURE RESEARCH

From the real data experiments with DEM by using different splitting schemes and comparing the results with data from the measurement stations (see Section 5) the following conclusions can be drawn:

- In general, the results, obtained with the weighted sequential splitting, are quite similar to those of the sequential splitting. This could be an indication for a small commutator of the two operators.
- For some pollutants (NO_2, SO_2) the symmetric splitting scheme (due to Marchuk and Strang) gives results, which are in most cases closer to the measurements than those, obtained with the other two splitting methods. More experiments are needed in order to investigate the consistency of such behavior.

- Although the weighted and the symmetric splitting schemes are of higher order of accuracy with respect to the splitting error, this cannot be clearly seen by the experimental results. This is probably because of the influence of other sources of error as well as the uncertainty of the measurements.

The examples given in the previous section show that the splitting procedure in general causes splitting errors. Therefore, it is worthwhile to carry out work in two important directions:

- attempt to avoid the use of splitting procedures when this is possible,
- finding reliable and robust methods for evaluating the splitting errors.

Both problems are still open when general large-scale air pollution problems are to be handled. Success in the solution of any of these two problems is a very desirable, but also a very challenging task. It should be noted that even a partial solution of any of the two problems will lead to a very considerable progress in the numerical treatment of large-scale air pollution models and many other large-scale computational problems arising in science and engineering.

Better understanding of the interference of errors due to the splitting procedures and errors that are caused by other reasons (numerical methods, uncertainties of the input data, etc.) is highly desirable. Research efforts in this direction are needed.

Acknowledgement: This research was supported in part by grant I-901/99 from the Bulgarian NSF, grant ICA1-CT-2000-70016 – Center of Excellence and by NATO grant NATO ARW project "Impact of future climate changes on pollution levels in Europe".

REFERENCES

Bagrinovskii, K. A. and Godunov, S. K., 1957, Difference schemes for multidimensional problems. *Dokl. Akad. Nauk USSR* **115**, pp. 431-433.

Csomós, P., Faragó, I. and Havasi, Á., 2003, Weighted sequential splittings and their analysis. *Comput. Math. Appl.* (to appear).

Dimov, I., Faragó, I. and Zlatev, Z., 1999, Commutativity of the operators in splitting methods for air pollution models, Report 04/99, Central Laboratory for Parallel Processing, Bulgarian Acad. Sci..

Dimov, I., Faragó, I., Havasi, Á. and Zlatev, Z., 2001, L-commutativity of the operators in splitting methods for air pollution models. Annalles Univ. Sci. Budapest, Sec. Math. **44**, pp. 127-148.

Dyakonov, E. G., 1962, Difference schemes with splitting operators for multidimensional stationary problems. *Journal of Computational Mathematics and Mathematical Physics* **2(4)**, pp. 57-79 (in Russian).

Faragó, I. and Havasi, Á., 2002, The mathematical background of operator splitting and the effect of non-commutativity. In: *Large Scale Scientific Computations III (Margenov, S., Yalamov P. and Wasniewski, J. eds.),* Springer, Berlin, pp. 264-271.

Faragó, I. and Havasi, Á., 2003, On the convergence and local splitting error of different splitting schemes. *Progress in Computational Fluid Dynamics,* (to appear).

Faragó, I., Havasi, Á. and Korotov, S., 2002, Mathematical analysis of the weighted splitting. Reports of the Department of Mathematical Information Technology, University of Jyväskylä, No. B/12.

Hunsdorfer, W. and Verwer, J. G., 2003, Numerical solution of time-dependent advection-diffusion-reaction equations. Springer, Berlin.

Lanser, D. and Verwer, J. G., 1999, Analysis of operators splitting in advection-diffusion-reaction problems in air pollution modelling. *Journal of Computational and Applied Mathematics* **111**, pp. 201-216.

Marchuk, G. I., 1968, Some application of splitting-up methods to the solution of mathematical physics problems. *Applik. Mat.* **13(2)**.

Marchuk, G. I., 1980, Methods of computational mathematics. Nauka, Moscow, 1980.

Marchuk, G. I., 1982, Mathematical Modeling for the Problem of the Environment. Nauka, Moscow (in Russian).

Marchuk, G. I., 1986, Mathematical modelling for the problem of the environment. *Studies in Mathematics and Applications,* No. 16, North-Holland, Amsterdam.

Marchuk, G. I., 1994, Applications of Adjoint Equations to Problems of Global Change and Environmental Protection. In: *Proceedings of HERMIS'94,* Vol. 1, (E.A. Lipitakis, ed.), Hellenic Mathematical Society, Athens, pp. 1-20.

McRae, G. J., Goodin, W. R. and Seinfeld, J. H., 1982, Numerical solution of the atmospheric diffusion equations for chemically reacting flows. *Journal of Computational Physics* **45**, pp. 1-42.

Penenko, V. V. and Obraztsov, N. N., 1976, A variational initialization model for the fields of meteorological elements. *Soviet Meteorol. Hydrol.* **11**, pp. 1-11.

Strang, G., 1968, On the construction and comparison of difference schemes. *SIAM J. Numer. Anal.* **5**, pp. 505-517.

Tikhonov, A. N. and Samarski, A. A., 1977, Equations of the Mathematical Physics. Nauka, Moscow.

Yanenko, N. N., 1962, On convergence of the splitting method for heat equation with variable coefficients. *Journal of Computational Mathematics and Mathematical Physics* **2(5)**, pp. 933-937 (in Russian).

Zlatev, Z., 1995, Computer treatment of large air pollution models, Kluwer.

SIMULATION OF LIBERATION AND DISPERSION OF RADON FROM A WASTE DISPOSAL

Maria de Lurdes Dinis and António Fiúza
Department of Mining Engineering, Geo-Environment and Resources Research Center (CIGAR), Engineering Faculty of Oporto University, Rua Dr. Roberto Frias, s/n 4200-465 Oporto, Portugal

Abstract: Radon emissions from a radioactive waste disposal may constitute a major source of environment contamination and consequently a potential health hazard to the nearby population. Gaseous Radon-222 is generated from the radioactive decay of Radium-226 present in the tails. When it is formed, radon is free to diffuse along the pores of the residues to the surface and escape to the atmosphere.

Waste management and long term stabilisation has a major concern in reducing radon emissions to near-background levels. The common theoretical approach is done by calculating the cover thickness that allows a radon flux inferior to a stipulated and accepted value. The fundamentals of the conceptual model are based in the principles of diffusion across a porous medium, which allows the mathematical description of the radon transport through the waste and the cover. The basic diffusion equations are used for estimating the theoretical values of the radon flux formed from the decay of the Radium-226 contained in the waste material. The algorithm incorporates the radon attenuation originated by an arbitrary cover system placed over the radioactive waste disposal.

Once the radon is released into the atmosphere, it is available for atmospheric transport by the wind. Radon atmospheric dispersion is modelled by a modified Gaussian plume equation, which estimates the average dispersion of radon released from a point source representative of one or several uniform area sources. The model considers the medium point release between all the areas contaminated. The dispersion can be simulated in different wind directions, with different wind velocities, as well as in the dominant wind direction.

Key words: Waste, disposal, radium, radon, flux, dispersion.

I. Farago et al. (eds.), Advances in Air Pollution Modeling for Environmental Security, 133–142.

1. INTRODUCTION

Uranium milling tailings represent the highest potential source of environmental contamination for the great majority of uranium mining activities. In the ore milling, the uranium ore is grinded promoting liberation and thus increasing the possibility of radon to escape to the environment. The milling process generates large volumes of tailings being generally disposed in piles. The radionuclides presents in the tailings, ^{226}Ra, ^{230}Th and ^{222}Rn, are a major concern to the human health and to the environment. They are not dissolved during the leaching process, which breaks down the equilibrium chains of the ^{238}U and ^{235}U decay families.

The principal radon isotope, ^{222}Rn, formed from the ^{226}Ra radioactive decay, has a half-life of 3.82 days, which allows a large period of time for migration before it decays to another nuclide. Radon generation may continue for thousand of years due to the long decay periods of ^{226}Ra and ^{230}Th, present in the uranium tailings. Radon is an inert gas, which emanates from the solid tailings particles and is free to diffuse to the surface of the pile, escaping to the atmosphere. Coming up from the ground, it may become locally hazardous or it may be transported by the wind into the surrounding area, dispersing the potential damages.

The work proposes a two-dimensional model for calculating the ^{222}Rn flux diffusion from a radioactive waste disposal, having as a result the ^{222}Rn concentrations at a defined mixing height which will be the starting point to the atmospheric dispersion, as well as the bidimensional dispersion in the prevalent wind direction.

2. METHODS AND RESULTS

2.1 The ^{222}Rn diffusion

The basic equations of diffusion may be used for estimating the theoretical values of the radon flux from the ^{226}Ra content in the waste material. Radon migration to the surface is a complex process controlled mainly by porosity (ε) and moisture (θ), leading the cover efficiency in attenuating the radon flux. This efficiency depends on the capacity of the cover material for keeping the diffusion so slow that radon decays to another non-gaseous nuclide, becoming trapped by the cover system.

The movement of radon in soil is characterised by the diffusion coefficient, D, which can be measured, either in laboratory or in field, or be estimated by empirical correlation. These have the advantage of being simple and easy to use with a minimal amount of information needed. A correlation using the fraction of saturation, m, is recommended (Rogers, 1984):

$$D\left(cm^2.s^{-1}\right) = 0{,}07\, e^{\left[-4\left(m - m\varepsilon^2 + m^5\right)\right]} . \qquad (1)$$

Values for radon diffusivity in porous media may vary over a wide range of several orders of magnitudes depending on the porous material and particularly on its degree of water saturation. The generic diffusion equation can be represented by:

$$D\frac{\partial^2 C}{\partial x^2} - \lambda C + \frac{R\rho\lambda E}{\varepsilon} = 0 . \qquad (2)$$

The diffusion process occurs in a multiphase system, where the porosity is either filled with air, or with water. If we apply the generic diffusion equation to each one of the phases (a-air, w-water, filled pore space):

$$D_a \frac{\partial^2 C_a}{\partial x^2} - \lambda C_a + \frac{R\rho\lambda E_a}{\varepsilon - \theta} + \frac{T_{wa}}{\varepsilon - \theta} = 0 \qquad (3)$$

$$D_w \frac{\partial^2 C_w}{\partial x^2} - \lambda C_w + \frac{R\rho\lambda E_w}{\theta} - \frac{T_{wa}}{\theta} = 0 \qquad (4)$$

In these equations, D ($m^2.s^{-1}$) represents the radon diffusivity, λ the radon decay constant (s^{-1}), C ($Bq.m^{-3}$) the radon concentration in the pore space, R ($Bq.kg^{-1}$) the radium concentration in the material, ρ ($kg.m^{-3}$) the bulk density of the dry material, E (dimensionless) the radon emanation power coefficient for the pore spaces, ε (dimensionless) the total porosity, θ (dimensionless) the moisture and T_{wa} ($Bq.m^{-3}.s^{-1}$) the radon transfer rate from water to the air.

The solution of the diffusion equation for a homogeneous medium represents the flux release from the tailings to the surface, J ($Bq.m^{-2}.s^{-1}$). For a system without cover we obtain (Rogers, 1984):

$$J_t = R\rho E\sqrt{\lambda D_t}\, \tanh\left(\sqrt{\frac{\lambda}{D_t}}\, x_t\right) . \qquad (5)$$

In this equation, x_t represents the tailings thickness. If we consider a two-medium problem represented by the tailings (t) and a homogeneous cover material (c), the solution of the diffusion equation can be represented by (Rogers, 1984):

$$J_c(x_c) = \frac{2J_t e^{-b_c x_c}}{\left[1+\sqrt{a_t/a_c}\,\tanh(b_t x_t)\right]+\left[1-\sqrt{a_t/a_c}\,\tanh(b_t x_t)\right]e^{-2b_c x_c}} \tag{6}$$

In this solution, $b_i = \sqrt{(\lambda/D_i)}$ (m) (i = c or t) and $a_i = \varepsilon_i^2 D_i[1-0.74m_i]^2$ ($m^2.s^{-1}$), where m is the degree of saturation of the soil.

From the precedent equations it is possible to obtain a generic solution for the radon released in three main situations: flux directly diverted to the atmosphere without cover trapping; radon flux through a homogeneous cover and radon flux through a multilayer cover system.

Another possible approach consists in calculating the thickness of the cover which allows a value stipulated and accepted for the radon flux. This can be done directly using the waste and cover parameters, mainly the ^{222}Rn coefficient diffusion, porosity and moisture, the ^{226}Ra content in the waste material and ^{222}Rn emanation, rearranging the last equation (Rogers, 1984):

$$x_c = \sqrt{\frac{D_c}{\lambda}}\ln\left[\frac{2J_t/J_c}{\left(1+\sqrt{a_t/a_c}\,\tanh(b_t x_t)\right)+\left(1-\sqrt{a_t/a_c}\,\tanh(b_t x_t)\right)\left(Jc/Jt\right)^2}\right]. \tag{7}$$

2.2 The 222Rn atmospheric dispersion

The model uses a Gaussian model of plume dispersion to account for the transportation of ^{222}Rn from the source area to a downwind receptor, and is represented by the equation of Pasquill as modified by Gifford (Chacki, 2000):

$$X = \frac{Q}{2\pi\mu\sigma_y\sigma_z} e^{\left[-1/2\left(\frac{y}{\sigma_y}\right)^2\right]}\left\{e^{\left[-\frac{1}{2}\left(\frac{z-H}{\sigma_z}\right)^2\right]}+e^{\left[-\frac{1}{2}\left(\frac{z+H}{\sigma_z}\right)^2\right]}\right\} \quad . \tag{8}$$

This equation represents a Gaussian distribution, where X ($Bq.m^{-3}$) represents the radionuclide concentration, Q ($Bq.s^{-1}$) the source strength, and H (m) the corrected source released height. Dispersion parameters, σ_y (m) and σ_z (m), are the standard deviations of the plume concentration in the horizontal and vertical directions, respectively. The standard deviations can be evaluated by the Pasquill-Gifford coefficients for flat rural areas or by Briggs method for urban areas. The atmospheric transport is done at wind speed (height-independent), u ($m.s^{-1}$), to a sampling position located at surface elevation, z, and transverse horizontal distance, y, from the plume centre.

The concentration available for the dispersion is calculated from the radon flux released. The contamination source is defined as an emission area, or the sum of several areas, where the radon is diluted directly in the ambient air-breathing zone above the contaminated source zone. The wind speed for ^{222}Rn dilution should be matched to the average annual value through the mixing zone.

The contaminated area can be compared to an environment compartment in which the ^{222}Rn emission is uniform in all the area, characterised by its length and width. These values and the mixing height define the volume in which the concentrations are spatially homogeneous and instantaneously mixed.

The modified Gaussian plume equation estimates the average dispersion of radon released from a point source representative of one or several uniform area sources. This point represents the medium point release between all the areas contaminated, with the same mass flux as the entire affected zones, and it should be located in the weighted centre of the total contaminated area.

The ^{222}Rn concentration dispersion is calculated from the release point, at the mixing height, for defined distances in each wind direction. The release point source is located in the centre of a conceptual regular polygon area defined as the contaminated site. The concentrations along each direction are evaluated taking into account the respective mean wind velocity and the frequency of the occurrence. This will lead to different ^{222}Rn concentrations in the sectors defined by each direction in the collateral wind rose.

When we consider exclusively the dominant wind direction, we assume that all the ^{222}Rn concentration is being distributed in that direction, being inexistent in all other directions. Another assumption is that ^{222}Rn concentration is uniform in the volume defined by the central point source and the boundaries of the release area, and by the mixing height. Dispersion occurs outside this area.

2.3 A case study

The model was applied to a specific contaminate site, the Urgeiriça uranium tailings. The Urgeiriça uranium mine is located in the north of Portugal near Nelas (Viseu). The mine is surrounded by small farms and country houses, with most of the local population living in the village Canas de Senhorim within about 2 km of the mine.

The exploitation of the mine began in 1913 for radium extraction. The activity of the Urgeiriça mine was maintained until 1944, then exclusively dedicated to the production of radium. In 1951, a chemical treatment unit for the production of low-grade U_3O_8 concentrates was built, and in 1967 was transformed into a modern unit (Bettencourt *et al*, 1990). In 1991, local mining stopped but the facilities were still used for the treatment of ores from other mines, in the same region, until 2000.

The extensive exploitation and treatment of the uranium ore in the Urgeiriça mine, has led to an accumulation of large amounts of solid wastes (tailings). About $4x10^9$ kg of rock material was routed into natural depressions confined by dams that cover an area of about 0.11 km^2. The solid wastes are composed mainly of sand and silt particles, transported as a pulp to the site from the counter-current decantation. This operation was proceeded by grinding and acid leaching. The radon exhalation from the tailings is potentially one of the main sources of contamination for the nearby areas.

We tested the model in simple situations varying some of the parameters involved. The necessary parameters were adopted from some measurements made by ITN (Nuclear and Technological Institute) in the Urgeiriça tailings piles (Reis *et al*, 2000). Local meteorological data, namely wind velocity and frequency, was used for simulating the dispersion. The unknown parameters were estimated from available data.

The calculating procedure uses the following steps: (i) estimative of the ^{222}Rn diffusivity in the waste and in the cover, (ii) estimate of the ^{222}Rn flux release to the air in the breathing zone and (iii) calculation of the ^{222}Rn concentration dispersed in the atmosphere along each wind direction, taking into account for its intensity and frequency of blow.

The contaminated site is composed by four different tailings with a total area of approximately 110 321 m^2. We considered the inexistence of a covering system in this area. The medium point for the global area was defined by the arithmetic average of the medium point for each singular area and has the following coordinates: $x = 20612$, $y = 93326$.

The ^{222}Rn flux was calculated for the global area, now conceptually defined as a regular octagon. Each sector has a characteristic average wind speed. The area contaminated is the same as the polygon area, and from it, one can estimate the length of the polygon sides and the radius of the circumscribed circumference. The air breathing or mixing height was defined as 1.7 m. The ^{222}Rn concentrations were calculated in each sector at this height. The total average flux estimated is 0.2042 $Bq.m^{-2}.s^{-1}$.

The wind speed data in each direction (N S E W NW NE SE SW) refers to the Nelas meteorological station. The values fit in the Pasquill stability class D (neutral) so this was chosen to estimate the typical dispersion coefficients. The dominant wind direction is NE.

The concentration at the breathing height in the dominant wind direction refers only to the polygon side that limits the respective sector.

The dispersed concentrations are different in each sector. We defined a 2-km distance from the centre for simulating the ^{222}Rn dispersion in each wind direction; the same distance was considered for the dominant wind direction (NE). The dispersion results can be seen in the figures below.

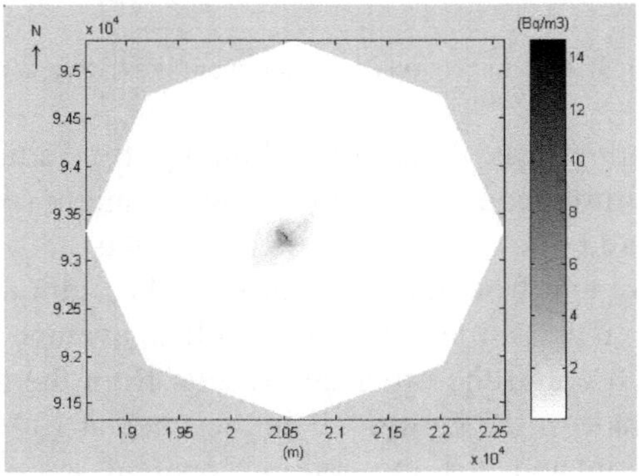

Figure 1: Radon dispersion in each wind direction, $Bq.m^{-3}$

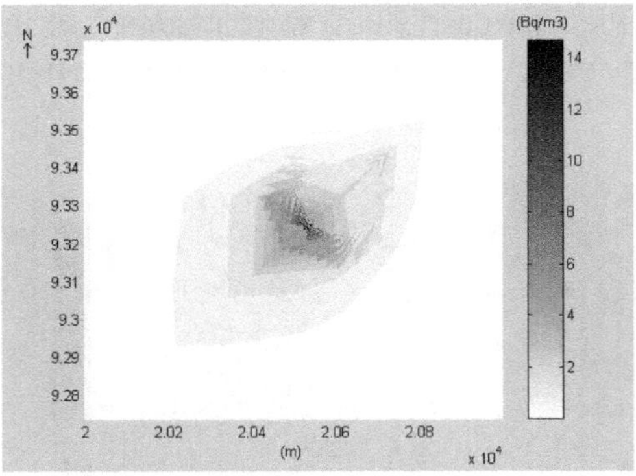

Figure 2: Radon dispersion in each wind direction, amplified, $Bq.m^{-3}$

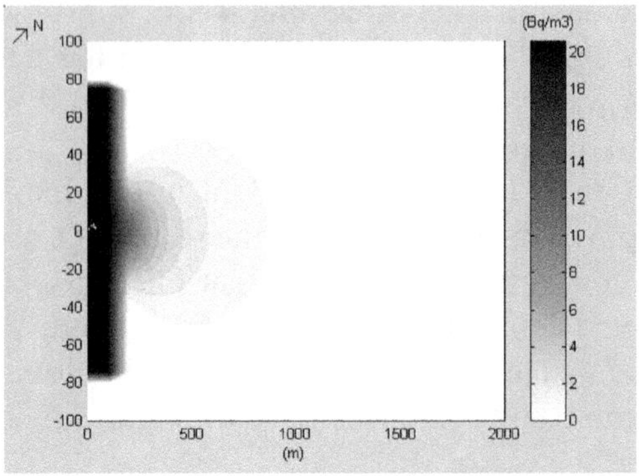

Figure 3: Radon dispersion in the dominant wind direction, Bq.m^{-3}.

A new simulation was done by estimating the radium content in the tailings assuming initial equilibrium in the two uranium series: ^{238}U and ^{235}U. We have considered the same four contaminated areas (dams) but with different radium content from the previous case. We admitted that three of the tailings result from the treatment of an ore with an average grade of 1 kg.ton^{-1}, being leached at 90% and the other one results from the treatment of an ore with an average grade of 0.2 kg.ton^{-1}, being leached at 100%.

The values estimated for the radium content are respectively 10376 Bq.kg^{-1} and 2075 Bq.kg^{-1}. The ^{222}Rn flux was estimated for the global area also without cover. The value obtained was 5.96 Bq.m^{-2}.s^{-1}. The average concentration obtained inside the compartmental box was 70.2 Bq m^{-3}. All the other values from the first case study were assumed in this simulation. The results of the dispersion for this case can be seen in the figures below.

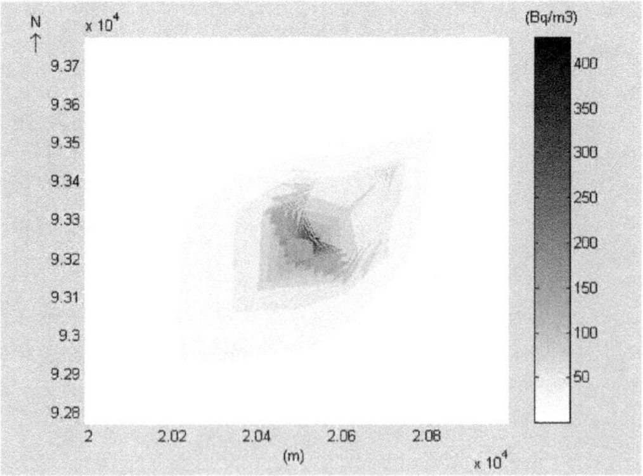

Figure 4: Radon dispersion in each wind direction, amplified (2nd case), Bq.m^{-3}.

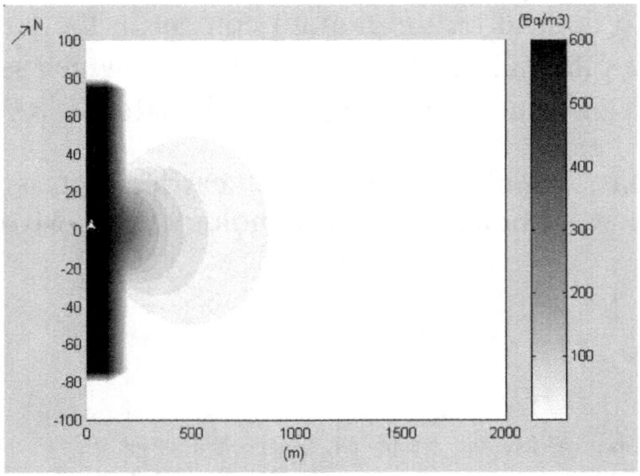

Figure 5: Radon dispersion in the dominant wind direction, (2nd case), Bq.m^{-3}.

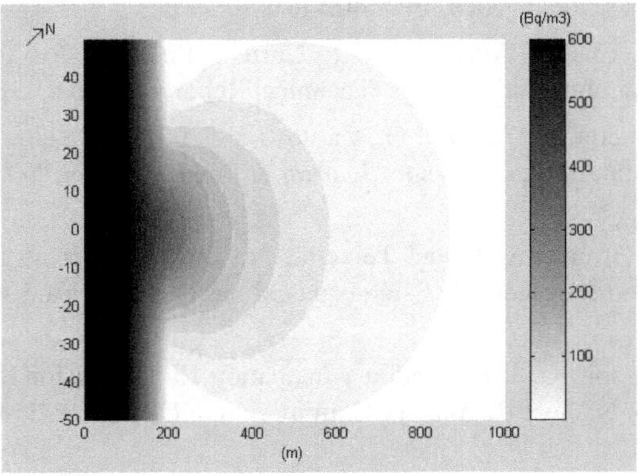

Figure 6: Radon dispersion in the dominant wind direction, amplified (2nd case), Bq m^{-3}.

3. CONCLUSIONS

The work presented may be considered a first approach in modelling the dispersion of ^{222}Rn, thus allowing the assessment of exposure to uranium tailing piles. The obvious limitations are mostly related to the reduction of the source to a point. In addition, variations due to complex terrain topography cannot be modulated; our model assumes implicitly a flat plain. The wind velocity is constant with height and the dispersion is only two-directional, although this direction rotates horizontally.

We should refer to the difficulties that may be found when trying to obtain the data needed to characterise the contaminated soil and also the variability of these data, both in space and time. We stress that the radon flux depends on the diffusion coefficient, which greatly varies with the material moisture and porosity. These parameters will vary over the year, due to the climatic changing. They also depend on the radium content in the tailing, which may be different in each pile and on the water retention, which is also a seasonal parameter, with spread values at the Urgeiriça site (Reis *et al*, 2000).

Acknowledgements: This work has been carried out with the financial support of Foundation for Science and Technology (FCT-MCES).

REFERENCES

Betteencourt, A. O., Teixeira, M. M., Elias, M. D., & Madruga, M. J., 1990, Environmental monitoring in uranium mining areas. In: Environmental behaviour of radium, vol. 2. Technical Reports Series n.º 310, Vienna, IAEA, pp. 281-294.

Chacki, S. and Parks, B., 2000, Update's User's Guide for Cap88-PC, Version 2.0. U. S. Environmental Protection Agency, EPA 402-R-00-004.

International Atomic Energy Agency, 1982, Current Practices for the Management and Confinement of Uranium Mill Tailings. Technical Reports Series n.º 335, Vienna, Austria.

Madruga, M. J., Brogueira, A., Alberto, G., Cardoso, F., 2001, ^{226}Ra bioavailability to plants at the Urgeiriça uranium mill tailings site. *Journal of Environmental Radioactivity* **54**, pp. 175-188.

Reis, M., Faísca, C., Brogueira, A. and Teixeira, M., 2000, Avaliação da Exalação de ^{222}Rn para a Atmosfera em Barragens de Estéreis das Minas da Urgeiriça. Report DPRSN serie A, n.º 2.

Rogers V. C. and Nielson K., 1984, Radon Attenuation Handbook for Uranium Mill Tailings Cover Design. U.S. Nuclear Regulatory Commission, NUREG/CR-3533.

METHODS OF EFFICIENT MODELING AND FORECASTING REGIONAL ATMOSPHERIC PROCESSES

Anatoliy Yu. Doroshenko[(1)], Vitaly A. Prusov[(2)]
[(1)]Institute of Software Systems of the National Academy of Sciences of Ukraine, Acad. Glushkov prosp., 40, block 5, 03187 Kiev, Ukraine; [(2)]Ukrainian Research Hydrometeorological Institute, Science prosp. 37, 03650, Kiev-28, Ukraine

Abstract: Regional forecasting is of great importance because of its potential for preventing economical and social consequences of natural and man-caused atmospheric pollution. A new non-standard approach and computational method is offered in this paper for the efficient solution for forecasting regional meteorological processes. The method replaces the Cauchy problem in the atmospheric model by a boundary-value problem and introduces a specific interpolation technique that has a number of advantages for the method to be computationally efficient. The model and method have been tested by the Hydrometeorological Centre of Ukraine and successfully applied in regional short- and middle-term weather forecasting for regions of Ukraine.

Key words: Regional weather forecast, mesoscale grids, boundary-value problem, interpolation.

1. INTRODUCTION

Mathematical modeling and forecasting atmospheric phenomena is of great importance because of its potential for preventing negative economical and social consequences of atmospheric pollution from natural and man-caused emergencies. These problems are being intensively discussed and investigated in recent years especially in the context of climate changes (Houghton et al., 2001), (Zlatev, 1995) as well as the regional prediction of atmospheric events (Xue et al., 2000). Forecasting regional atmospheric processes is of ever growing interest due to, on one hand, huge detriment caused by mesoscale weather events (floods, tornadoes, hail etc.) and, on

I. Farago et al. (eds.), Advances in Air Pollution Modeling for Environmental Security, 143–152.

another hand, due to obvious benefits in saving costs from using environmental modeling instruments and large-scale computing. For example, in the USA annual economic losses caused by regional atmospheric events are greater than 13 billion US dollars (Xue et al., 2000), and at the same time the total expenses on forecasting can be up to 15 times less than the expense of covering possible damages.

For Ukraine as a big industrial country in Eastern Europe regional weather modeling and forecasting can have critical significance for a number of reasons, among which the most important ones are the following. Ukraine has a well-developed chemistry, metallurgy, oil and gas industry with extremely high concentrations of huge industrial complexes in single places. About 1800 of the Ukrainian enterprises use toxic substances; and the total area of zones subject to chemical pollution due to emissions into the atmosphere is 81500 km^2 with 40% of the population. Moreover, Ukraine has five nuclear power stations with 16 working reactors. In overall 6.34 megatons of pollutants hit the atmosphere annually in Ukraine, which means 126 kg of pollutants per person.

Regional atmospheric processes are influenced by macroscale atmospheric circulation, so modeling meteorological values over a restricted area is to be considered as a task with transitional boundary conditions. To achieve a prescribed level of accuracy of the solutions for a model in places of heavy gradients of the related functions it is often necessary to apply a numerical method with variable grid steps for restricted terrains. However the common techniques of mathematical physics (Tikhonov and Samarskiy, 1977) cannot often satisfy these requirements because of low accuracy, slow divergence and stability problems, so some dedicated numerical methods are needed to make the computation more time- and cost-effective.

In this paper we follow the approach of "unilateral influence" to combine macro- and mesoscale models (Phillips and Shukla, 1973), (Miyakoda and Rosati, 1977) and use a non-standard technique for modeling and forecasting atmospheric processes over a region. The technique replaces the Cauchy problem in the atmospheric model by a boundary-value problem and introduces a specific interpolation method that has a number of advantages concerning computational efficiency. Our methodology is well tested and approved in complex regional ecological-meteorological modeling in Ukraine (Prusov and Doroshenko, 2003; Prusov et al., 2004; Dovgyi et al., 2000).

2. A MOTIVATING EXAMPLE

Usually, the numerical solution of a regional weather forecast problem as a classical initial and boundary value problem is difficult as all explicit difference schemes of numerical solution of the equations of hydrodynamics, heat and mass transmission are conditionally stable, i.e., they have an upper bound for the size of the time integration step. The main idea behind our method is to replace the problem by a similar one but satisfying special requirements following from the concept of "unilateral influence" (Miyakoda and Rosati, 1977). To give the flavor of the approach, we will consider a simple example.

Assume that we have to find a function $f(\eta)$ satisfying the differential equation $\frac{df}{d\eta} = R(f)$ on a segment $\eta_{i-1} \le \eta \le \eta_i$ with additional conditions of prehistory (background) $f(\eta_i) = f_i$, $i = 1,2,...,N$. Here R is generally a nonlinear operator, f_i are discrete values of function $f(\eta)$ in the nodes $\eta = \eta_i$ $(i = 1,2,...,N)$ of a large-scale grid ϖ_h with grid step $h_i = \eta_i - \eta_{i-1}$. For ordinary differential equations R is algebraic, and for partial differential equations it is represented by a spatial differential operator.

Now let us confine the problem above to an ordinary differential equation. The natural classical way of the numerical solution would be a difference scheme like the following. On a segment $[\eta_{i-1}, \eta_i]$ a small-scale grid $\overline{\varpi}_\tau$ of points ζ_k $(k = 0,1,...,M)$ with a step $\tau_{k-1} = \zeta_k - \zeta_{k-1}$ is introduced, provided that $\zeta_0 = \eta_{i-1}$, $\zeta_M = \eta_i$ and the differential equation is considered only on the segment $[\eta_{i-1}, \eta_i]$. Replacing derivative $f^I(\zeta_k)$ by its approximation, for example, $f^I(\zeta_k) \approx [f(\zeta_{k+1}) - f(\zeta_k)]/\tau_k$ we are led to the Cauchy difference scheme $\frac{1}{\tau_k}[f(\zeta_{k+1}) - f(\zeta_k)] = L(f(\zeta_k))$, $k = 1,2,...,M-1$. where initial values are defined as $f(\zeta_1) = f(\eta_{i-1}) = f_{i-1}$.

However we should not necessarily solve this Cauchy problem. Instead we propose an alternative way to find function value $f(\eta)$ in the segment $[\eta_{i-1}, \eta_i]$ – by means of Hermite polynomials interpolation. Having known values of function $f_i = f(\eta_i)$ and its first derivative $L(f(\eta_i)) = df/d\eta\big|_{\eta=\eta_i}$ at points $\eta = \eta_i$ $(i = 1,2,...,N)$ we define the interpolated function values in other points with the help of polynomial $P_{2N}(\eta)$ where terms of the polynomial are determined by $\frac{d^\alpha P(\eta)}{d\eta^\alpha} = \frac{d^\alpha f(\eta)}{d\eta^\alpha}\bigg|_{\eta=\eta_i}$, $i = 1,2,...,N$, $\alpha = 0,1$ and the polynomial itself is given in the following form:

$$P_{2N}(\eta) = \sum_{i=1}^{N} \frac{\omega^2(\eta)}{[(\eta - \eta_i)\omega^I(\eta_i)]^2} \left\{ \left[1 - (\eta - \eta_i)\frac{\omega^{II}(\eta_i)}{\omega^I(\eta_i)} \right] f(\eta_i) + (\eta - \eta_i) f^I(\eta_i) \right\},$$

$$\omega_n(\eta) = (\eta - \eta_1)(\eta - \eta_2)...(\eta - \eta_N)$$

If the background is particularly given in two-point units (N = 2), the interpolating polynomial has the form:

$$P_4(\eta)=\left(\frac{\eta_2-\eta}{\eta_2-\eta_1}\right)^2\left[\left(1+2\frac{\eta-\eta_1}{\eta_2-\eta_1}\right)f_1+(\eta-\eta_1)f_1^I\right]+$$

$$+\left(\frac{\eta-\eta_1}{\eta_2-\eta_1}\right)^2\left[\left(1+2\frac{\eta_2-\eta}{\eta_2-\eta_1}\right)f_2+(\eta_2-\eta)f_2^I\right].$$

It is easy to see that the last formula gives correct results at $f=1,\eta,\eta^2$ and η^3. It is easy to prove also its uniqueness and to estimate the interpolation error. At f = η^4 the maximal error on an interval $(0,1)$ is equal to $-0{,}0625$, while for example Bessel's formula yields $-0{,}5625$. So, such a solution of the considered problem can be sufficiently accurate and efficiently carried out.

The major advantages of the constructed interpolation formulae involving a function and its derivative $f^{(\alpha)}(\eta_i)$, $i=1,2,\dots,N$, $\alpha=0,1$, are the following:

- they have greater accuracy than any of the formulae using only function values $f(\eta_i)$;
- no data are required on the right border of the interpolation interval, and so the formulae can also be used for the rightmost interval;
- the values of function $f(\eta_i)$ and its derivatives $f^{(\alpha)}(\eta_i)$ can be given through unequal intervals.

3. PROBLEM STATEMENT AND A METHOD OF ITS NUMERICAL SOLUTION

For forecasting values of meteorological quantities (components v_1, v_2, v_3 of velocity $\mathbf{V}$, pressure, temperature, specific humidity, specific liquid water content, concentration of pollutants and others) in the atmosphere in a bounded territory $\overline{G}$ we will follow the basics of the method of "unilateral influence" (Miyakoda and Rosati, 1977), where results of analyses and forecasts received from a macroscale (hemisphere or global) model are used as boundary conditions in a regional model.

Let the state of the atmosphere at spatial point $r=(\lambda,\varphi,\sigma)$ of the macrospace area G be defined by a vector of meteorological quantities $\Re(r,t)$ of discrete values of the analysis and, similarly, forecast $\Re(r,t^{m+1})=\Re^{m+1}(r)$ received from a macroscale model at time $t=t^{m+1}$ $(m=0,1,\dots,M)$ with a step $\tau=t^{m+1}-t^m$.

Then for determining the atmospheric state in the bounded domain $\overline{G}(r)\subset G(r)$ at $\forall t\in\left[t^m,t^{m+1}\right]$ we will solve a task of the following kind in vector representation:

$$\frac{\partial\Re}{\partial t}=D\Re,\ \forall t\in\left[t^m,t^{m+1}\right],\forall r\in\overline{G} \tag{1}$$

$$\Re\left(r,t^{m+1}\right)=\Re^{m+1}(r),\quad m=0,\,1,\ldots,M\,,$$

where

$$D\Re=\frac{1}{r\cos\varphi}\frac{\partial}{\partial\lambda}\left(\frac{v_1}{r\cos\varphi}\frac{\partial\Re}{\partial\lambda}\right)+\frac{1}{r}\frac{\partial}{\partial\varphi}\left(\frac{v_2}{r}\frac{\partial\Re}{\partial\varphi}\right)+\frac{\partial}{\partial r}\left(v_3\frac{\partial\Re}{\partial r}\right)-$$

$$-\frac{v_1}{r\cos\varphi}\frac{\partial\Re}{\partial\lambda}-\frac{v_2}{r}\frac{\partial\Re}{\partial\varphi}-v_3\frac{\partial\Re}{\partial r}+F$$

is the right-hand side function describing the momentum, heat and mass transmission in spherical coordinates with sink/source term F.

Now replace continuum $\overline{G}$ by a spatial grid of points obtained by discretization of the domain $\overline{G}$ with a set of $J-1$ elements $\Delta\lambda_j$, $K-1$ elements $\Delta\varphi_k$ and $L-1$ elements $\Delta\sigma_l$. Let us construct a vector $\{r_{jkl}\}$, defining the continuous variable r only at points j $(1\le j\le J)$, k $(1\le k\le K)$, l $(1\le l\le L)$. As a result we will have

$$\lambda_J=\lambda_1+\sum_{\mu=2}^{J-1}\Delta\lambda_\mu\quad \varphi_K=\varphi_1+\sum_{\mu=2}^{K-1}\Delta\varphi_\mu\quad \sigma_L=\sigma_1+\sum_{\mu=2}^{L-1}\Delta\sigma_\mu$$

In the domain $\overline{G}$ instead of function $\Re(r,t)$ defined on a macroscale grid, we will construct below a function of discrete argument on a regional grid in the nodes $\left(\lambda_j,\varphi_k,\sigma_l,t^m\right)\in R$, $1\le j\le J$, $1\le k\le K$, $1\le l\le L$, $1\le l\le L$. Our aim is to put in correspondence the differential operator D in (1) and the grid operator Λ (see the next section). After filling up function $\Re_{jkl}\left(t^{m+1}\right)=\Re_{jkl}^{m+1}$ in the nodes of the regional grid and computing the right parts $f\left(t^{m+1}\right)=f^{m+1}=\Lambda\Re^{m+1}$, $m=1,2,\ldots,M$, in all nodes of the grid $\left(\lambda_j,\varphi_k,\sigma_l\right)$, $1\le j\le J$, $1\le k\le K$, $1\le l\le L$, we will search for a solution of the problem (1) for $\forall t\in\left[t^m,t^{m+1}\right]$ with the help of a Hermite polynomial like above for number of points M = 3:

$$\begin{aligned}\Re(t)=\Re^m+\frac{t-t^m}{\tau}\Bigg[\,\tau f^m+\frac{t-t^m}{4\tau}\Bigg[\;&4\left(\Re^{m+1}-2\Re^m+\Re^{m-1}\right)-\tau\left(f^{m+1}-f^{m-1}\right)+\\ +\frac{t-t^m}{4\tau}\Bigg[\;&5\left(\Re^{m+1}-\Re^{m-1}\right)-\tau\left(f^{m+1}+8f^m+f^{m-1}\right)-\\ -\frac{t-t^m}{4\tau}\Bigg[\;&2\left(\Re^{m+1}-2\Re^m+\Re^{m-1}\right)-\tau\left(f^{m+1}-f^{m-1}\right)+\\ +\frac{t-t^m}{4\tau}\Bigg[\;&3\left(\Re^{m+1}-\Re^{m-1}\right)-\tau\left(f^{m+1}+4f^m+f^{m-1}\right)\Bigg]\Bigg]\Bigg]\Bigg]\Bigg]\end{aligned}\qquad(2)$$

for each node of the grid $\left(\lambda_j,\varphi_k,\sigma_l\right)$, $1\le j\le J$, $1\le k\le K$, $1\le l\le L$.

It is easy to check up that the scheme (2) has interpolation properties, i.e., at $t=t^m$ or $\left(\tau=t-t^m=0\right)$ and $t=t^{m+1}$ or $\left(\tau=t^{m+1}-t=0\right)$ the equalities $\Re\left(t^m\right)=\Re^m$ and $\Re\left(t^{m+1}\right)=\Re^{m+1}$ hold, respectively. So the maximal error of the solution of problem (1) with the help of (2) is inside the interval $t^m\le t\le t^{m+1}$ and it has an order of approximation $O\left[(\tau)^4\right]$.

4. SMOOTH FILLING UP AND APPROXIMATION OF DIFFERENTIAL OPERATORS

To provide a fourth-order approximation of a differential operator D in (1) by a grid operator Λ we need to guarantee the accuracy of the same order in the interpolation method for smooth filling up of the given discrete function in the nodes of the regional grid. To this aim we propose in this section the following computational scheme.

Designate with η one of the horizontal axes of the system of coordinates $r=(\lambda,\varphi,\sigma)$ and with interval $a \le \eta \le b$ the linear size of the area of the solutions of the macroscale model along this axis. Let any points $a<\eta_1<\eta_2<...<\eta_{N-1}<b$, form a non-uniform macroscale grid $\varpi_h[a,b]$ with grid step $h_{i-1}=\eta_i-\eta_{i-1}$. Let us enumerate all nodes in some order $\eta_0,\eta_1,\eta_2,...,\eta_N$ and consider the values of macroscale function $\Re(\eta_i,t^m)$ in the nodes as components of a vector $\Re=\{\Re_i(t^m),\ i=0,1,...,N\}$.

The task of filling up values of a function defined on a macroscale grid in nodes of a regional grid on each interval $[\eta_i,\eta_{i+1}]$ will be performed with the help of a polynomial of the fifth degree:

$$Q_i(\eta)=a_0+a_1(\eta-\eta_i)+a_2(\eta-\eta_i)^2+a_3(\eta-\eta_i)^3+a_4(\eta-\eta_i)^4+a_5(\eta-\eta_i)^5, \quad (3)$$

where $a_0=\Re_i$, $a_1=\dfrac{h_{i-1}}{h_i(h_i+h_{i-1})}\left[\Re_{i+1}-\left(1-\dfrac{h_i^2}{h_{i-1}^2}\right)\Re_i-\dfrac{h_i^2}{h_{i-1}^2}\Re_{i-1}\right]$,

$$a_2=\frac{1}{h_i(h_i+h_{i-1})}\left[\Re_{i+1}-\left(1+\frac{h_i}{h_{i-1}}\right)\Re_i+\frac{h_i}{h_{i-1}}\Re_{i-1}\right],$$

$$a_3=-h_i(a_4+h_ia_5),\quad a_4=-\frac{5}{2}h_ia_5,$$

$$a_5=\frac{2}{h_i^3}\left[a_2-\frac{1}{h_{i+1}(h_i+h_{i+1})}\left(\Re_{i+2}-\left(1-\frac{h_{i+1}}{h_i}\right)\Re_{i+1}+\frac{h_{i+1}}{h_i}\Re_i\right)\right],$$

As for vertical changes of the meteorological values near the underlying surface, where they have the heaviest gradients, it is needed to use grids with small steps. On the other hand, to save computer memory and time it is expedient to make use of a rough grid far from the land surface. So, irregular grids are needed for solving mesoscale problems. However macroscale models are usually determined on the standard levels of pressure σ $(z_0, 850, 700, 500, ...\ \text{hPa})$ where z_0 stands for sea level. Evidently, there is no unique interpolating formula which provides necessary accuracy of interpolation in the segment $[z_0, 850]$ of the atmospheric boundary layer.

Let us divide the domain height $\sigma=H$ into two pieces: $0\le\sigma\le h$ and $h\le\sigma\le H$, where h is the 850 hPa pressure level. Values of the meteorological quantities in the nodes of the vertical grid $h\le\sigma\le H$ will be filled in with an interpolation polynomial spline like (3) above, and values on another layer $0 \le \sigma \le h$ will be based on the commonly known theory of the turbulent atmospheric boundary layer (Dovgyi et al., 2000).

We will adopt the conditions of horizontal homogeneity of the meteorological fields, the absence of heating or chilling effects and other factors except turbulent exchange in the atmosphere. Then a system of equations for the mesoscale processes in the layer $0 \le \sigma \le h$ can be written as follows:

$$\frac{\partial v_1}{\partial t} = \frac{\partial}{\partial z}\left[\nu_T \frac{\partial v_1}{\partial z}\right] + \ell\left(v_2 - v_{2g}\right), \tag{4}$$

$$\frac{\partial v_2}{\partial t} = \frac{\partial}{\partial z}\left[\nu_T \frac{\partial v_2}{\partial z}\right] - \ell\left(v_1 - v_{1g}\right),$$

$$\frac{\partial \theta}{\partial t} = \frac{\partial}{\partial z}\left[\frac{\nu_T}{\Pr}\frac{\partial \theta}{\partial z}\right] + S_\theta,$$

$$\frac{\partial q}{\partial t} = \frac{\partial}{\partial z}\left[\frac{\nu_T}{Sc}\frac{\partial q}{\partial z}\right] + S_q,$$

$$\frac{\partial p}{\partial z} = -g\rho,$$

$$\theta = T(1 + 0.608q),$$

$$\rho = \frac{p}{R\theta},$$

where t is time, playing the role of the iteration parameter; v_1 and v_2 are the components of the wind velocity; v_{1g} and v_{2g} are those at the height $\sigma = h$ (geostrophic wind); S_θ and S_q the sources and outflows of enthalpy and humidity, respectively; ν_T is the turbulent viscosity; $\Pr$ is the Prandtl number; Sc is the Schmidt number, ℓ is a Coriolis parameter. The further designations are commonly known.

We construct a vertical grid of M levels with uneven grid steps estimated as

$$z = 1 - \frac{\ln\{[\beta + 1 - (\sigma/h)]/[\beta - 1 + (\sigma/h)]\}}{\ln[(\beta+1)/(\beta-1)]} \tag{5}$$

where $1 < \beta < \infty$ should hold and the closer parameter β to 1 the more nodes are collected near the level $z = 0$.

The formulated nonlinear problem (4) has a numerical solution on the grid (5) (Dovgyi et al., 2000).

Equation system (4) concerns all internal points of the whole layer $z_0 < \sigma < H$. Particularly, for the sub-domain $h < \sigma < H$, where the turbulence viscosity coefficient can be considered as constant, system (4) has an analytical solution (Dovgyi et al, 2000). Combining the numerical solution on the segment $z_0 < \sigma < h$ with the analytical solution on the other segment $h < \sigma < H$ and imposing respective boundary conditions, one can define a divergent iterative process to reproduce the vertical profiles of the meteorological fields based on their known values on the standard levels $(z_0, 850, 700, 500, \ldots \text{hPa})$.

The offered method of filling up the vertical grid allows us to take into account the heterogeneity of the underlying surface, which can disturb the macroscale flow.

Now the computation of the grid values of the partial derivatives of the first order $\psi_i = (\partial\mathfrak{R}/\partial\eta)_i$ and of the second order $\zeta_i = \left(\partial^2\mathfrak{R}/\partial\eta^2\right)_i$ included in $f_{jkl}^m = \Lambda\mathfrak{R}_{jkl}^m$, will be performed on the basis of the following relations:

$$\psi_{i+1} + 2\left(1+\frac{h_i}{h_{i-1}}\right)\psi_i + \frac{h_i}{h_{i-1}}\psi_{i-1} = \frac{3}{h_i}\left\{\mathfrak{R}_{i+1} - \left[1-\left(\frac{h_i}{h_{i-1}}\right)^2\right]\mathfrak{R}_i - \left(\frac{h_i}{h_{i-1}}\right)^2\mathfrak{R}_{i-1}\right\} -$$
$$-\frac{h_i h_{i-1}^2}{24}\left[1-\left(\frac{h_i}{h_{i-1}}\right)^2\right]\frac{\partial^4\mathfrak{R}}{\partial\eta^4}, \tag{6}$$

$$\frac{h_{i-1}}{h_i}\left[\frac{h_{i-1}}{h_i}\left(1-\frac{h_{i-1}}{h_i}\right)+1\right]\xi_{i+1} + \left(1+\frac{h_{i-1}}{h_i}\right)\left[\frac{h_{i-1}}{h_i}\left(3+\frac{h_{i-1}}{h_i}\right)+1\right]\xi_i + \left[\frac{h_{i-1}}{h_i}\left(1+\frac{h_{i-1}}{h_i}\right)-1\right]\xi_{i-1} =$$
$$= \frac{12}{h_i^2}\left[\frac{h_{i-1}}{h_i}\mathfrak{R}_{i+1} - \left(1+\frac{h_{i-1}}{h_i}\right)\mathfrak{R}_i + \mathfrak{R}_{i-1}\right] +$$
$$+\frac{h_i^2 h_{i-1}}{360}\left[1-\left(\frac{h_{i-1}}{h_i}\right)^2\right]\left\{5\frac{h_{i-1}}{h_i} + 2\left[1-\left(\frac{h_{i-1}}{h_i}\right)^2\right]\right\}\frac{\partial^5\mathfrak{R}}{\partial\eta^5}. \tag{7}$$

It is obvious that the relations (6-7) have third order at $h_i \neq h_{i-1}$ and fourth order at $h_i = h_{i-1}$. Derivatives $\psi_i = (\partial\mathfrak{R}/\partial\eta)_i$ and $\xi_i = \left(\partial^2\mathfrak{R}/\partial\eta^2\right)_i$ belong to (4-5) implicitly. But these are systems of algebraic equations with tridiagonal matrices, so solutions can be found effectively with the help of the sweep method (Godunov and Riabenkiy, 1973) with the boundary conditions

$$-\frac{h_1}{6}(\xi_2 - \xi_1) + \psi_1 + \psi_2 = 2\frac{\mathfrak{R}_2 - \mathfrak{R}_1}{h_1} + O\left[h_1^4\right], \tag{8}$$

$$-\frac{h_{N-1}}{6}(\xi_N - \xi_{N-1}) + \psi_{N-1} + \psi_N = 2\frac{\mathfrak{R}_N - \mathfrak{R}_{N-1}}{h_{N-1}} + O\left[h_{N-1}^4\right]. \tag{9}$$

The main advantage of the offered method for the approximation of derivatives is that the solution of the system of algebraic equations (6)-(7) at all points depends on values at other points, i.e., it depends on $\mathfrak{R}_i$ globally, which means smooth filling up and approximation of the differential operators by the grid operators.

The model has been tested on real weather data obtained from the Deutscher Wetterdienst (DWD), Offenbach. Numerical experiments were carried out for the territory of Ukraine and its administrative districts.

5. CONCLUSION

We have presented a new non-standard computational method for the efficient solution of the complex problem of forecasting regional meteorological processes. Our method follows the approach of "unilateral influence" to combine macro- and mesoscale models (Phillips and Shukla, 1973; Miyakoda and Rosati, 1977). It gives opportunity to replace the Cauchy problem in the atmospheric model (1) by a boundary-value problem and introduces a specific interpolation method (2) that has a number of advantages:

- the time step in getting macroscale information for regional forecasting can be significantly increased and reach $\tau = 12$ hours (Prusov et al., 2004);
- as opposed to classical numerical methods for solving the equations of mathematical physics, the offered method is deprived of instability problems;
- the accuracy of the offered method has fourth order and is determined by the same order of accuracy of the following constituent methods: smooth filling up of macroscale values into mesoscale grid nodes (3)–(4), approximating differential operators by grid operators (6)–(9) and interpolation method (2) for solving the boundary-valued problem based on the approach of "unilateral influence".

The model and method have been implemented in a software package and tested for a one-year period by the Hydrometeorological Center of Ukraine. The comparison with actual wheather cards has shown that the numerical forecasts qualitatively and quantitatively well coordinate with real observed data. The model and method have been successfully applied in regional short- and middle-term weather forecasting for districts of Ukraine.

Results on the implementation of the computational scheme for solving problems in regional meteorological forecasting over Ukraine show its good computational efficiency, scalability and applicability for parallel computation.

Acknowledgement: The work is partially supported by NATO Collaborative Linkage Grant 980505.

REFERENCES

Dovgyi, S. A., Prusov, V. A., Kopeika, O. B., 2000, Mathematical modeling of man-caused environmental pollution. Kiev, Naukova Dumka, 247 p. (in Russian).

Godunov, S. K., Riabenkiy, V. C., 1973, Difference schemes (introduction to the theory). Moscow, Science, 400 p. (in Russian).

Houghton, J. T., Ding, Y., Griggs, D. J., Noguer, M., van der Linden, P. J., Dai, X., Maskell, K., and Johnson, C. A. (eds.), 2001, Climate Change 2001: The Scientific Basis. Cambridge University Press.

Miyakoda K., Rosati A., 1977, One-way nested grid models: The interface condition and the numerical accuracy. *Mon. Weather Review* **105**, pp. 1092–1107.

Phillips, N., Shukla, J., 1973, On the strategy of combining coarse and fine grid meshes in numerical weather prediction. *J. Appl. Meteor.* **12**, pp. 763-770.

Prusov, V., Doroshenko, A., 2003, Modeling and Forecasting Atmospheric Pollution over Region. *Annales Univ. Sci. Budapest* **46**, pp. 47-64.

Prusov, V., Doroshenko, A., Prikhodko, S., Yu. Tyrchak, Chernysh, R., 2004, Methods for efficient solution of problems in modeling and predicting regional atmospheric processes. *Problems in Programming* **3-4**, pp. 556-569 (in Russian).

Tikhonov A. N., Samarskiy A. A., 1977, The equations of mathematical physics. Moscow, Science, 735 p. (in Russian).

Xue, M, Droegemeier, K. K., and Wong, V., 2000, The Advanced Regional Prediction System (ARPS) – A multi-scale nonhydrostatic atmospheric simulation and prediction model. Part I: Model dynamics and verification. *Meteorol. Atmos. Phys.* **75**, pp. 161-193.

Zlatev, Z., 1995, Computer treatment of large air pollution models. Environmental Science and Technology Library, Vol. 2. Kluwer Academic Publishers, Dordrecht-Boston-London.

NUMERICAL FORECAST OF AIR POLLUTION – ADVANCES AND PROBLEMS

Adolf Ebel, Hermann J. Jakobs, Michael Memmesheimer, Hendrik Elbern, and Hendrik Feldmann
Rhenish Institute for Environmental Research at the University of Cologne (RIU), Aachener Str. 201 – 209, 50931 Cologne, Germany

Abstract: Application of tropospheric chemical transport models (CTMs) to chemical weather forecast on the global and regional scale is growing. In this paper advances and problems of numerical forecasts with Eulerian regional models are treated. After reviewing the progress of computer performance since the appearance of first complex CTMs, selected models are briefly introduced in order to demonstrate the spectrum of forecast activities. Ozone and particulate matter are chosen as examples of air quality forecasts on continental (European) and smaller spatial scales. The question of forecast and model evaluation is discussed and the potential of chemical data assimilation for the upcoming fourth generation of air quality forecast models emphasized.

Key words: Air quality, chemical weather, numerical forecast, atmospheric chemistry, atmospheric transport, ozone, aerosol, PM10, PM2.5, evaluation, chemical data assimilation, accuracy of data, reliability of predictions, fourth-generation air quality model.

1. INTRODUCTION

Regarding the negative impact of poor air quality on human health, economy, cultural heritage and ecology it is evident that chemical weather forecasts can significantly contribute to the alleviation of harmful effects of air pollution and thus to environmental security. The demand of accurate and early information of the public and significant increase of computer power allowing fast execution of chemical transport algorithms have lead to growing activity in the field of short- to medium range chemical weather forecast. The

I. Farago et al. (eds.), Advances in Air Pollution Modeling for Environmental Security, 153–163.

chemical transport system of the atmosphere exhibits a high degree of complexity so that, despite impressive performance characteristics of present day computers, various limitations for the treatment of numerical models of the system still exist. Attempts to deal with them and to overcome them have lead to different strategies of forecast model design. The model structure also depends on different goals of prediction activities, e. g. focusing on local, regional or global air quality phenomena.

This paper is emphasising forecasts employing Eulerian chemical transport models (CTMs) on regional to local scales (alpha- to gamma-meso-scale). Also, it does not consider statistical approaches which are sometimes integrated in numerical models. Though an attempt is made to generalise the analysis of advances and problems of numerical forecasts, it is admitted that the authors are guided in their views by experience with their own modelling activities based on the EURAD system (European Air Pollution Dispersion model system, Ebel, 2002).

The main forecast parameters are surface ozone concentrations and particulate matter (PM) in most cases. They are also preferably used in this study though the range of predictable quantities is much broader in all advanced chemical weather models. After a brief review of recent development of computer performance and a short overview of ongoing numerical forecast activities (as information has become available so far) examples of ozone and PM forecasts are discussed in order to demonstrate the state of the art. Results of forecast evaluation are presented and the future role of chemical data assimilation for the improvement of chemical weather prediction is discussed. Finally, the question of systematic adaptation of numerical forecast models to existing limitations of computer power is addressed. It seems that such adaptation is mainly governed by intuition instead of cost/benefit assessment till now.

2. PROGRESS OF COMPUTER PERFORMANCE

Around 1985, when the first advanced complex chemistry transport models (second generation models, Peters et al., 1995) came into operation (e. g. RADM, Chang et al., 1987), the peak performance of the most powerful computers amounted to roughly 2 Gflops with main memory near 5 Gbyte. Roughly 10 years later first computers with a peak performance in the range of 2 Tflops were constructed. Presently (2004), the most powerful system is the NEC Earth Simulator from 2002 in Japan, with 40 Tflops peak performance and an aggregated memory of 10 Tbyte. The most efficient system in Europe with a peak performance of 8.9 Tflops was recently installed at the Research Centre Juelich, Germany.

Though it cannot be expected that the most advanced computers will always and exclusively be available for routine chemical weather predictions, i.e., complex chemical transport simulations, the numbers characterizing the progress of computer power give a general impression about the possible growth of CTM complexity and performance in general and of chemical weather forecast in particular. Yet it should be noted that not only the complexity of models, but also their efficiency through structural changes and intelligent adaptation to the task of prediction exhibits a considerable increase. Through such advancement of model design in accordance with the ongoing growth of PC performance it is now possible to perform operational chemical weather forecasts on smaller PC clusters with a peak performance around 10 Gflops.

3. FORECAST ACTIVITIES BASED ON EULERIAN MODELS

Chemical weather forecast has become a popular issue and activities in this field are growing. It is difficult to prepare a state-of-the-art overview of existing models. Therefore, activities mentioned in this section should only be regarded as examples demonstrating various approaches to the task.

An attempt to establish a forecast network combining operational weather forecast carried out with the "Lokal-Modell" (LM) by the German Weather Service, EURAD-CTM simulations, advanced emission modelling at the University of Stuttgart and high resolution regional-to-local modelling at the Research Centre Karlsruhe yielded promising results (Jakobs et al., 2002a), yet failed to become operational due to unexpected discontinuation of funding by the sponsoring German institution. Nevertheless, it stimulated the development of an operational forecast version of the EURAD system (Jakobs et al., 2002b). A comparison of the performance of five CTMs used for chemical weather forecast (EURAD-CTM with DWD-LM of the mentioned German network system, REM3-FU-Berlin/Germany, MATCH/Sweden, THOR/Denmark, NILU/Norway) was carried out by Tilmes et al. (2002) who also provide short descriptions of the participating model systems. The study revealed clear, but unexplained differences of forecast accuracy. Two differing prediction systems have been developed in France (MOCAGE, CHIMERE). Whereas MOCAGE starts with global simulations scaling them down to regional dimensions with a resolution of 10 km, CHIMERE offers predictions for Europe, France and smaller areas (Ille de France). Forecasts of air quality on local scales are the main product of OPANA developed at the University of Madrid (Spain). Forecasts combined with integral (impact and assessment) modelling are available from NILU/Norway (EMEP/AIRQuiz system) and NERI/Denmark (THOR system). THOR can also be applied to air quality forecasts in street canyons.

In North America various pilot studies aiming at regional numerical forcasts of air quality have been carried out, e. g. employing a combination of the meteorological meso-scale model MM5 with a chemistry model (MM5-CHEM). The Meteorological Service of Canada (MSC) initiated a forecast model, named CHRONOS, in 1999 for ozone and PM to be applied to the eastern provinces. Later it was extended to the major part of all Canada (Pudykiewicz et al., 2003). In 2002, congressional interest has stimulated increased activities regarding the development of a National Air Quality Forecasting System for the US with NOAA and EPA as the main collaborating institutions (Davidson, 2003; Schere, 2003).

A convenient way to more information about most existing chemical weather forecast activities is "Bernd Krüger's Linklist" (Krueger, 2000). The European Environmental Agency also offers a collection of links to air quality forecast activities (not all of them based on numerical models) on its web sites (EEA, 2004).

4. FORECASTING

Daily air quality forecasts provided by various operational numerical models (REM-3, THOR, CHIMERE, EURAD and others) for EUROPE or parts of it may be obtained from the internet. All models follow a similar strategy for preparation of forecasts. Input data like those for land use, orography and emissions are chosen from available sources. Meteorological input to the employed CTM is either chosen from external regional weather forecasts or generated internally with initial and boundary conditions taken from global forecasts (e. g. NCEP, ECMWF) in the case of regional models. Chemistry is treated off-line (i.e., without feed back to meteorology) in today complex air quality forecast systems in order to avoid delays of weather forecasts due to considerable computational demands of chemical transport mechanisms. Ideally, emission data used in the models should also be forecasted which is not yet possible for the anthropogenic component. Natural emissions are usually parameterised employing meteorological and land use parameters and are thus included in the forecast process.

Accuracy and reliability of air quality predictions significantly depend on the weather forecasts. Based on them medium range chemical weather forecasts are possible up to about five days where the reliability is decreasing with increasing time span of the predictions.

Most users of air quality predictions have a strong interest in the small-scale or local state of air pollution. To meet their needs forecast models enable downscaling of regional forecasts which generally cover larger domains comprising all relevant sources of anthropogenic emissions (e. g. the European continent) in a first approach. Usually, the technique of one-way or two-way sequential nesting is employed. Figure 1 exhibits an example of ozone and PM10 forecasts with the EURAD system scaled down from the European domain (with a resolution of 125 km) to Central Europe and North-Rhine Westphalia in Germany (with a resolution of 25 and 5 km, respectively). Further downscaling may require adaptation of parameterizations of sub-grid processes or the use of specific models, e. g. in the case of micro-scale forecasts for street canyons with the THOR system.

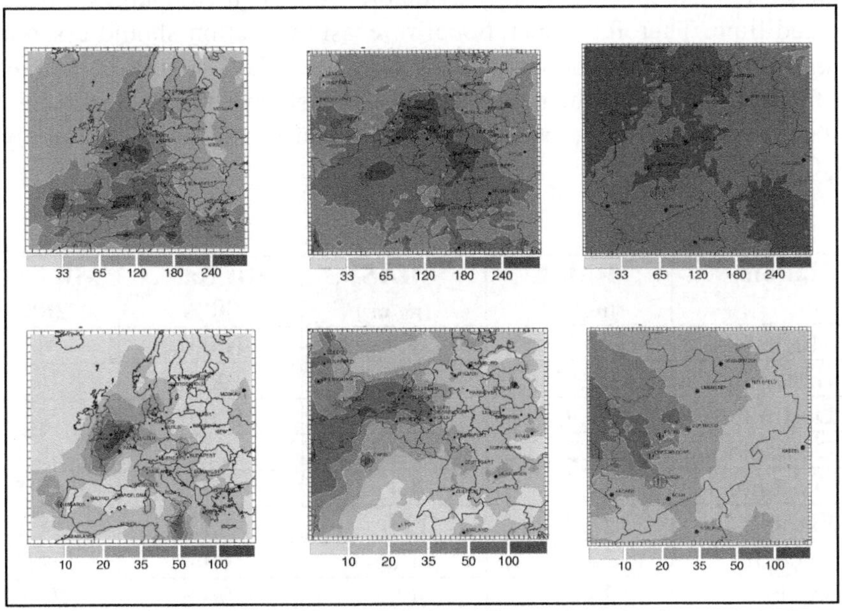

Figure 1: Nested forecast of maximum concentrations of surface ozone (hourly mean, in μg/m³, upper panels) and daily maximum of PM10 (μg/m³, lower panels) employing the EURAD model system. Horizontal resolution from left to right: 125 (Europe), 25 Central Europe), 5 km (Northrhine-Westphalia). Height of the model surface layer 36 m. 12 August 2003.

5. FORECAST AND MODEL EVALUATION

Evaluation of forecasts is an inherent and indispensable task of chemical weather prediction. In general, methods of evaluation are not much different from those applied to usual weather forecasts. Yet availability of relevant and reliable chemical data for this task is a significant limiting factor. Quality control and assurance (QA, QC) require much more tedious and time consuming procedures than for meteorological observations. This is especially true for measurements which cannot regularly be carried out in operational networks. For instance, split concentrations of VOCs, essential for photo-chemical processes, or size distributions and composition of aerosols needed for impact assessments are usually available from field campaigns limited in space and time. Therefore, operational forecast evaluation should essentially be based on operational network measurements. Ways should be sought to make them accessible to chemical weather forecasters as fast and as widely (e. g. on a European scale) as possible as preliminary data sets before applying QA/QC measures.

Station	**RMSE ($\mu g/m^3$)**	**BIAS ($\mu g/m^3$)**	**Hit Rate, 50%**	**Hit Rate, 20%**
Muenster	22.7	-5.7	100.0	74.2
Bielefeld	21.5	-2.2	96.8	74.2
Dortmund	21.9	-3.6	100.0	77.4
Duesseldorf	26.5	9.8	90.3	58.1
Duisburg	24.1	4.4	90.3	64.5
Essen	23.8	12.9	83.9	58.1
Aachen	20.7	-1.2	100.0	74.2
Koeln	24.8	4.9	96.8	71.0
M'gladbach	24.1	9.6	80.6	58.1
Wuppertal	30.2	-15.1	96.8	58.1
Eggegebirge	27.4	11.5	100.0	71.0
Eifel	27.4	0.9	96.8	71.0

Table 1: Evaluation of surface ozone forecasts employing the EURAD model system, August 2003. RMSE and bias of daily maxima.

Useful data gathered in larger networks are surface concentrations of ozone, NO_x, SO_2, and PM. In addition to on-line checks of predictions, evaluation of the forecast model performance over longer past periods is necessary (Tilmes et al., 2002) and can advantageously be used for the improvement of air quality forecast systems.

An example is shown in Table 1 containing results for the statistical evaluation parameters RMSE, bias and 50% and 20% hit rates, which were derived from forecasts of surface ozone concentrations at selected stations in Northrhine-Westphalia with the EURAD system in August 2003. What is evident from this special case is the fact that the quality of ozone (and other forecasts) may considerably vary in smaller domains. This may partly be due to reduced representativeness of some stations. Treatment of this important aspect of model evaluation is out of the scope of this paper.

With regard to the hit rates they are representative for most ozone simulations and remain in a range which is needed for useful forecast applications (above approximately 80 and 60% for deviations from observation up to 50 and 20%, respectively). In Germany it is possible to use daily analyses of maximum hourly concentrations published by the federal environmental agency UBA (Umweltbundesamt) in the internet (http://www.env-it.de/luftdaten/map.fwd?measComp=O3) for a quick visual check of the predicted structure of ozone maximum fields.

Though input data for PM predictions are comparatively vague comparisons with available measurements (usually PM10 in Europe) yield quite promising results. What seems to be a problem for most models is correct prediction of the composition of aerosol mass (if aerosol chemistry is treated at all). The EURAD system using the Modal Aerosol Dynamic Module for Europe (MADE, Ackermann et al., 1998) has been applied to the analysis of long-term air quality changes. The simulations for 1997 have been used for model evaluation. Some of the results which are also pertinent to forecast reliability are shown in Figure 2. Requesting 50% accuracy, a hit rate of PM10 calculations between 50 and 90% is found for individual months, when 125 and 25 km horizontal resolution is used. The results for 5-km horizontal resolution are better in autumn and winter, as it is expected for finer grids. But they become considerably worse in summer, in contrast to expectation. This can be explained by the use of a specific emission inventory for the small domain (Northrhine-Westphalia) evidently missing efficient sources during the summer months. The lower part of Fig. 2 demonstrates that this is not the case for NO_2 exhibiting the expected improvement of predictions with increasing resolution and thus giving more trust to the NO_x part of the fine grid emission inventory.

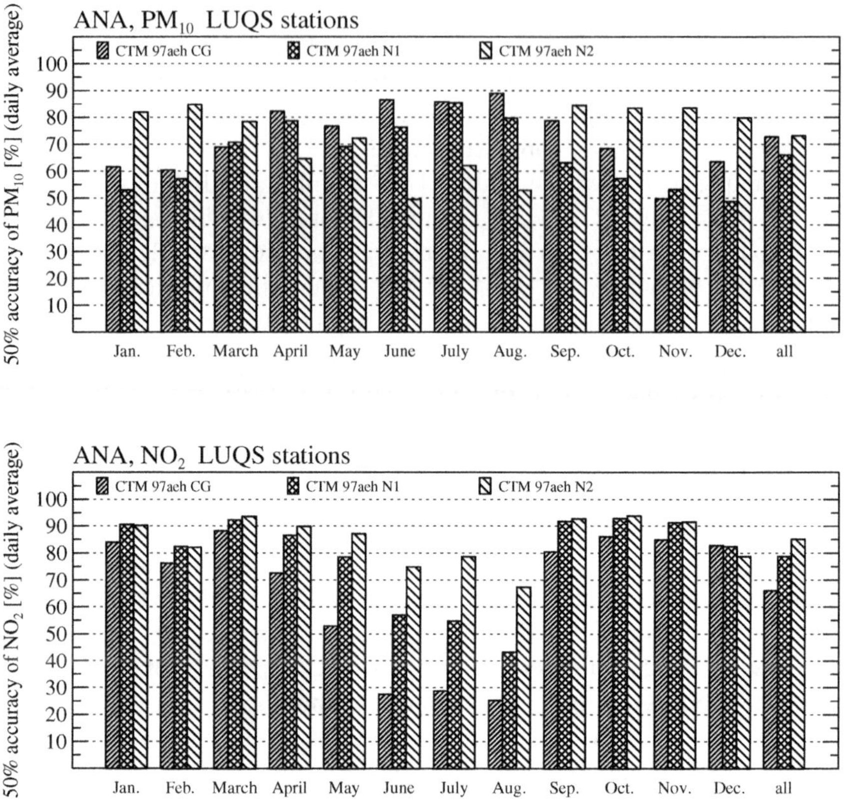

Figure 2: 50% hit rates of PM10 (upper panel) and NO_2 (lower panel) obtained for long-term simulations (1997) with the EURAD system. Comparison of coarse grid (125 km, left columns, Europe) and nested calculations (resolution 25 and 5 km, middle and right columns, respectively; Central Europe and Northrhine-Westphalia).

6. CHEMICAL DATA ASSIMILATION

The development of assimilation schemes for chemical data has made considerable progress during recent years and is already successfully applied in episodic simulations of lower atmosphere composition. For instance, van Loon et al. (2000) used the method of Kalman filtering, whereas Elbern et al. (1997) and Elbern and Schmidt (2001) demonstrated the applicability of 4-dimensional variational data assimilation to chemistry transport modelling. Since immense computer resources are needed for this technique it will still last some time till chemical data assimilation can fully be integrated into chemical weather forecast models. Yet for the improvement of the predictions it is absolutely necessary to use this technique in future. Applied as inverse modelling it also has the potential to solve the problem of short- and medium-range emission analysis (Elbern and Schmidt, 2002) which is needed for the increase of reliability of chemical weather forecasts.

7. CONCLUDING REMARKS

Evidently, quality and reliability of chemical weather prediction essentially depend on three factors, namely availability of sufficient computer power, accuracy of input data, i. e. initial and boundary conditions and meteorological forecasts, and sufficient complexity of accurately defined and executed chemical and physical processes. Yet it should be noted in the latter case, that increasing complexity which in principle leads to an increase of reliability needs more model parameters and other input data which are only known in a certain range of accuracy and thus may increase the uncertainty of model calculations. Therefore, an optimum may be expected for the design of CTMs regarding their complexity and reliable information content (Irwin, 2000; Ebel, 2001). Such optimum may be used to determine the computer resources needed for reliable forecasting. Using moderate resolution for regional applications without applying assimilation methods, i.e., using so-called third-generation models (Peters et al., 1995) on grids of medium size, optimum reliability may well be reached with quite moderate computational power. Furthermore, it is not reasonable to increase the performance of air quality prediction models beyond the range of accuracy which is given by measurements available for forecast evaluation.

It would be advantageous to develop a commonly accepted systematic strategy for air quality forecast model improvement. Such strategy does not yet exist. All model elements are usually mentioned as being worthy of improvement. Yet the ranking of problems by individual modellers differs considerably. The meteorological part being the basis of reliable air quality forecasts causes concerns mainly through often inadequate representation of the atmospheric boundary layer, local circulations and clouds including precipitation. Knowing that fine particles are a major cause of human health effects, strongest emphasis should be put on the improvement of aerosol chemistry and dynamics in CTMs and their numerical treatment. Also wet deposition and actinic flux estimation should be mentioned as problems which need to be solved with higher priority. With regard to the input data needed for CTMs surface parameters, lower boundary and initial conditions seem to cause the most bothering problems. Surprisingly, free troposphere/lower stratosphere conditions are ranking rather low on the priority list though episodic studies show that downward fluxes of ozone and trans-continental and –oceanic transport of pollutants may have a significant impact on the composition and chemistry of the lower troposphere. On the numerical side, problems originating from advection algorithms should be mentioned.

Initialisation of chemical fields and emission estimation as specific important aspects of chemical transport modelling will gain from ongoing chemical data assimilation activities which among other observations will also exploit remote sensing data of the troposphere from satellites. Such computationally demanding "fourth-generation models" are evolving now and will bring significant advancements of chemical weather forecast on the regional as well as global scale.

Acknowledgement: The figures and the table shown in this paper are taken from projects funded by the Environmental Agency of Northrhine-Westphalia and the German Federal Ministry for Education and Research.

REFERENCES

Ackermann, I. J., Hass, H., Memmesheimer, M., Ebel, A., Binkowski, F. B., Shankar, U., 1998, Modal Aerosol dynamics model for Europe: Development and first applications. *Atmos. Environm.* **32**, pp. 2891-2999.

Chang, J. S., Brost, R. A., Isaksen, I. S. A., Madronich, S., Middleton, P., Stockwell, W. R., Walcek, C. J., 1987, A three-dimensional Eulerian acid deposition model: Physical concepts and formulation, *J. Geophys. Res.* **92**, pp. 14681-14700.

Davidson, P., 2003, National air quality forecast capability: first steps toward implementation, US EOA 2004 National Air Quality Conference, 22 – 25 Febr. 2004, Baltimore, MD, USA; http://www.epa.gov/airnow/2004conference/tuesday/davidson.pdf

Ebel, A., 2001, Evaluation and reliability of meso-scale air pollution simulations, in: *Large-Scale Scientific Computing*, Margenov, S., Wasniewski, J., Yalamov, P., eds., Springer-Verlag Berlin Heidelberg, pp. 255-263.

Ebel, A., 2002, Changing atmospheric environment, changing views – and an air quality model's response on the regional scale, in: *Air Pollution Modelling and Its Application XV*, C. Borrego, G. Schayes, eds., Kluwer Academic/Plenum Publ., New York, Boston, Dordrecht, London, Moscow, pp. 25-36.

EEA, 2004, Forecast web sites of ground-level ozone for regions in Europe; http//www.eea.eu.int/Highlights/20030819134711/ozone

Elbern, H., Schmidt, H., 2001, Ozone episode analysis by four-dimensional variational chemistry data assimilation. *J. Geophys. Res.* **106**, pp. 3569-3690.

Elbern, H., Schmidt, H., 2002, Chemical 4-D variational data assimilation and its numerical implications for case study analyses, in: *Atmospheric Modeling*, Chock, D. P., Carmichael, G. R., eds., IMA Volumes in Mathematics and its Applications, Volume 130, pp. 165 – 184.

Elbern. H., Schmidt, H., Ebel, A., 1997, Variational data assimilation for tropospheric chemistry modeling, *J. Geophys. Res.* **102**, D13, pp. 15967-15985.

Irwin, J. S., 2000, Modeling air quality pollutant impacts, *EURASAP Newsletter,* 40, pp. 2-28.

Jakobs, H. J., Tilmes, S., Heidegger, A., Nester, K., Smiatek, G., 2002a, Short-term ozone forecasting with a network model system during summer 1999. *J. Atmos. Chem.* **42**, pp. 23-40.

Jakobs, H. J., Friese, E., Memmesheimer, M., Ebel, A., 2002b, A real-time forecast system for air pollution concentrations. in: *Proceedings from the EUROTRAC-2 Symposium 2002*, Midgley, P. M., Reuther, M., eds.; CD-ROM; Margraf Verlag, Weikersheim.

Krueger, B., 2000, Bernd Krüger's Linklist; http://homepage.boku.ac.at/krueger/ozontrop.htm

Peters, L. K., Berkowitz, C. M., Carmichael, G. R., Easter, R. C., Fairweather, G., Ghan, S. J., Hales, J. M., Leung, L. R., Pennell, W. R., Potra, F. A., Saylor, R. D., Tsang, T. T., 1995, The current state and the future direction of Eulerian models in simulating the tropospheric chemistry and transport of species: a review, *Atmos. Environm.* **29**, pp. 189-222.

Pudykiewicz, J. A., Kallaur, A., Moffet, R., Bouchet, V. S., Jean, M., Makar, P. A., Moran, M. D., Gong, W., Venkatesh, S., 2003, Operational air quality forecasting in Canada: numerical model guidance for ground-level ozone and particulate matter, 5th Conference on Atmos. Chem. Gases, Aerosols, and Clouds, 83rd Annual, AMS, 10 – 13 Febr. 2003, Long Beach, CA, USA; http://ams.confex.com/ams/pdfpapers/54490.pdf

Schere, K., 2003, NOAA air quality forecast modelling program, 2003 National Air Quality Conference, 2 – 5 Febr. 2003, San Antonio, Texas, USA; http://www.epa.gov/airnow/2003 conference/presentations2003/KenSchere.pdf

Tilmes, S., Brandt, J., Flatoy, F., Bergström, R., Flemming, J., Langner, J., Christensen, J. H., Frohn, L. M., Hov, O., Jacobsen, I., Reimer, E., Stern, R., Zimmermann, J., 2002, Comparison of five Eulerian Air Pollution Forecasting Systems for the summer of 1999 using the German ozone monitoring data, *J. Atmos. Chem.* **42**, pp. 91-121.

van Loon, M., Builtjes, P. J. H., Segers, A. J., 2000, Data assimilation of ozone in the atmospheric transport chemistry model LOTOS, *Environm. Mod. Software* **15**, pp. 603-609.

ALTERNATIVE TECHNIQUES FOR STUDYING / MODELING THE AIR POLLUTION LEVEL

Liviu – Daniel Galatchi[(1)]
[(1)] *Ovidius University, Department of Ecology and Environmental Protection, Mamaia Boulevard 124, 900241 Constanta - 3, Romania*

Abstract: There are various aspects of the air pollution, its measurement, modeling, control and effects. They might be considered to be ones of the most effective means of informing and educating university students in modeling problems of the atmospheric environment and air quality protection. Four experiments which might be carried out are described: collection and identification of biogenic matter in air, determination of the effect of ozone on rubber under laboratory conditions and evaluation of ozone in ambient air using the effect of rubber cracking, lichens as indicators of air pollution, and how an aerosol diffuses and how its diffusion is affected by temperature inversion.

Key words: Alternative techniques, air quality, air pollution modeling, ecology, practical training.

1. INTRODUCTION

There are plenty of opportunities to develop experiments which might be carried out as alternative techniques for studying and modeling the air pollution level at universities, particularly in the bio-ecological studies. The performance of the experiments requires a certain amount of the basic scientific background usually taught at universities. It is thus felt that performance of the problems of air pollution should contribute to a practical training in scientific disciplines.

Four experiments that might be carried out are proposed, related to the atmosphere:

a) collection and identification of biogenic matter in air;

I. Farago et al. (eds.), Advances in Air Pollution Modeling for Environmental Security, 165–174.

b) determination of the effect of ozone on rubber under laboratory conditions and evaluation of ozone in ambient air using the effect of rubber cracking;

c) lichens as indicators of air pollution;

d) how an aerosol diffuses and how its diffusion is affected by temperature inversion.

2. COLLECTION AND IDENTIFICATION OF BIOGENIC MATTER

As well as man-made pollutants, air is likely to be carrying what can be called "*natural pollutants*": dust, pollen, spores. The latter may be easily collected and inspected. In order to do that, a microscope slide is thinly smeared with petroleum jelly and left exposed to air for a few hours. Pollen grains are likely to adhere to the jelly, and can be inspected under a microscope.

The jelly should be covered with a cover slip, and examined under low power. Pollen grains vary enormously in size (from 0.0025 mm to 0.25 mm in diameter) and in color (and this can be distinguished under low power). Under high power (as x 400) the external structure of the grains is easily visible.

The concentration of pollen in the air is dramatically increasing in early summer. Figure 1 shows a chart of pollen concentrations from measurements made in the period May – September 2003 in Constanta, Romania. The pollen count is a figure referring to the average number of pollen grains in a cubic meter of air. It is extremely interesting to plot the pollen count obtained from local authority sources against meteorological data (humidity etc.) to see if there is any correlation.

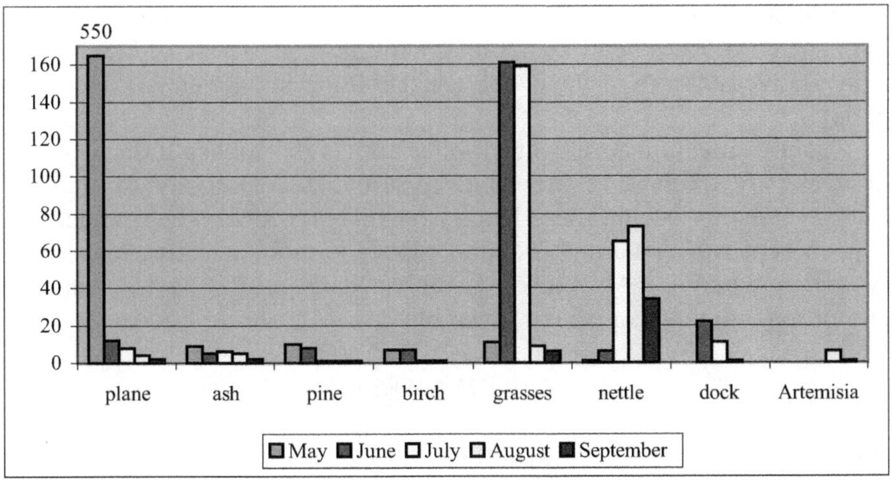

Figure 1: Pollen in Constanta, Romania, 2003 (average monthly concentration in a cubic meter of air).

3. DETERMINATION OF THE EFFECT OF OZONE ON RUBBER UNDER LABORATORY CONDITIONS AND EVALUATION OF OZONE IN AMBIENT AIR USING THE EFFECT OF RUBBER CRACKING

One of the most important oxidants, which appear in photochemical smog, is ozone (O_3). It rapidly oxidizes most organic materials, and seriously affects health and damages plants by itself or via its highly toxic reaction products. The rupture of double bonds in rubber resulting in so-called "rubber cracking" is mainly due to ozone. Ozone is formed in the atmosphere by the addition of oxygen atoms to molecular oxygen. In polluted air the oxygen atoms are produced by the photolysis of nitrogen dioxide (NO_2).

From the natural ozone layer in the stratosphere a small fraction of ozone reaches the lower regions of the atmosphere by vertical mixing, establishing occasionally a natural background ozone concentration of at most 0.03 ppm near ground level. Higher concentrations, decreasing with height in contrast to the profile of the natural ozone concentration, which increases with height, must be considered as secondary pollutants due to the photolysis of NO_2, which is directly emitted from combustion processes or is an atmospheric reaction product of nitrogen monoxide (NO). Certain chain reactions, especially in the presence of hydrocarbons, ensure a rapid transformation of NO into NO_2. Therefore, nitrogen oxides and hydrocarbons are called the precursors of ozone, the major component of photo-oxidants.

The object of the proposed experiments is the determination of the effect of ozone on rubber under laboratory conditions and the use of these effects as a qualitative measure of the ozone concentration in ambient air as a function of day.

Light to medium ozone attack only affects the rubber surface producing little cracks, whereas severe attack destroys the elasticity of the rubber. Rubber cracking occurs more readily on rubber which is held under tension (e.g. in bent rubber tubing). Natural rubber is more sensitive to ozone than synthetic materials. As expected, rubber damage RD is to some extent proportional to the ozone dose, given by:

$$RD = [O_3] \cdot t$$

where $[O_3]$ = ozone concentration (ppm), and t = time for which the rubber material was exposed to ozone (hr) (Hunter and Wohlers, 2000).

In the laboratory, irradiating ambient air with ultraviolet light produces ozone. As a light source the mercury resonance line at 1.849 Å can be used. At this wavelength molecular oxygen is photolysed (1), and the oxygen atoms recombine with other oxygen molecules in a three-body reaction to ozone (2):

(1) $O_2 + h_u \rightarrow 2\ O$

(2) $O + O_2 + air \rightarrow O_3 + air$

Two experiments have been developed:

a) Rubber cracking by an ozonized air stream;

b) Evaluation of ozone in ambient air by means of rubber cracking.

3.1 Rubber cracking by an ozonized air stream

The following instruments and materials are required:

- Flow meter for measuring the flow rate of the air stream.
- A quartz tube 16 cm in length and 1.1 cm in outer diameter.
- A glass tube about 16 cm in length and 1.5 cm in outer diameter, which can be disconnected by two standard glass joints.
- A mercury resonance lamp with an appropriate power source.
- Natural rubber tubing pieces.

- Needle valve or a simple valve for regulating the airflow roughly (a stop-cock can be used).
- A mechanical pump with a pumping speed of about 70 $dm^3 * h^{-1}$ or higher.

The ambient air is pumped through a flow system in the experiment. A pumping speed of about 70 $dm^3 * h^{-1}$ should be obtained by a normal laboratory backing pump. Any valve between flow tube and pump attenuates the flow. The gas flow can be controlled at the inlet of the flow system by a "rota meter" or in the exhaust line of the backing pump by a "bubble meter".

Care should be taken that the flow rate remains constant during the experiment. The air first passes a quartz tube 16 cm in length and 1.1 cm in outer diameter parallel to a mercury lamp. Thereafter, connected by plastic tubing, the gas stream enters a glass tube 22 cm in length and 5.5 cm in outer diameter that is finally connected by another plastic tube to the regulating valve and the backing pump. It should be possible to open the latter glass tube easily by disconnecting two standard glass joints, in order to introduce pieces of rubber tube about 16 cm in length and 1.5 cm in outer diameter.

The pieces of rubber tube have first to be bent and fixed in that position by an insulated wire before introduction (there must be no metal surfaces in the gas stream because ozone is rapidly destroyed at metal surfaces). The ozone is produced by photo dissociation of molecular oxygen with the ultraviolet light of a mercury resonance lamp. An ORIEL spectral calibration lamp of the "pen-ray" type can be used as a source (the researcher should not look directly into such a lamp). This lamp has the shape of a capillary, 5 cm in length and 0.6 in outer diameter.

An appropriate transformer as power supply operates the mercury lamp. Mercury resonance lamps emit an Hg spectral line at 2.537 Å, which is transmitted through any quartz glass but not through normal glass, and a line at 1.849 Å, which is transmitted only through clean quartz windows. The mercury lamp is set parallel to the quartz glass tube of the flow system at a distance of 0.4 cm. By variation of the lamp length (the lamp is covered partially with a thin movable metal foil), the light intensity can be changed.

In the experiment described, the steady state concentration of ozone, which builds up in the flow tube downstream of the photolysis lamp, is determined spectroscopically by measuring the absorption of the Hg 2.537 Å line intensity by ozone in a cuvette of known length. For such an ozone measurement, an additional mercury resonance lamp, an interference filter or a monochromator for selecting the Hg 2.537 Å line, and a photoelectric light detector are necessary. The relative amount $\Delta I * I_o^{-1}$ of 2.537 Å radiation absorbed by a concentration c ppm of ozone of path length $*$ cm is given by the formula:

$$\Delta I * I_o^{-1} = 1 - 10^{-kxc}$$

$$c = -\ 8{,}000 * \log (1 - \Delta I * I_o^{-1}) * x^{-1} \text{ (ppm)}.$$

I_o is the intensity of the 2.537 Å radiation at zero ozone concentration and k = $1.25 * 10^{-4}$ $ppm^{-1} * cm^{-1}$ is the decadic absorption coefficient at 23 °C, x is the path length of the cuvette in cm, and $\Delta I = I_o - I$ the absorbed light intensity (Keith, 1996).

Any other ozone meter can also be used. Relative measurement on rubber cracking can also be done without any ozone meter. With the lamp in full operation at atmospheric pressure in the flow tube, a steady state concentration of about 0.6 ppm O_3 can be obtained, which depends on the airflow rate. If only a fraction of the lamp is used, the ozone concentration can easily be reduced to about 0.05 ppm. Provided the lamp – quartz tube distance is not changed, the ozone concentration should always be proportional to the lamp surface, which is allowed to illuminate the quartz tube.

Table 1 lists the rubber tube probes, which were exposed to different ozone concentrations in the glass tube of the flow system for different lengths of time at a flow rate of 70 $dm^3 * h^{-1}$.

Probe no.	Exposure Time (hr)	O_3 Concentration (ppm)	O_3 Dose (ppm * hr)
1.	0.3	0.6	0.18
2.	0.5	0.6	0.30
3.	1.0	0.6	0.60
4.	3.5	0.6	2.10
5.	18.0	0.6	10.80
6.	0.3	0.05	.0.015
7.	2.0	0.05	0.10
8.	4.0	0.05	0.20
9.	6.0	0.05	0.30
10.	22.0	0.05	1.10

Table 1: Exposure time, O_3 concentration, and O_3 dose on a rubber cracking experiment by an oxidized air stream.

By eye inspection, aided by a pocket lens, rubber damage can be observed from a dose value of 0.10 ppm * h upwards. A dose of 10 ppm * h shows severe destruction of the rubber tube. Comparison of probe no. 2 with probe no. 9, which were both exposed to the same dose, indicates a slightly stronger damage of the rubber surface with lower ozone concentration but longer exposure times. In ambient air O_3 concentrations may range from 0.01 ppm (background) to 0.2 ppm in photochemical smog. During heavy smog periods O_3 concentrations reach values of 0.8 ppm. Therefore, in photochemical smog, exposure of bent rubber tubes for about 2 hours should always show a remarkable rubber cracking effect.

3.2 Evaluation of ozone in ambient air by means of rubber cracking

10 pieces from a rubber tube all about 15 cm in length and 1.5 cm in outer diameter should be prepared. The rubber pieces should be fixed in a bent position by string or insulated metal wire and stored in a dark box.

In the experiment, every hour from about 8 a.m. until 8 p.m., a probe is exposed for duration of 2 hours. One probe should be kept in the ambient air for 24 hours. One probe should be exposed free in the outdoor air not too close to the walls of buildings.

After collecting the probes, the degree of rubber cracking can be checked and related to the time of day exposure and the duration of exposure.

By comparing the different probes, the time of the day of most severe rubber cracking, which is related to the highest daily concentration, can possibly be determined. If such a maximum in rubber damage does not correlate with the daily change of the sun intensity, the findings indicate that ozone has been transported in from other urban regions, which might be identified on a map by taking wind directions and sites with large industries or dense traffic into account.

By comparison of the degree of rubber cracking on the outdoor probes with the cracking of the probes exposed in the flow apparatus, a rough estimate might be possible of the ozone doses and the related concentrations in the ambient air.

4. LICHENS AS INDICATORS OF AIR POLLUTION

Lichens are sensitive to sulphur dioxide in the atmosphere, and air pollution maps can be based on lichen distribution maps.

The fact that lichens are sensitive to air pollution has been realized for over 100 years. Although the metabolism of lichens is not yet fully understood, it is most likely that sulphur dioxide kills them by reducing the rates of photosynthesis and respiration. Fluorides also affect lichens, but the effect of fluorides is usually limited to the neighborhood of aluminum smelters and similar industrial sites. Sulphur dioxide travels from its source at considerable distances in the atmosphere, and lichen populations are thought to reflect the average levels of this gas in the air, rather than isolated peaks of higher concentrations. Smoke does not seem to affect the growth or vitality of lichens.

Lichens are slowly growing and long-lived. They are often abundant in unpolluted conditions. There are many lichen species and the different levels of tolerance of these species to sulphur dioxide in the air means that lichen populations and distribution maps can be drawn, which show dramatically the mean levels of this pollutant. Lichen tolerance seems to be affected by the substrate, and this factor must be taken into account when lichen distribution maps are made.

Lichens are difficult to identify. Although they exhibit a great variety of form, three main types can be distinguished by means of habit of growth and attachment to the substrate:

1. Crustose (crustaceous) lichens – form roundish or irregular crusty patches on the substrate, up to 10 cm or more across. They are closely attached to the substrate and are impossible to separate from it without crumbling the lichen into small pieces.

2. Foliose lichens – often look like many-lobed "leaves" or scales, lying flat on the substrate and attached to it by root-like threads.

3. Fruticose lichens – either erect or bush-like, or hanging tassel-like, are attached only at the base.

Just what type of lichen and substrate are chosen will, of course, depend on what range is available locally, but the example we give illustrate the approach. They are examples of simple studies, which will give evidence of the local air pollution situation and which easily could be used as air pollution modeling methods.

A. Map common wall top species, such as *Parmelia sexatilis* or *Leconora muralis* down a suspected pollution gradient, for example, along a line from a factory town out to seemingly unpolluted countryside. Then map common epiphytic lichens such as *Parmelia sulcata* or *Evernia prunastri* down the same line. In both cases, it is important to specify the substrate.

The limits of growth of *Parmelia sp.* and *Evernia sp.*, for example, will establish the principle of lichen "zones".

B. Maps can be made of lichens zones by plotting the occurrence of easily recognizable types of different substrates, for example the bright orange-yellow crustose *Xanthoria sp.* on asbestos roofs and lime-stone walls, the grayish foliose lichens *Parmelia sp.* on specific trees, the yellow-green fruticose lichens *Evernia*, *Ramalina* and *Usnea sp.* on trees.

Further work can be suggested: does the pH of bark (the pH may be determined by pressing a narrow range indicator paper against the bark when it is wet with rain) or type of tree affect survival? Does "shelter" from other trees or from walls affect survival?

Simple mapping as described above, using mapping pins on a large-scale map, will show the air pollution situation around a town or industrial complex.

5. EFFECT OF TEMPERATURE INVERSION ON POLLUTANTS DIFFUSION

The diffusion and dispersion of pollutants is strongly influenced by meteorological conditions such as temperature inversion. The object of the experiment is to show in a laboratory assembly how an aerosol diffuses and how its diffusion is affected by temperature inversion.

Air temperature decreases with height. Usually the decrease is approximately 10 °C/km. This value corresponds to the change of temperature a gas would undergo if expanded as a result of a pressure change due to an increase if 1 km in height. Sometimes, however, the temperature increases with height. The temperature gradient is in this case reversed and this condition is termed temperature inversion.

When a temperature inversion occurs, it means that a layer of warmer air sits on top of a cooler air mass, preventing the rise of the air mass. Polluted air is thus trapped below the inversion layer. If emissions are continuing, exceedingly high concentrations of pollutants can occur, which may have noxious effects.

The experiment consists of production of smoke in a suitable box in which a positive and a negative temperature gradient can be set up, and shows how the smoke rise is affected by temperature inversion.

A cabinet termed an inversion box is required to obtain a temperature gradient. It is separated into two halves by a horizontal wooden slide. The size (70 * 35 * 35 cm) is not, however, critical and any similar arrangement can be used. The front of the box is a glass plate. By using a source of warm air, a temperature gradient can be produced as described below (Liberti, 1995).

Two thermometers (0 – 50 °C) and a flashlight are needed. Black smoke can be obtained by burning any carbon-containing material (a piece of rubber, rag soaked with kerosene etc.). By blowing air into two bottles containing concentrated hydrochloric acid and ammonia, a white smoke of ammonium chloride is obtained. The smoke will be sent into the box from the opening at the bottom. If ammonium chloride or other white smokes are used, the inside of the box has to be painted to have a better view of the experiment.

Remove the slide and with all holes open insert the tubing for smoke emission at the bottom of the box and gently blow air through the bottles to obtain ammonium chloride smoke. The smoke diffuses and in a short time will fill the box.

After blowing clean air through the box to get rid of the smoke, set the slide in place and close the holes in the lower half of the box.

Create an increase in temperature in the upper half of the box by inserting the source of warm air into one of the holes. Check the temperature with a thermometer and record it. Close all holes and remove the slide and gently introduce some smoke from the bottom. Observe the position of the smoke layer.

Heat the metal surface, which covers the bottom of the box. A draught forms, which will break the inversion layer, and diffusion of the smoke into the upper part of the box will occur.

The formation of the inversion layer can be seen in daylight. A better view is obtained in the dark by shining the flashlight through the hole at the top of the box, producing a thin light beam.

Experiments can be carried out with different temperature gradients. It can be shown that the higher the temperature, the more stable the inversion layer.

6. CONCLUDING REMARKS

The described experiments fulfill a double aim: a social one as the students evaluate the air they breathe and its alterations as a consequence of human activities, and a technical one as students acquire experience, through the practical work, on scientific principles upon which experiments are based.

As air pollution affects the environment from various aspects, it is felt that experiments of an interdisciplinary nature might have a certain influence on the students in biology and ecology themselves to make them aware of the complex nature of environmental problems. The equipment required for the experiments is simple and inexpensive, and there are also given directions for the construction of some equipment by the students.

In spite of the simplicity of most experiments, the results that can be obtained are quite significant and can supply useful information to local authorities engaged in the evaluation of air quality.

Acknowledgement: We thank Prof. Stoica – Preda Godeanu for helpful suggestions. Our thought and gratitude are going to Professor Florian Vasiliu, who will remain forever in our memories.

REFERENCES

Hunter, D. C., Wohlers, H. C., 2000, Methods of Air Sampling Analysis. American Public Health Association, Washington, D. C.

Keith, L. H., 1996, Principles of Environmental Sampling, American Chemical Society, Washington, D. C.

Liberti, A., 1995, Manual of Experiments for Science Students. Council of Europe.

APPLICATION OF FUNCTIONS OF INFLUENCE IN AIR POLLUTION PROBLEMS

Kostadin Ganev[(1)], Nikolai Miloshev[(1)], and Dimitrios Melas[(2)]
[(1)]Institute of Geophysics, Bulgarian Academy of SciencesAcad. G. Bonchev Str., block.3, Sofia 1113, Bulgaria; [(2)]Laboratory of Atmospheric physics, Aristotle University of Thessaloniki Campus Box 149, 54006 Thessaloniki, Greece

Abstract: The present paper aims at giving a general idea of the functions of influence and their application to air pollution and environmental problems, introducing a rather general formulation of the problem; application of the functions of influence in pollution abatement strategies and model sensitivity studies. The functions of influence had been originally defined for the case of linear chemistry as solutions of an equation conjugated to the pollution transport one. The problem of functions of influence in case of small disturbances of model parameters is introduced in the present paper, which makes it possible to apply the approach of functions of influence for the case of nonlinear chemistry. Some applications of the functions of influence in pollution abatement strategies, model sensitivity studies and model parameters optimization are discussed.

Key words: Air pollution, pollution criteria, conjugated equations, small parameter variations, functions of influence.

1. INTRODUCTION

The functions of influence are solutions of an equation conjugated to the pollution transport one. They depend on the meteorological conditions and the domain where pollution has to be estimated, and on the choice of the air pollution characteristic (for example pollution quantity, concentration, deposition, etc) but not on the air pollution sources. Thus if the functions of influence are known for a chosen domain and pollution characteristic, the pollution characteristic value can be easily obtained for arbitrary pollution

I. Farago et al. (eds.), Advances in Air Pollution Modeling for Environmental Security, 175–184.

sources. That is why they have many interesting and useful applications in environmental problems.

2. THE BASIC POLLUTION TRANSPORT AND TRANSFORMATION PROBLEM

The evolution of N interacting compounds in a region $D=\{(x,y)\in S, 0<z<1\}$ is described by the system of equations

$$\frac{\partial c_i}{\partial t}+L_{ij}c_j+\alpha_{ij}c_j+A_i+\alpha_{wij}c_j+A_{wi}=Q_i\,, \qquad i=1,...,N\,, \tag{1}$$

where $Q_i(x,y,z,t)$ is the volume source and $c_i(x,y,z,t)$ is the concentration, $L_{ij}=\frac{\partial}{\partial x}(u-k\frac{\partial}{\partial x})+\frac{\partial}{\partial y}(v-k\frac{\partial}{\partial y})+\frac{\partial}{\partial z}(w_{ij}-k_z\frac{\partial}{\partial z})$ - the transport operator; $\{w_{ij}\}$ - a diagonal matrix whose main diagonal elements are $w_{ij}=w+w_{gi}$, w - the vertical component of the velocity of atmospheric motions, w_{gi} - gravitational deposition velocity of the i th admixture; $\{\alpha_{ij}\}$ - the NxN square matrix of the coefficients of linear chemical transformations and $\{\alpha_{wij}\}$ - the NxN square matrix (usually diagonal) of the wash-out coefficients; A_i - the sources and sinks of the i th compound, due to nonlinear chemical transformations, A_{wi} - the term describing nonlinear wash-out processes for of the i th compound.

Summation according to repeated indexes is inferred in (1) and further in the text.

Let (1) be solved under the following initial conditions

$$c_i=c_i^0 \text{ at } t=0\,,\ i=1,...,N\,, \tag{2}$$

and with boundary conditions

$$k_z\frac{\partial c_i}{\partial z}-w_{ij}c_j=\beta_{ij}c_j-q_i \text{ at } z=0\,,\ i=1,...,N\,, \tag{3}$$

$$k_z \frac{\partial c_i}{\partial z} - w_{ij} c_j = \gamma_{ij} c_j \qquad \text{at } z = 1,\ i = 1,...,N, \tag{4}$$

$$a_u \frac{\partial c_i}{\partial n} = b_u (c_i - c_{iB}) \ \text{ at } (x, y) \in \sigma,\ i = 1,...,N, \tag{5}$$

where $\{\beta_{ij}\}$ and $\{\gamma_{ij}\}$ are matrixes, usually diagonal, with elements describing the absorption by the surface of the earth and the penetration through the upper boundary, $q_i(x, y, t)$ are the pollution sources at the earth's surface, $\sigma = \{ x, y) \in l_S, 0 < z < 1\}$ - the lateral boundary of D, l_S - the contour of S, $\frac{\partial}{\partial n}$ is the derivative in direction normal to σ (the normal vector assumed to be positive if directed inward), $a_u = 0$ or $a_u = 1$. Quite often $a_u = 0$ is chosen when the normal wind component u_n is positive (inflow) and $a_u = 1$ when the normal wind component u_n is non-positive (outflow) and $b_u = 1 - a_u$. In such a case $c_i = c_{iB}$ at the inflow and $\frac{\partial c_i}{\partial n} = 0$ at the outflow parts of the lateral boundary, which seems to be a realistic boundary condition.

Some integral air pollution characteristics may be of practical interest together with the concentration field. A rather general definition of such characteristics is the linear functional[1]

$$K = \iiint_D \left(m_i c_i \big|_{t=T} + m_i^0 c_i^0 \right) dD +$$
$$\int_0^T \left[\iiint_D \left(p_i c_i + r_i Q_i \right) dD + \iint_S \left(d_i c_i \big|_{z=1} + g_i c_i \big|_{z=0} + o_i q_i \right) dS \right] dt. \tag{6}$$

It can be easily seen that, depending on the choice of the parameters $m_i(x, y, z)$, $m_i^0(x, y, z)$, $p_i(x, y, z, t)$, $r_i(x, y, z, t)$, $g_i(x, y, t)$, $d_i(x, y, t)$ and $o_i(x, y, t)$, relation (4) can express also the most popular integral pollution characteristics, such as: time/space averaged concentration in a chosen sub-region of D and in particular different countries; wet and dry depositions in a chosen sub-region of D and in particular different countries; pollution fluxes through the side boundaries of a chosen sub-region of D and in particular through the different country borders; any kind of loads on soils or vegetation; etc. Apparently, depending on the explicit form of the coefficients, the functional K may concern any single compound, or arbitrary compound combinations.

3. THE CONJUGATED POLLUTION TRANSPORT PROBLEM

The functional K can be calculated when the problem (1)-(5) is solved for the time period $[0-T]$, but there is also another way to obtain it. Let, for the moment, the case of linear chemistry be assumed, i.e., $A_i = 0$ and $A_{wi} = 0$. The problem conjugated to (1)-(3) is (Marchuk, 1976, 1977, 1982, Penenko and Aloian, 1985, Ganev, 1991):

$$-\frac{\partial c_i^*}{\partial t} + L_{ij}^* c_j^* + \alpha_{ij}^* c_j^* + \alpha_{wij}^* c_j^* = p_i \tag{7}$$

$$c_i^* = m_i \quad \text{at } t = T,\ i = 1,\dots,N, \tag{8}$$

$$k_z \frac{\partial c_i^*}{\partial z} = \beta_{ij}^* c_j^* - g_i \qquad \text{at } z = 0,\ i = 1,\dots,N,$$

$$k_z \frac{\partial c_i^*}{\partial z} = \gamma_{ij}^* c_j^* + d_i \qquad \text{at } z = 1,\ i = 1,\dots,N, \tag{9}$$

and at $(x, y) \in \sigma$

$$k \frac{\partial c_i^*}{\partial n} = -(u_n - kb_u) c^* \text{ when } a_u \neq 0,\ i = 1,\dots,N, \tag{10}$$

$$c^* = 0, \text{ when } a_u = 0,\ i = 1,\dots,N. \tag{11}$$

Here $\overset{*}{c_i}(x,y,z,t)$ are the functions of influence and $L_{ij}^{*}=-(u\frac{\partial}{\partial x}+\frac{\partial}{\partial x}k\frac{\partial}{\partial x})-(v\frac{\partial}{\partial y}+\frac{\partial}{\partial y}k\frac{\partial}{\partial y})-(w_{ij}\frac{\partial}{\partial z}+\frac{\partial}{\partial z}k_z\frac{\partial}{\partial z})$.

It is relatively easy to prove, that if $\alpha_{ij}^{\ *}=\alpha_{ji}$, $\alpha_{wij}^{\ *}=\alpha_{wji}$, $\beta_{ij}^{\ *}=\beta_{ji}$, $\gamma_{ij}^{\ *}=\gamma_{ji}$, the functional (6) can be written also in the form:

$$K=\iiint\limits_D\left(c_i^{\ *}\Big|_{t=0}+m_i^0\right)c_i^0dD+\int\limits_0^T\left[\iiint\limits_D\left(c^*{}_i+r_i\right)Q_idD+\iint\limits_S\left(c_i^{\ *}\Big|_{z=0}+o_i\right)q_idS\right]dt+$$
$$\int\limits_0^T\left\{\oint\limits_{l_S}[a_ub_uc_Bc^*+(1-a_u)c_Bk\frac{\partial c^*}{\partial n}]dl\right\}dt \qquad (12)$$

The form (12) reveals the physical meaning of the functions of influence as time/space weight of the contribution of the volume and surface pollution sources, the initial pollution field and the pollution concentration at the lateral boundary to the formation of the integral pollution characteristic K. Knowing the functions of influence, one knows the contribution of each separate air pollution source. An important application of the functions of influence is also for optimization of the location and working regime of existing or planned new sources.

Some results (Ganev et al., 2002) of the studies of the pollution origin for cases of extremely large mean sulfur concentrations in Bulgaria or Northern Greece are given below as an example of the application of the method of functions of influence. Several plots of the time integrated functions of influence are shown in Figure 1, together with the relative contribution (in %) of the SO_2 sources in each of the cells of the EMEP grid.

First of all it should be noted that the direct estimations agree very well with these obtained by using the method of the functions of influence, which may be treated as an indirect evidence of the consistency of the evaluations of both models. The time integrated functions of influence reflect the wind field pattern during the whole integration period, including the space heterogeneity and the time evolution. That is why in some of the cases they are rather complex.

The study results can be briefly summarized as follows: in most of the cases more than 50% of extreme surface SO_2 pollution in Northern Greece (up to 82% in some cases) originates from Bulgarian sources, with a particularly large contribution of the "Maritza" power plants (in one of the episodes nearly 80%). There are cases, however, when the SO_2 pollution in Northern Greece is formed mostly (more then 60%) from Greek sources. In all the cases studied, more than 70% of the extreme SO_2 pollution in Bulgaria is caused by Bulgarian sources.

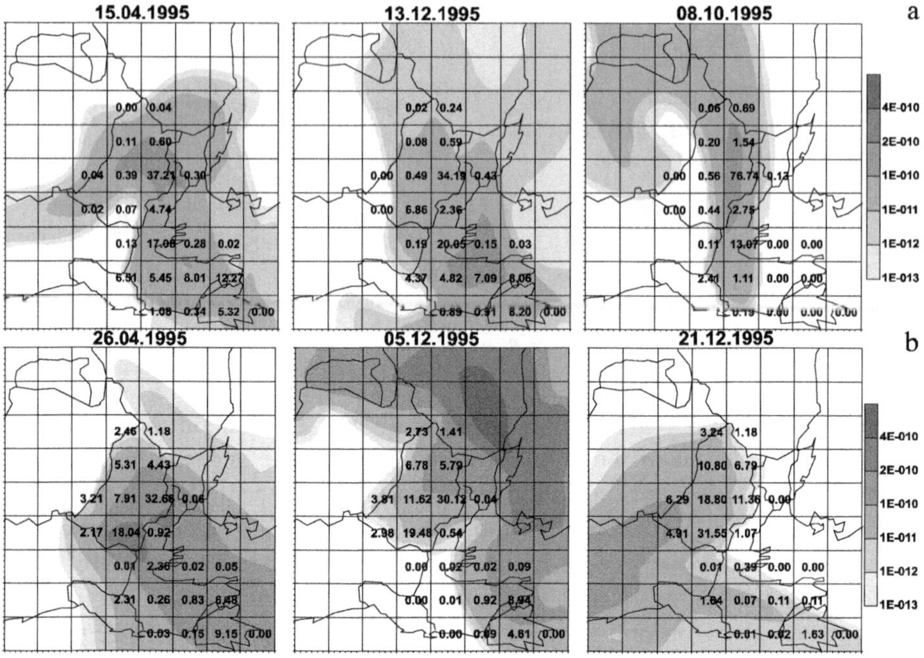

Figure 1: Horizontal cross-sections of the time integrated functions of influence [s] at z = 200 m and the contribution [%] of the sources in the corresponding EMEP cell to the 24h mean, spatially averaged surface SO_2 concentration in Northern Greece (a) or Bulgaria (b) for the respective dates.

4. FUNCTIONS OF INFLUENCE AND SMALL DISTURBANCES OF MODEL PARAMETERS

The set of parameters in (1 – 5) defines the solution:

$$u,\ v,\ w, k_z, \ldots\ldots, A_i(c_1, c_2, \ldots\ldots, c_N), \ldots, Q, q, \ldots \ \rightarrow\ c_1, c_2, \ldots\ldots, c_N$$

Some small disturbances in the parameters lead to another solution:

$$u+\delta u,\ v+\delta v,\ w+\delta w, k_z+\delta k_z, \ldots, A_i(c_1+\delta c_1, c_2+\delta c_2, \ldots, c_N+\delta c_N), \ldots,$$
$$Q+\delta Q, q+\delta q, \ldots \rightarrow c_1+\delta c_1, c_2+\delta c_2, \ldots\ldots, c_N+\delta c_N$$

If the parameter disturbances are small enough, so that $\delta L_{ij}\delta c_j << L_{ij}c_j$ and

$$A_i(c_1+\delta c_1, c_2+\delta c_2, \ldots\ldots, c_N+\delta c_N) \approx A_i(c_1, c_2, \ldots\ldots, c_N)+\frac{\partial A_i}{\partial c_j}\delta c_j$$

$$A_{wi}(c_1+\delta c_1, c_2+\delta c_2, \ldots\ldots, c_N+\delta c_N) \approx A_{wi}(c_1, c_2, \ldots\ldots, c_N)+\frac{\partial A_{wi}}{\partial c_j}\delta c_j,$$

the problem of the model sensitivity to parameter disturbances can be defined (Ganev, 2004) as

$$\frac{\partial \delta c_i}{\partial t}+L_{ij}\delta c_j+(\alpha_{ij}+\frac{\partial A_i}{\partial c_j})\delta c_j+(\alpha_{wij}+\frac{\partial A_{wi}}{\partial c_j})\delta c_j = \delta Q_i-\delta L_{ij}c_j-\delta\alpha_{ij}c_j-\delta\alpha_{wij}c_j, \quad i=1,\ldots,N, \tag{13}$$

where $\delta c_i(x,y,z,t)$ are the concentration disturbances;

$$\delta L_{ij}=\frac{\partial}{\partial x}(\delta u-\delta k\frac{\partial}{\partial x})+\frac{\partial}{\partial y}(\delta v-\delta k\frac{\partial}{\partial y})+\frac{\partial}{\partial z}(\delta w_{ij}-\delta k_z\frac{\partial}{\partial z})$$

$$\delta c_i=\delta c_i^0 \text{ at } t=0,\ i=1,\ldots,N \quad ,$$

$$k_z\frac{\partial \delta c_i}{\partial z}-w_{ij}\delta c_j=\beta_{ij}\delta c_j-(\delta q_i+\delta k_z\frac{\partial c_i}{\partial z}-\delta w_{ij}c_j-\delta\beta_{ij}c_j) \quad \text{at} \quad z=0, \quad i=1,\ldots,N, \tag{14}$$

$$k_z\frac{\partial \delta c_i}{\partial z}-w_{ij}\delta c_j=\gamma_{ij}\delta c_j-(\delta k_z\frac{\partial c_i}{\partial z}-\delta w_{ij}c_j-\delta\gamma_{ij}c_j) \quad \text{at} \quad z=1, \quad i=1,\ldots,N, \tag{15}$$

and

$$a_u \frac{\partial \delta c_i}{\partial n} = b_u (\delta c_i - \delta c_{iB}) \text{ at } (x,y) \in \sigma \text{ , } i=1,...,N \text{ .} \quad (16)$$

The problem (13 – 16) is linear, so the approach of functions of influence is applicable to it (apparently in this case $\alpha_{ij}^* = \alpha_{ji} + \frac{\partial A_j}{\partial c_i}$, $\alpha_{wij}^* = \alpha_{wji} + \frac{\partial A_{wj}}{\partial c_i}$). Thus a dual presentation of the variation $\delta K = K(c_1 + \delta c_1, c_2 + \delta c_2, ..., c_N + \delta c_N) - K(c_1, c_2, ..., c_N)$ of the pollution criteria is possible[6]:

$$\delta K = \iiint_D \left(m_i \delta c_i \big|_{t=T} + m_i^0 \delta c_i^0 \right) dD +$$
$$\int_0^T \left[\iiint_D (p_i \delta c_i + r_i \delta Q_i) dD + \iint_S \left(d_i \delta c_i \big|_{z=1} + g_i \delta c_i \big|_{z=0} + o_i \delta q_i \right) dS \right] dt \text{ ,} \quad (17)$$

and

$$\delta K = \iiint_D \left(c_i^* \big|_{t=0} + m_i^0 \right) \delta c_i^0 dD +$$
$$\int_0^T \left[\iiint_D (c^*{}_i + r_i) \widetilde{Q}_i dD + \iint_S \left(c_i^* \big|_{z=0} + o_i \right) \widetilde{q}_i dS - \iint_S c_i^* \big|_{z=1} \widetilde{\widetilde{q}}_i dS \right] dt +$$
$$\int_0^T \left\{ \oint_{l_S} [a_u b_u \delta c_B c^* + (1 - a_u) \delta c_B k \frac{\partial c^*}{\partial n}] dl \right\} dt \text{ ,} \quad (18)$$

where

$$\widetilde{Q}_i = \delta Q_i - \delta L_{ij} c_j - \delta \alpha_{ij} c_j - \delta \alpha_{wij} c_j \text{ ,}$$
$$\widetilde{q}_i = \delta q_i + (\delta k_z \frac{\partial c_i}{\partial z} - \delta w_{ij} c_j - \delta \beta_{ij} c_j)_{z=0} \text{ ,}$$
$$\widetilde{\widetilde{q}}_i = (\delta k_z \frac{\partial c_i}{\partial z} - \delta w_{ij} c_j - \delta \gamma_{ij} c_j)_{z=1} \text{ .}$$

The dual presentation allows many interesting applications, like evaluation of the contribution of each separate air pollution source to the functional K, optimization of the location and working regime of existing or planned new sources, model tuning and validation.

Assume that measurements are carried out in M different points with coordinates (x_m, y_m, z_m) and the respective measured concentrations are $c_i^m(t),\ m=1,2,...,M, i=1,2,....,N$. Then a set of MxN functionals $C_k^m = \int_0^T c_k^m\, dt,\quad m=1,..,M,\ k=1,...,N$ can be calculated using the measured concentrations.

Let $c_k(x_m, y_m, z_m, t)$ be the simulated concentrations at the same points. Then the same functionals K_k^m may be constructed using (6). In this case the parameters in (6) will be:

$$m_i = m_i^0 = r_i = d_i = g_i = o_i = 0;\ p_i = \delta_{ki}\delta(x - x_m)\delta(y - y_m)\delta(z - z_m).$$

Small variations δK_k^m of K_k^m can be defined by solving the MxN conjugated problems (7-11) and obtaining a set of functions of influence $\overset{*m}{c}_k$. Thus a set of MxN functionals δK_k^m, calculated according to (18) can be obtained. Then the system:

$$K_k^m + \delta K_k^m = C_k^m,\ m = 1,2,...,M,\ k=1,...,N, \qquad (19)$$

where K_k^m is calculated from the model solution by using (6) and δK_k^m is presented in the form (18), defines the problem of optimization of the model parameters, so that the model solution becomes as close as possible to the measured concentrations.

If the parameter variations are assumed constant with time and space, or linear combinations of defined functions with unknown coefficients, the system (19) becomes a system of linear algebraic equations for the parameter variations or the unknown coefficients respectively.

This is just an example of how the functions of influence can be used for validation of the model parameters. Obviously the functionals C_k^m may be defined in many different ways, which will result in different sets of parameters for K_k^m, different parameters for the conjugated equations, different sets of functions of influence and thus different explicit forms of the optimization problem (19).

5. CONCLUSIONS

The idea of applying functions of influence in the air pollution problems is not a new one. The aim of the present paper is just to give a more detailed and explicit formulation of some not so popular and interesting applications like studying the model sensitivity to small parameter variations, using the functions of influence in case of nonlinear chemistry, etc. These applications make the functions of influence a useful tool for air pollution and environment studies even nowadays, when the limitations of the computer resources are not so severe.

Acknowledgement: The present work is supported by the NATO Collaborative Linkage Grant EST.CLG 979794, by the BULAIR project (EU 5FP Contract EVK2-CT-2002-80024) as well as by the Greek and Bulgarian ministries of education and science in the framework of an agreement for bilateral cooperation (Contract No. ES-1212-Gr/02, NScC) and by the Bulgarian National Science Council (Contract No. ES-1002/00).

REFERENCES

Ganev, K., 1991, Formulation of the influence function problem for a complex of interacting admixtures. *Compt. Rend. Acad. bulg. Sci.* **45** (3), pp. 29-32.

Ganev, K., Syrakov, D., Zerefos, Ch., Prodanova, M., Dimitrova, R., Vasaras, A., 2002, On some cases of extreme sulfur pollution in Bulgaria or Northern Greece, *Bulg. Geoph. J.* **28**(1-4), pp. 15-22.

Ganev, K., 2004, Functions of influence and air pollution models sensitivity. *Compt. Rend. Acad. bulg. Sci.*, (accepted for publishing).

Marchuk, G. I., 1976, The environment and some optimization problems (in Russian), *Docl. AN USSR* **226**, pp. 1056-1059.

Marchuk, G. I., 1977, Methods of Computational Mathematics (in Russian). Science, Moscow.

Marchuk, G. I., 1982, Mathematical Modeling in Environmental Problems (in Russian). Science, Moscow.

Penenko, V., Aloian, A., 1985, Models and Methods for the Environmental Protection Problems (in Russian). Science, Novosibirsk.

LONG-TERM CALCULATIONS WITH A COMPREHENSIVE NESTED HEMISPHERIC AIR POLLUTION TRANSPORT MODEL

Camilla Geels, Jør gen Brandt, Jesper H. Christensen, Lise M. Frohn and Kaj M. Hansen
National Environmental Research Institute, Department of Atmospheric Environment, Frederiksborgvej 399, DK- 4000 Roskilde, Denmark

Abstract: The Danish Eulerian Hemispheric Model (DEHM) developed at the National Environmental Research Institute (NERI) in Denmark has been applied for numerous studies of air pollution. DEHM was originally developed with the main purpose of investigating the atmospheric transport of pollutants to the Arctic region, but is now also applied for more local scale environmental problems. The DEHM model has therefore recently been developed with nesting capabilities, which allows for a large mother domain (the Northern Hemisphere) as well as one or two nested domains each with a three times higher horizontal resolution over a limited area. The chemistry version of the model (DEHM-REGINA) has recently been run for a period of 14 years with a nested domain covering Europe and driven by meteorological data from the nested MM5 model. A chemical scheme with 60 species is included and the long-term calculations were made possible due to the application of a new low-cost Linux cluster system recently implemented at NERI. Examples from the validation of the model show that the observed variations and tendencies on both daily to seasonal and longer time scales are well captured for both primary and secondary air pollution components.

Key words: Transport-chemistry modeling, nesting, long-term trends.

1. INTRODUCTION

Many air pollution problems are influenced by various processes like synoptic-scale variability in the transport, heterogeneity in emissions, atmospheric chemistry, deposition processes, local variations in the boundary

I. Farago et al. (eds.), Advances in Air Pollution Modeling for Environmental Security, 185–196.

layer processes as well as long range transport. These processes all act in concert and their relative importance will depend on time/location and e.g. on the chemical species involved. In order to describe these complex problems comprehensive air pollution models have been developed through the years (Peters et al., 1995, Russell and Dennis, 2000). The models are based on our current understanding of the relevant processes and these are described by complex non-linear partial differential equations that make the models computationally intensive. This is especially true for mesoscale models where a high model resolution is included in order to represent e.g. the urban air pollution plume. The use of such models has, therefore, typically been concentrated on shorter time periods (Bauer and Langmann, 2002, Tang, 2001, Walker et al., 1999, Becker et al., 2002, Hass et al., 1997, Lu et al., 1997) due to computer limitations. Running the models for longer time periods (years to decades) can, however, give us important knowledge about the variations and trends observed in long time series of air pollution components. Model simulations provide better insight into the impact of emission reductions and changes in atmospheric transport and have e.g. shown that long-range transport of pollution from North America to Europe is strongly correlated with variations in the North Atlantic Oscillation Index (Li et al., 2002).

At NERI the Danish Eulerian Hemispheric Model (DEHM) has been developed through the last decade. DEHM is a full 3D atmospheric transport model covering the majority of the Northern Hemisphere and including multiple two-way nesting capabilities. The model thereby covers a large domain as well as the possibility to perform high-resolution simulations for smaller sub-regions. Both the long-range and local to regional scale processes are hence included in the model, which makes DEHM a useful model tool suitable for many applications. The model has through the years been extended to include different chemical species and tracers and the following model versions are currently applied:

- DEHM-Sulphur: SO_2, SO_4 and lead (Christensen, 1997)
- DEHM-Mercury: including 13 mercury species (Christensen et al., 2003)
- DEHM-CO_2 (Geels, 2003)
- DEHM-POP: Persistent Organic Pollutants (Hansen et al., 2004)
- DEHM-REGINA including 60 chemical species e.g. O_3, NH_3 and NO_x (Frohn et al. 2003; Frohn, 2004)

The different versions of DEHM have all been run for longer time periods covering several years in a setup with the meteorological model MM5.

In this paper the general setup of DEHM-REGINA and the applied meteorological models are described. The model has been validated against EMEP observations for the years 1989-2001 and examples of the model performance will be given.

2. THE DEHM MODEL SYSTEM

The Danish Eulerian Hemispheric Model is currently run in two versions in which the required high-resolution meteorological input data are obtained from the mesocale numerical weather prediction models MM5 (Grell et al. 1995) or Eta (Nickovic, 1998). As seen in Figure 1 the MM5 DEHM version is set up with a hemispheric mother domain (150 km x 150 km resolution), a nest over Europe or Greenland (50 km x 50 km resolution) and a second nest covering Denmark and near surroundings (16.67 km x 16.67 km resolution). All domains are defined on a regular 96 x 96 grid and the model can be run without nests or with one or two nests. The Eta DEHM version covers Europe with a resolution of 50 km x 50 km. In the vertical the model is divided into 20 sigma levels extending up to ca. 15 km in the atmosphere. The MM5 and Eta models are run in a corresponding setup with meteorological input data from ECMWF and NCEP, respectively.

The basic continuity equation for each chemical component is divided into sub-models. It is thereby possible to apply the most appropriate numerical technique for each sub-model separately at each time step during the integration. The horizontal and vertical advection is solved by a modified accurate space derivative scheme and a finite element scheme, respectively. For the time integration of the advection a Taylor series expansion is applied. The three-dimensional diffusion is solved by applying a finite elements scheme for the spatial discretizations and the Crank-Nicolson method for the integration in time. The time step in the integration is controlled by the Courant-Fridriechs-Lewy stability criterion and will for the mother domain typically be between 10 to 20 minutes. A detailed description of the numerical methods as well as the testing of the methods can be found in Christensen (1997), Frohn et al. (2002), and references herein.

The model system is currently run at the recently purchased cluster of nine 244 Opteron computers with dual processors. The meteorological models are parallelized over nodes and processors using Message Passing Interface, while the transport model is parallelized over the dual processors using Open Multi Processing. The computing time is thereby reduced and calculations for one year with DEHM-REGINA including one nest can now be done within 10 days on one node.

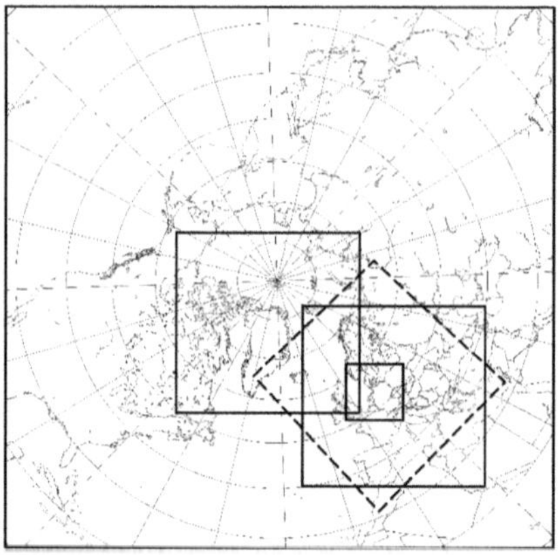

Figure 1: The DEHM domains.

2.1 DEHM-REGINA

The original DEHM model has been further developed to include a comprehensive chemistry scheme, a process that has resulted in a new 3-D nested Eulerian transport-chemistry model: DEHM-REGINA (REGIonal high resolutioN Air pollution model) (Frohn, 2004). The new chemical scheme covers 60 different species and is based on the scheme by Strand and Hov (1994). The original scheme has been modified in order to improve the description of e.g. the transformations of nitrogen containing compounds. The chemical scheme is hence extended with a detailed description of the ammonia chemistry through the inclusion of ammonia (NH_3) and related species, ammonium nitrate (NH_4NO_3), ammonium bisulphate (NH_4HSO_4), ammonium sulphate ($(NH_4)_2SO_4$) and particulate nitrate (NO_3^-) formed from nitric acid (HNO_3).

The emission of the primary pollutants is based on a combination of global data from the Global Emissions Inventory Activity (GEIA, Graedel et al., 1993) and the Emission Database for Global Atmospheric Research (EDGAR, Olivier et al., 1996) as well as European data from the European Monitoring and Evaluation Programme (EMEP, Vestreng, 2001). The implemented emission data is then composed of the EMEP data when available and either GEIA or EDGAR emissions for the rest of the domain (see Table 1). Thereby a high resolution of the emission data within the nested European domain is secured. When the second nest is included additional emission data from the GENEMIS project (Generation and Evaluation of Emission data, Schwarz et al., 2000, Wickert et al., 2001) are incorporated in the resulting input data.

Information on the seasonal variation of the NO_x emissions from lightning and soil as well as the isoprene emissions is included, but the emissions are otherwise assumed to be constant throughout the year. The emission data covers different years, see Table 1. In the European domain the emissions for 1993 have been used also in the years before 1993, while the emissions for 2000 have been used in the years after 2000.

The MM5 DEHM-REGINA version has been run for the period 1989-2002 with a nested domain over Europe and with the second nest for selected months in 1998. The Eta version has been run for the years 1999-2003 and is now running operationally four times each day and used for a 72 hour air pollution forecast in the THOR system operated at NERI.

Compound	Emission data
NO_x	EDGAR (1995), EMEP (1993-2000), GENEMIS (1994)
NO_x, lightning and soil	GEIA (1990)
SO_x	EDGAR (1995), EMEP (1993-2000), GENEMIS (1994)
NH_3	EDGAR (1990), EMEP (1993-2000), GENEMIS (1994)
CO	EDGAR (1995), EMEP (1993-2000), GENEMIS (1994)
CH_4	EDGAR (1995)
NMVOC	EDGAR (1990, 1995), EMEP (1993-2000), GENEMIS (1994)
Isoprene	GEIA (1990)

Table 1: The applied emission data. The original GEIA and EDGAR data have a horizontal resolution of 1°x1°, while the resolution of the EMEP data is 50 km. The data have been interpolated to the horizontal grid in the hemispheric and nested European domains. The original emissions are compiled for different years, which are given in the parentheses.

3. RESULTS

DEHM-REGINA has been validated against measurements from the available EMEP monitoring stations (see www.emep.int) across Europe for the period 1989-2001. The validation includes air concentrations of NO_2, O_3, SO_2, SO_4, HNO_3, NO_3, NH_3 and NH_4, wet deposition of SO_4, NO_3 and NH_4 as well as accumulated precipitation (Frohn, 2004). In the present paper the overall performance of the model will be discussed by comparing measured and simulated values of selected chemical species averaged over all the European stations. It should be noted that the number and quality of monitoring sites and data vary throughout the period in focus here. No quality criterions have been applied for the EMEP data included in the validation.

3.1 Daily time series for 1998

The observed and simulated daily time series of a few air pollution components in Figure 2 gives a view of the model's ability to simulate gas-phase transport and chemistry on daily to seasonal time scales. The correlation coefficients for the atmospheric concentration of NO_2, O_3, SO_2, SO_4 and the wet deposition of SO_4 are seen to be in the range 0.61 to 0.86 in 1998, when averaged over all the European stations.

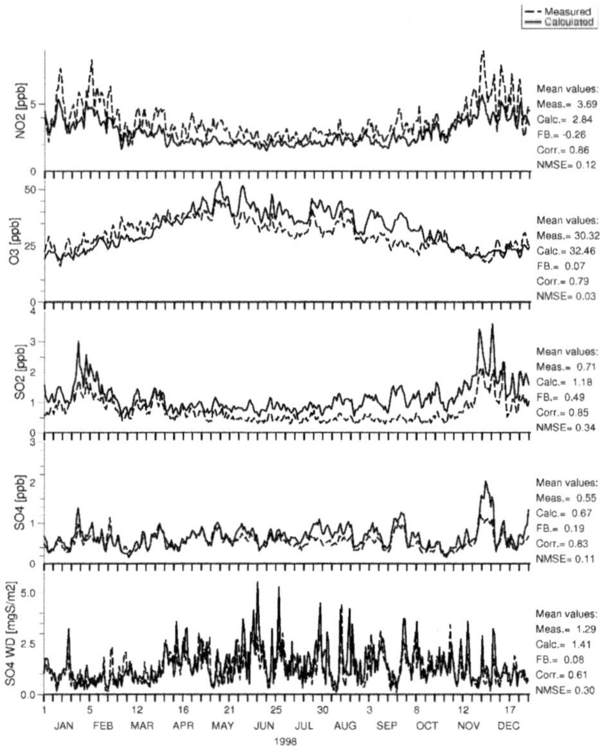

Figure 2: Daily time series of observed and simulated concentrations of NO_2, O_3, SO_2, SO_4 and wet deposited SO_4 for the year of 1998 and averaged over all available measurements. from the EMEP network. A few statistical parameters are given to the right of each plot: measured and calculated mean, fractional bias, the correlation coefficient and normalized mean square error (NMSE).

The overall seasonal cycle of higher and lower summer concentrations of O_3 and NO_2 respectively, is seen to be captured by the model.

The primary component NO_2 is an important precursor for photochemical oxidant formation, but the overall level tends to be underestimated by the model throughout most years. Such a bias is likely caused by local sources (e.g. busy roads) which will turn up in the measured time series, but will be underestimated in the simulations with the current model resolution. It should also be noted that the quality of the EMEP NO_2 observations are varying due to the application of various measuring techniques that are more or less accurate (Fagerli et al., 2003). The O_3 concentration is slightly underestimated during winter and early spring due to a general underprediction of the background level of O_3 in the model. The observed daily maximum O_3 (not shown) with a mean of 40.13 ppb is therefore underpredicted by about 6 ppb, but the general spatial and temporal variations in the daily maximum is well simulated with a significant correlation coefficient of 0.72 in 1998. The day to day variations in concentration and deposition due mainly to synoptic scale variations in atmospheric transport are also seen to be well simulated for the primary and secondary component SO_2 and SO_4, respectively.

3.2 Yearly time series for 1989 – 2001

The long-term trends in the observed and simulated concentrations of five pollutants in the period 1989 to 2001 are discussed in the following. The long-term development of the concentration of the sulphur components (SO_2, SO_4 and SO_4 in wet deposition) is to a large degree driven by the temporal and spatial changes in the emission of SO_x, see Figure 3. The total yearly emission within the EMEP domain decreases almost continuously from about 20000 ktonnes SO_x in 1989 to about 9000 ktonnes SO_x in 2000. In the same period the decrease in the concentration of SO_2 and SO_4 is, however, more pronounced in both the observed and calculated time series.

This deviation between emission and concentrations probably reflects the fact that the majority of the included EMEP monitoring sites are located in Western Europe where the cut in emissions throughout this decade is larger than the average decrease for Europe. DEHM-REGINA is generally seen to predict a higher concentration of the sulphur components than observed. This bias is nevertheless seen to decrease in the period, which indicates that the quality of the applied emission inventories increases throughout the nineties.

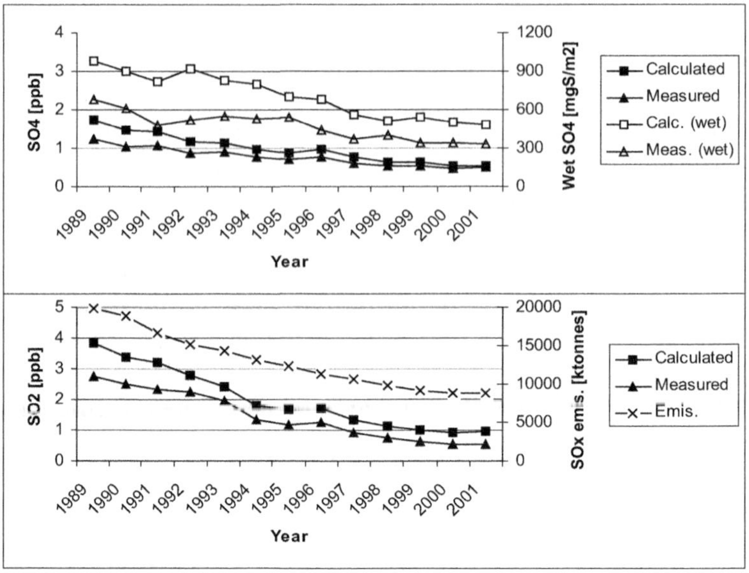

Figure 3: Examples of yearly calculated and measured time series for the period 1989-2001. Averaged over all available measurements from the EMEP network.

The overall year to year variations in the air components are generally captured by the model, while larger deviations as expected are seen in wet SO_4 due to the more complicated processes related to precipitation and the following wet scavenging.

The time series for ozone only covers 1990 and the period 1992-2001 due to limited availability of observations prior to that, see Figure 4. The model is seen to agree well with the observations both in regards to the level (with a maximum bias of 2.3 ppb in 1990) and in regards to the concentration tendencies and variations throughout the period. The observed and predicted high level of O_3 in 1999 coincide with a minimum in the accumulated precipitation and can hence be related to an enhanced production of ozone during relatively cloud free conditions this year.

As previously described, the new developments of DEHM-REGINA have been focused on an improved description of the nitrogen-containing components. Time series for two of these components (NO_3 and NH_4) are shown in Figure 4. The model is again seen to reproduce the observed time series for both nitrate and ammonium with a general decrease in the concentration from 1989 to 2001. The largest deviation is seen for nitrate where the model systematically overestimates the level with up to 0.3 ppb, while the year to year variations are well captured. The long-range transported pollutant ammonium is easier to model than other components related to nitrogen chemistry. An example is the air concentration of ammonia, which due to the local characteristics of the emission regions and the complex emission rates driven by agricultural activity and local-scale meteorological processes is difficult to simulate with a mesoscale model. Sensitivity studies with the DEHM-REGINA models have, however, shown that the observed pronounced variability in the ammonia concentration is better captured when the second nest (16.67 km x 16.67 km resolution) is included in the simulations (Frohn, 2004).

4. CONCLUSIONS

A full 3-D Eulerian transport-chemistry model covering the Northern Hemisphere and including nesting capabilities has been developed. The model, DEHM-REGINA, has been run for the period 1989-2002 on a cluster of nine 244 Opteron computers with dual processors. The model was run in a setup with the meteorological MM5 and with a nested domain over Europe. The overall performance of the model for Europe has been studied by analyzing the observed and simulated data averaged over all the available EMEP monitoring stations. Thereby site specific characteristics are averaged out and regional differences are not in focus here.

Examples from the model validation against daily EMEP data for 1998 demonstrate that the model captures the main daily to seasonal variations with a correlation coefficient above 0.79 for the air concentration of NO_2, O_3, SO_2 and SO_4. For the period 1989 to 2001 observed and simulated yearly time series of a few primary and secondary components have been compared in order to study the ability of the model to simulate long-term trends and year to year variations over Europe. Emissions of the primary pollutants have been constructed by combining the global GEIA/EDGAR data for 1990 or 1995 with EMEP emissions covering the period 1993-2000. The final emission data set thereby includes a higher resolution of 50 km x 50 km in the nested domain over Europe as well as the spatiotemporal changes of the emissions in the 1990s. The decreasing tendencies in the European emissions are clearly displayed in the time series for SO_2, SO_4, NO_3 and NH_4 in both the observed and simulated time series. A systematic overestimation is seen for the calculated sulphur components, but the observed concentration level and the year to year variations are generally well captured by DEHM-REGINA.

In this paper the first and encouraging results from a long model run with one nest have been discussed. The next step will be to analyze the results in more details at the specific EMEP stations whereby e.g. differences due to the various climate and pollution regimes across Europe can be studied. Likewise, new model calculations including the second nest over Denmark are currently running on the Linux cluster system. The high-resolution simulations are necessary when the processes related to e.g. atmospheric deposition of nitrogen and coastal eutrophication are to be studied.

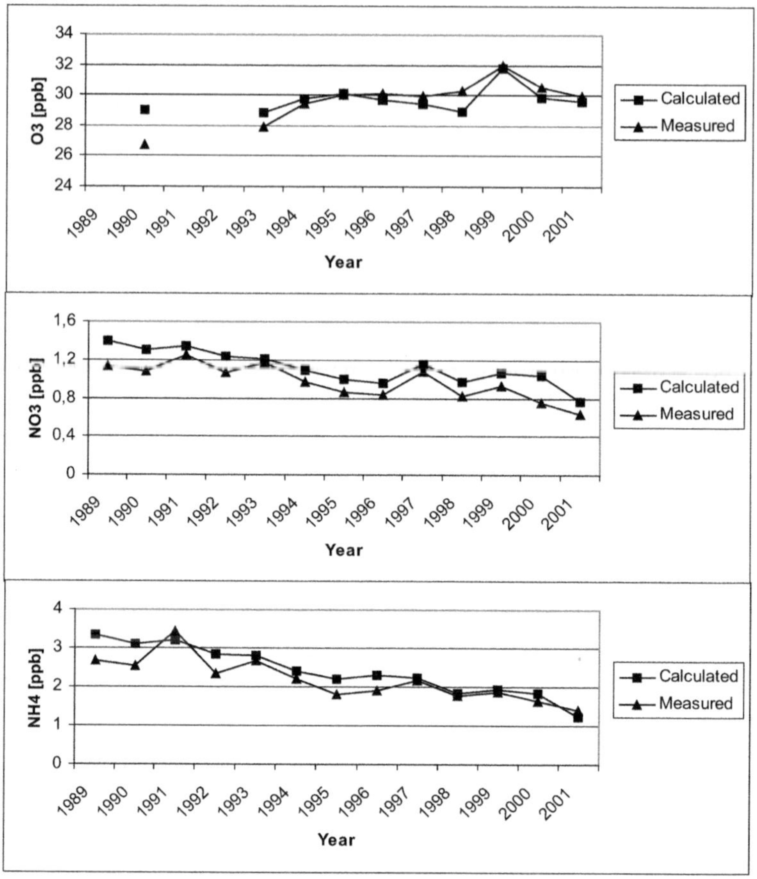

Figure 4: Examples of yearly calculated and measured time series for the period 1989-2001. Averaged over all available measurements from the EMEP network.

Acknowledgements: The Arctic Monitoring and Assessment Program supported part of the developments of DEHM-REGINA.

REFERENCES

Bauer, S. and Langmann, B., 2002, An atmosphere-chemistry model on the meso-γ scale: model description and evaluation. *Atmospheric Environment* **36**, pp. 2187-2199.

Becker, A., Scherer, B., Memmesheimer, M. and Geiß, H., 2002, Studying the City Plume of Berlin on 20 July 1998 with Three Different Modelling Approaches. *Journal of Atmospheric Chemistry* **42**, pp. 41-70.

Christensen, J. H., 1997, The Danish Eulerian Hemispheric Model – a three-dimensional air pollution model used for the Arctic. *Atm. Env.* **31**, pp. 4169–4191.

Christensen, J., Brandt, J., Frohn, L. and Skov, H., 2003, Modelling of mercury with the Danish Eulerian Hemispheric Model. *Atm. Chem. Phys. Disc.* **3**, pp. 3525-3541.

Fagerli, H., Simpson, D., Tsyro, S., Solberg, S., Aas, W., EMEP report 1/2003, Norwegian Meteorological Institute.

Frohn, L. M., Christensen, J. H. and Brandt, J., 2002, Development of a regional high resolution air pollution model – the numerical approach. *J. Comp. Phys.* **179**, No. 1, pp. 68-94.

Frohn, L. M., Christensen, J. H., Brandt, J., Geels, C. and Hansen, K. M., 2003, Validation of a 3-D hemispheric nested air pollution model. *Atm. Chem. Phys. Disc.* **3**, pp. 3543-3588.

Frohn, L. M., 2004, A study of long-term high-resolution air pollution modelling. PhD thesis. National Environmental Research Institute, Roskilde, Denmark. 444 p.

Geels, C., 2003, Simulating the current CO_2 content of the atmosphere: Including surface fluxes and transport across the Northern Hemisphere. PhD thesis, National Environmental Research Institute, Roskilde, Denmark, 238 p.

Graedel, T. E., Bates, T. S., Bouwman, A. F., Cunnold, D., Dignon, J., Fung, I., Jacob, D. J., Lamb, B. K., Logan, J. A., Marland, G., Middleton, P., Pacyna, J. M., Placet, M., and Veldt, C., 1993, A Compilation of Inventories of Emissions to the Atmosphere. *Global Biogeochem. Cycl.* **7**, pp. 1–26.

Grell, G. A., Dudhia J. and Stauffer, D. R., 1995, A Description of the Fifth-Generation PennState/NCAR Mesoscale Model (MM5). NCAR/TN-398+STR. NCAR Technical Note, p. 122. Mesoscale and Microscale Meteorology Division. National Center for Atmospheric Research. Boulder, Colorado, June 1995.

Hansen, K. M., Christensen, J. H., Brandt, J., Frohn, L. M., and Geels, C., 2004, Modelling atmospheric transport of α-hexachlorocyclohexane in the Northern Hemisphere with a 3-D dynamical transport model: DEHM-POP. *Atm. Chem. Phys.* **4**, pp. 1125-1137.

Hass, H., Builtjes, P. J. H., Simpson, D. and Stern, R., 1997, Comparison of model results obtained with several European regional air quality models. *Atm. Env.* **31**, No. 19, pp. 3259-3279.

Li, Q, Jacob, D. J., Bey, I., Palmer, P., Duncan, B. N., Field, B. D., Martin, R. V., Fiore, A. M., Yantosca, R. M., 2002, Transatlantic Transport of Pollution and its Effects on Surface Ozone in Europe and North America. *J. Geophys. Res.* **107**(D13), pp. ACH 4-1.

Lu, R., Turco, R. P. and Jacobsen, M. Z., 1997, An integrated air pollution modelling system for urban and regional scales: 1. Structure and performance. *J. Geophys. Res.* **102**, No. D5, pp. 6063-6079.

Nickovic S., Mihailovic, D., Rajkovic, B. and Papadopoulos, A., 1998, The weather forecasting system SKIRON, Vol. II, description of the model. Athens, p. 228.

Olivier, J. G. J., Bouwman, A. F., Van der Maas, C. W. M., Berdowski, J. J. M., Veldt,C., Bloos, J. P. J., Visschedijk, A. J. H., Zandveld, P. Y. J., and Haverlag, J. L., 1996, Description of EDGAR Version 2.0. A set of global emission inventories of greenhouse gases and ozone-depleting substances for all anthropogenic and most natural sources on a per country basis and on 1x 1grid, RIVM, Bilthoven, RIVM report no. 771060 002, TNO MEP report nr. R96/119.

Peters, L. K., Berkowitz, C. M., Carmichael, G. R., Easter, R. C., Fairweather, G., Ghan, S. J., Hales, J. M., Leung, L. R., Pennell, W. R., Potra, F. A., Saylor, R. D. and Tsang, T. T., 1995, The current state and future direction of Eulerian models in simulating the tropospheric chemistry and transport of trace species: A review. *Atm. Env.* **29**, No. 2, pp. 189-222.

Russell, A. and Dennis, R., 2000, NARSTO critical review of photochemical models and modeling, *Atm. Env.* **34**, pp. 2283-2324.

Schwarz, U., Wickert, B., Obermeier, A. and Friedrich, R., 2000, Generation of Atmospheric Emission Inventories in Europe with High Spatial and Temporal Resolution. Proceeding of EUROTRAC Symposium '98, Garmisch-Partenkirchen, WITpress, Southampton.

Strand, A. and Hov, Ø., 1994, A 2-dimensional global study of tropospheric ozone production. *J. Geophys. Res.* **99**(D11), pp. 22877-22895.

Tang, Y., 2002, A case study of nesting simulation for the Southern Oxidants Study 1999 at Nashville. *Atm. Env.* **36**, pp. 1691-1705.

Vestreng, V., 2001, Emission data reported to UNECE/EMEP: Evaluation of the spatial distributions of emission. EMEP/MSC-W Note 1/01, July 2001.

Walker, S.-E., Slørdal, L. H., Guerreiro, C., Gram, F. and Grønskei, K. E., 1999, Air pollution exposure monitoring and estimation. Part II. Model evaluation and population exposure. *J. Env. Monit.* **1**, pp. 321-326.

Wickert, B., Heidegger, A. and Friedrich, R., 2001, Calculations of Emissions in Europe with CAREAIR. Transport and Chemical Transformation in the Troposphere. Proceedings of EUROTRAC Symposium 2000, Garmisch-Partenkirchen, Springer.

DISPERSION MODELLING FOR ENVIRONMENTAL SECURITY: PRINCIPLES AND THEIR APPLICATION IN THE RUSSIAN REGULATORY GUIDELINE ON ACCIDENTAL RELEASES

Eugene Genikhovich
Voeikov Main Geophysical Observatory, Karbyshev Street 7, 194021 St. Petersburg, Russia

Abstract: A methodology of the use of the majorant (upper-percentile) concentration fields in the dispersion modelling for environmental security is discussed in this paper. General principles of constructing these fields are discussed, and an analytical expression for the majorant concentrations from the ground-level point source is presented here. A Russian regulatory dispersion model used for estimations of the air pollution from industrial sources had been formulated using these principles. Another Russian regulatory model dealing with accidental releases is described in this paper. This model is based on the dose approach and calculations of conditional majorants. It is used to determine the scale of contamination resulting from the accidental release of harmful chemical pollutants. Some results of validation of the models are also presented in this paper.

Key words: Air quality, concentration, dose, majorant, extremes, accidental release, atmospheric transport.

I. Farago et al. (eds.), Advances in Air Pollution Modeling for Environmental Security, 197–207.

1. INTRODUCTION

One of the main goals of the dispersion modelling is to provide a tool for supporting the policy- and decision-making. In practical applications, modelling results should be compared with officially established criteria to draw certain conclusions about safety of humans and environment. Short-term criteria are widely used in security applications, especially in those dealing with accidents. That is why this paper is focused on modelling of the short-term averaged characteristics of the air pollution.

Dispersion models usually predict mean values ("mathematical expectations") of short-time characteristics of impact of atmospheric pollutants, e.g., mean concentrations corresponding to a certain averaging time. It means, however, that such models underestimate observed concentrations in approximately 50% of the observations (assuming an approximately symmetrical distribution of individual short-term concentrations relative to their mathematical expectations). The models do not account for the influence of stochastic fluctuations, either, responsible, for example, for the plume meandering, which could enhance the magnitude of underestimation. That is why the upper-limit ("majorant") characteristics of the impact seem to be more appropriate when dealing with security issues.

A methodology of dispersion modelling with the use of the majorant concentrations had been introduced by Berlyand (1975), and its further development was discussed by Berlyand (1982), Genikhovich (1998) and Genikhovich et al. (1999). Since the 1960ies, Russian regulatory dispersion models, including their latest version presented by Berlyand et al. (1987), have exploited the majorant approach. These models can be used to calculate the "global" majorant concentration fields corresponding at each receptor point to 98th to 99th percentile of the annual PDF of short-term averaged concentrations. A somehow similar approach described by Brode (1988) was adopted by the US EPA for screening procedures. Both Russian and US dispersion models can be used to generate "global" majorant concentration fields that correspond to the "worst-case" combination of governing meteorological parameters "observed" during a year. Such characteristics are appropriate if the source operates more or less continuously in course of the year. In the case of accidental releases, however, which frequently could be most critical for the environmental security, the discharge of noxious pollutants occurs at certain "actual" combinations of governing meteorological parameters that could significantly differ from the worst-case scenario. Thus, the global majorant could noticeably overestimate the actual impact of accidental releases or any other emissions that occur comparatively rarely. A problem of constructing more efficient characteristics of the impact that are statistically stable is discussed in this paper. In this context, statistically stable ("robust") characteristics of the air pollution are introduced here too.

2. ERRORS IN DISPERSION MODELLING AND NATURAL VARIABILITY OF METEOROLOGICAL PARAMETERS

Short-term averaged concentrations predicted with a deterministic dispersion model usually correspond to the averaging time varying from 20 – 30 min in Russia to 1 hour in the US and EU countries. At given initial and boundary conditions as well as a given set of parameters determining meteorological conditions and governing physical processes under consideration, such a model should generate some values of concentration at certain receptor points. To validate this model, one should be sure that the same concentrations are well defined in the real atmosphere, which is characterized by the broad spectrum of turbulent motions. It was shown by Gifford (1959), however, that short-term averaged concentrations are stochastic variables. Genikhovich and Filatova (2001) have demonstrated that even the axial ground-level concentrations (GLCs) from a point source are heavily contaminated with noise and that the standard deviations of their stochastical fluctuations are approximately equal to their mean values. It means that the errors of predictions of axial GLCs with a deterministic model resulting from the "noisiness" of data of measurements, E_n, are equal at least about 100% (Genikhovich, 2003); correspondingly, the errors of prediction of concentrations at a given receptor point are of order of 1000% or even more. Let us compare these figures with other errors that could be expected in dispersion modelling.

Errors of predictions with atmospheric models are influenced by the following factors:

1. Modelling-, parameterization- and numerical errors;
2. Errors in input data.

Let us denote these errors as E_m and E_ε, correspondingly. The errors of the first type, E_m, are model-specific and vary very much from one dispersion model to another. At the moment it is hardly possible to characterize all the variety of the existing models with one value of E_m but the models seem to be accepted by the scientific community if their results are in agreement with data of measurements within a factor of two. So, one can use the value $E_m \sim$ 100% as a rough estimate. One can also expect that this error will be further reduced in the near future as a result of the improved understanding of physical processes governing dispersion of pollutants in the atmosphere and the increasing power of computational platforms.

The second type error, E_ε, results from uncertainties in the wind speed and wind direction, eddy diffusivities, mixing height and other parameters governing the atmospheric diffusion. Each of these components could be estimated using a standard sensitivity analysis. The following expression for the relative error of concentration in the Gaussian plume, $\delta c/c$, as a function of the relative error in the wind speed, $\delta U/U$, and absolute error in the wind direction, $\delta\varphi$, could be considered as a result of such an analysis:

$$\frac{\delta c}{c} = -\frac{\delta U}{U} + [a_1 + (a_2 \frac{y}{r} + a_3 \frac{y^3}{r^3}) r^{2(1-\alpha)}]\delta\varphi \,. \tag{1}$$

Here, r and y are along- and across-wind-direction coordinates of the receptor point; a_1, a_2, and a_3 are known coefficients; α is a constant depending on the speed of expansion of the plume in y- and z-directions (at neutral thermal stratification α is close to 1). Similarly, one can derive the expression for the relative error in the calculated concentration due to uncertainties in the mixing height and other governing parameters. For more complex models than the Gaussian one the error E_ε can be estimated using more sophisticated methods (see for example Kiselev and Gorelova, 1979, where the adjoint operator technique was used for deriving the corresponding expressions).

The errors δU and $\delta\varphi$ are influenced by the errors of measurements of these parameters. The corresponding uncertainties, however, are significantly less than the natural variability of the governing parameters in time and space. The last ones could be estimated, for example using the following empirical expressions (Nieuwstadt, Van Dop, 1982):

$$\sigma_{\varphi,r} = k_1 + k_2 \ln r; \qquad \sigma_{U,r} = k_3 + k_4 \ln r;$$
$$\sigma^2_{\varphi,U} = m_1^2 + (m_2/U)^2; \quad (\sigma_{U,U}/U)^2 = m_3^2 + (m_4/U)^2. \tag{2}$$

Here, $\sigma_{\varphi,r}$ and $\sigma_{\varphi,U}$ are standard deviations (estimated with the use of the temporal averaging) of the wind direction and wind speed at two points on the distance r (measured in km) one from another; $\sigma_{\varphi,U}$ and $\sigma_{U,U}$ are standard deviations of the wind direction and wind speed estimated with the use of the spatial averaging as functions of the mean wind speed. Empirical constants k_1, k_2, k_3, k_4, m_1, m_2, m_3, m_4 were determined using the data of experiments in the St. Louis region as follows: $k_1 = 15°$; $k_2 = 5.7°$; $k_3 = 0.47$ ms^{-1}; $k_4 = 0.24$ ms^{-1}; $m_1 = 5°$; $m_2 = 60°$; $m_3 = 0.15$; $m_4 = 0.6$ ms^{-1}.

It is evident from this analysis that the noisiness of the short-term averaged concentrations plays a dominant role in the estimates of performance of dispersion models and that corresponding errors cannot be reduced by any improvements of the model in use. It seems to be sensible in such a situation to replace dispersion modelling of "actual" (corresponding to the given set of governing meteorological parameters) concentrations with their "robust" (i.e., statistically stable) characteristics. In connection with the environmental and public security, such majorant characteristics like upper certain percentiles and maxima seem to be the most relevant characteristics of the air pollution. In such a case the modelling results could be used, for example, as a proof that the majorant characteristics are below the safety threshold. The majorants or maxima are determined in relation to the long-term (say, annual) sample of the short-term characteristics. Thus, the noise corresponding to time periods less than one year is filtered out there increasing the robustness of these majorants and maxima.

3. EXTREMES AND MAJORANTS

A well-know maximum principle for parabolic-type partial differential equations could be used as a proof that the field of the ground-level concentrations (GLCs) is limited. In the case of the finite-time single point source an analytical expression for the envelope of GLCs, c^{ENV}, was derived by Genikhovich using dimensional analysis (see Berlyand et al., 1977):

$$c^{ENV} = \frac{GM[t]_T}{r^3}. \tag{3}$$

Here M is the emission rate, and r is the distance to the receptor point from the source, G is a constant characterizing the worst-case meteorological and dispersion condition in the domain of interest, t is the diffusion time (from the moment when the source was switched on), T is the total duration of emission, and function $[t]_T$ is equal to t, if $t < T$, and equal to T, if $t > T$. The global envelope, however, seems to be a too conservative estimate for the expected air pollution; the fields of upper percentiles seem to be more informative for decision-makers.

For the sake of simplicity, let us assume that the process of the atmospheric diffusion from a stationary point source is a quasi-stationary one; it means that the short-term averaged concentration field can be found as a solution of the stationary advection-diffusion equation and that all dependence of the solution from the time is via the governing meteorological parameters $\Omega = (\omega_1, \omega_2, \omega_3 \ldots\ldots)$. In other words, the concentration of the pollutant, c, can be represented as

$$c = M \cdot F(x_1, x_2, x_3, \Omega); \tag{4}$$

where the function F implicitly depends also on technical parameters of the source (the stack height and diameter, effluent temperature and volume rate and so on). Assuming also that $\Phi(\Omega)$ is a joint PDF of governing parameters

ω_i with the density $\varphi(\Omega)$, one can find PDF of modelled concentrations, Pm(c), using the following expression (Pugachev, 1979):

$$Pm(c, x_1, x_2, x_3) = \int_{M \cdot F(x_1, x_2, x_3, \Omega) < c} \varphi(\Omega) d\Omega; \tag{5}$$

where $d\Omega = d\omega_1 d\omega_2 d\omega_3 \ldots$. The upper (1-δ) quantile of modelled concentrations, $c_{1-\delta}$, can be found as a solution of the equation $Pm(c, x_1, x_2, x_3) = 1-\delta$. It follows from Eq. (4) that $c_{1-\delta} = M \cdot \chi(x_1, x_2, x_3, \delta)$ where function χ does not depend on actual meteorological parameters but rather is governed by characteristics of their regime "hidden" in $\Phi(\Omega)$. The field $c_{1-\delta}$ is referred to as the "majorant" or "upper-limit" concentration field corresponding to the upper (1-δ) quantile. It should be noted that Pm(c) does not reproduce the influence of stochastic fluctuations of concentrations described in Eq. (1) so that Pm(c), generally speaking, should be different from the PDF of the experimentally measured concentrations, P(c). However, meandering of the plume, which is mainly responsible for these stochastic fluctuations, should not reduce highest measured concentrations because their occurrence corresponds to "direct hit" of the monitoring point by the plume. That is why one can expect a reasonably good agreement between measured and calculated majorant concentration fields.

The same majorant concentrations $c_{1-\delta}$ could be found as a solution of the minimax problem:

$$c_{1-\delta} = \underbrace{\min}_{\{S_\delta\}} \underbrace{\max}_{\{S/S_\delta\}} c(x_i, \Omega); \tag{6}$$

where S is the phase space {Ω} and S_δ is the subspace of S with the probabilistic measure equal to δ. It follows from (5) and (6) that $c_{1-\delta}$ could be estimated from the PDF of extremes of concentrations. It is well known from the general theory, that this PDF is described by the double exponential distribution (DED) $Pe(x) = \exp(-\exp(-a(x-u)))$, where "a" and "u" are parameters and x is an "individual" extremum of the short-term concentrations over some time interval (a day or a week, for example). It is obvious that $c_{1-\delta}$ can be found from the equation $Pe(c_{1-\delta}) = 1-\varepsilon$, where $\varepsilon = \varepsilon(\delta)$. Thus, $c_{1-\delta}$, in fact, depends only on two parameters, "a" and "u", determining DED.

An explicit description of S_δ was used, in particular, to derive formulae of the Russian national regulatory dispersion model (Berlyand et al., 1987). It is a multiple-source model, which is used for calculation of dispersion from point, line and area sources. The model accounts for technical parameters of the sources, characteristics of effluents, terrain and building features, and so on. It should be noted that this model can be used to generate the annual majorant of the short-term concentrations corresponding to the 98th percentile with the error inside 25% (see for example Genikhovich, 2004).

4. DISPERSION MODELLING OF ACCIDENTAL RELEASES

Accidental releases of noxious chemicals in the atmosphere unfortunately happen frequently enough to be considered as a serious problem for modern civilization. The number of victims of these accidents could be as high as several thousand like after the Bhopal tragedy. The main sources of accidents are industrial enterprises and storage facilities where poisonous chemicals are generated or stored or used as a part of technological processes (transportation of the noxious chemicals is also a possible reason for accidental releases). For these enterprises and facilities, the potential sources of accidental releases are usually well localized and possible amount of pollutants, which could be discharged in the case of accidents, is known in advance. For the emergency preparation, it is required in Russia to estimate the size of the zone (also called "the scale of contamination") around an enterprise, which could be affected by harmful pollutants in the case of accidental release at such level that certain precautions are to be taken (for example, people could be evacuated from this zone in the case of emergency). A corresponding regulatory model was introduced in Russia in the beginning of the 1990ies (Berlyand et al., 1990), and its description was given by Berlyand et al. (1994). It is based on the "dose approach" (it is often referred to as the "exposure approach"), i.e., on calculation of the doses, D, of different noxious pollutants, corresponding to 1-hour exposure time, τ, and comparison those with criteria limiting the permissible level of doses (PLD). It should be noted here that doses are monotonous functions of τ. Therefore, if the 1-hour dose, D(1 hour), corresponding to a certain hour is less than PLD, one can be sure that during this hour all doses $D(\tau)$ corresponding to exposure times $\tau < 1$ hour are limited with the same value of PLD. In other words, limiting 1-hour doses with a certain criterion means that all doses corresponding to lesser exposure times will also be limited with the same criterion.

An equation for doses was derived by integration over time of the advection-diffusion equation (ADE). It was assumed that the nonstationarity in ADE results from the nonstationary accidental discharge of pollutants into the atmosphere. The resulting equation, similar to those for the stationary concentration distribution, was solved numerically and the numerical solution, analytically approximated, is used in further computations. The model could be used for different kinds of sources (discharge of gases from tanks and pipes, evaporation of liquid spills and so on) as well as for preparation of emergency preparedness scenarios ("prognostic mode") and for calculations in actual emergency response situations ("diagnostic mode"). For the sake of simplicity, in this paper let us limit ourselves to the case of the gaseous emission.

The characteristic parameters of the accidental release like its height, orientation of impulse, and volume rate of the source are usually not known well, especially in the cases when the model works in the prognostic mode. It seems to be natural in this situation to use those combinations of these parameters that result in the highest level of impact of the release. The worst orientation of the jet emitted from the source, in particular, could be directed downward, and this situation could be simulated with the ground-level source of emission. The spatial distribution of pollutants from the ground source, which is an obvious majorant to those from the elevated source, can be given in the following form:

$$D = \frac{0.94\psi Q}{\lambda^{3/2} U x^2} \cdot e^{-\frac{1.8 y^2}{\lambda x^2}}, \tag{7}$$

where U is the wind speed at the vane level, λ is the stratification parameter, and ψ is a function of $(x/U\tau_0)$ with the time scale τ_0 expressed via the Coriolis parameter. Here, ψ is normalized with a condition $\psi(0) = 1$ and $\psi \sim (x/U\tau_0)^{1/2}$ when x tends to infinity. The parameter λ is determined as a ratio of the vertical eddy diffusivity, K_z, to the product of the wind speed U(z) and corresponding height z (this ratio should be calculated at the height $z = z_1$ with $z_1 = 1$ m). It is closely correlated with the Richardson number Ri, the Monin-Obukhov length scale L and other quantitative characteristics of the atmospheric stability. It should be stressed out that at large enough distances from the source, the influence of the source height could become negligible; thus, Eq. (7) could be also considered as an asymptotic of the corresponding distributions for elevated sources.

Eq. (7) is used for constructing a majorant dose field. An accidental release, however, usually does not continue for the whole year; that is why the approach aimed to determining the extremum of concentrations over the whole range of variations of thermal stratifications cannot be directly transferred here. It seems to be possible, instead, to introduce several "characteristic modes" of the thermal stratification of the atmosphere and derive formulae for majorant concentration fields corresponding to these modes and given values of the wind speed. Critical conditions here obviously correspond to low winds and most stable stratifications. That is why these conditions are assumed when running the model in the prognostic mode. In diagnostic mode, however, the actual meteorological conditions during the accidental release should be taken into account. In doing so, the value of the stratification parameter λ and the wind speed should be found as a certain low quantile of the probability distribution function of λ corresponding to the chosen stratification mode.

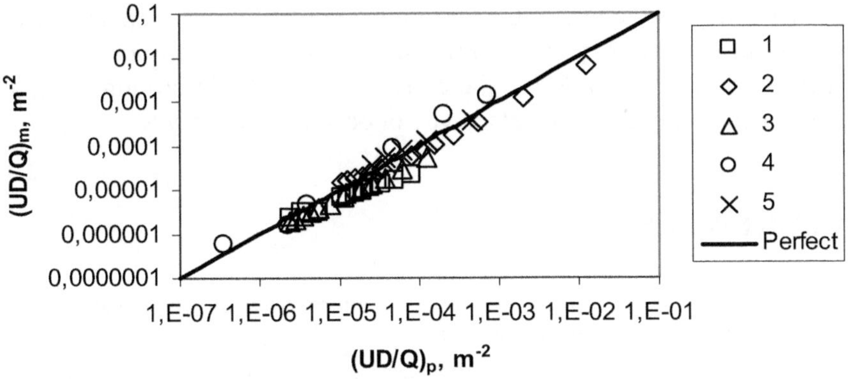

Figure 1: Comparison of predicted (p) doses with measurements (m) in the Mubarek field tests (1 – stable stratification), US field tracer experiments (2 – stable, 3 – neutral and 4 – unstable stratifications) and French wind-tunnel tests (5); the solid line corresponds to the perfect agreement.

The results of validation of this model are shown in Figure 1. Here, the marker (1) corresponds to the field tests carried out by the Main Geophysical Observatory at the Mubarek test site (Central Asia), (2), (3) and (4) represent results of validation upon experimental data collected in the well-know US field tests (Prairie Grass, Green Glow, Ocean Breeze, and Dry Gulch), and (5) is based on data of the wind-tunnel tests carried out by Solaile (1972).

5. CONCLUDING REMARKS

The existence of the stochastic components influencing meteorological variables was discussed by meteorologists for several decades. As a way to handle corresponding problems, the ensemble forecasts were introduced in the weather predicting practice. Probabilistic approaches are widely used in forecasting precipitations and other "spotty" meteorological events.

It seems to be reasonable to account for the stochastic nature of concentration field in modelling the air pollution too. The methodology in use in Russia gives an example of direct computation of the majorant concentration fields for regulatory purposes. The computed results are widely used in engineering and public practice including decision making, designing, and so on. Majorant concentration fields could be used in solution of different environmental problems where the maximal level of impact of pollutants should be limited.

The techniques for computation of majorant concentration or dose fields are still under development. This development is especially important for numerical dispersion models. When running these models, the majorants can be generated with the use of the ensemble simulation but such an approach is time-consuming and requires significant computational resources. It seems to be prospective to try to derive the equations directly describing the majorant fields but corresponding models have not been developed yet.

REFERENCES

Berlyand, M. E., 1975, Modern problems of atmospheric diffusion and air pollution, Hydrometeoizdat Publishers (in Russian).

Berlyand, M. E., 1982, Moderne Problemen der atmosphärische Diffusion und der Verschmutzung der Atmosphäre, Academie-Verlag (in German).

Berlyand, M. E., Genikhovich, E. L., Onikul, R. I., 1977, To setting the emission standards for ground-level sources. *Trudy GGO (Transactions of the Main Geophys. Observ.)* **387**, pp. 3-12 (in Russian).

Berlyand, M. E., Gasilina, N. K., Genikhovich, E. L., Onikul, R. I., Glukharev V. A. (Ed), 1987, Method for calculation of concentrations of air pollutants that industrial emissions contain. USSR Regulatory Document OND - 86. Hydrometeorological Publishers (in Russian).

Berlyand, M. E., Genikhovich, E. L., Gracheva, I. G., 1994, Modeling and forecast of air pollution under accidents. Proc. 4th Intern. conference on atmospheric sciences and applications to air quality, Seoul, Korea, pp. 91-93.

Berlyand, M. E., Suldin, Yu. I., Genikhovich, E. L., Gracheva, I. G., Malyshev, V. P., Isaev, V. S., Chicherin, S. S., Onikul, R. I., Eliseyev, V. S., Zachek, V. S., Korzunov, S. N., Semenov, V. I., 1990, Method for Forecasting of Scales of the Contamination With Poisonous Toxic Pollutant by Accidents (Destructions) at Chemically Dangerous Units and Transport. Regulatory Document RD 52.04.253-90. Hydrometeorological Publishers (in Russian).

Brode, R. W., 1988, Screening procedures for estimating the air quality impact of stationary sources – draft for public comments. Rpt. EPA-450/4-88-010, US EPA.

Genikhovich, E. L., 1998, Russian regulatory dispersion models: status, results of validation and international intercomparisons, In: Air pollution in the Ural Mountains (Ed. Linkov, I. and Wilson, R.), NATO ASI Series 2, v. 40, pp. 75-80.

Genikhovich, E. L., 2003, Indicators of performance of dispersion models and their reference values. *Int. J. Environment and Pollution* **20**, Nos. 1-6, pp. 321-329.

Genikhovich, E., 2004, Statistical approach to deterministic modelling of air pollution and its applications in Russia. In: Air pollution modelling and its application XVI (Ed. C. Borrego and S. Incecik). Kluwer Academic / Plenum Publishers, pp. 3-15.

Genikhovich E., Filatova E., 2001 A PDF approach to processing the data of tracer experiments for validation the dispersion models. In Harmonization within Atmospheric Dispersion Modelling for Regulatory Purposes, Belgirate, Italy, May 28-31. Proc. of the 7th Intern. Conf., pp. 33 - 36

Genikhovich, E. L, Berlyand, M. E., Onikul, R. I, 1999, Progress in the theory of atmospheric diffusion as a basis for development of the air pollution prevention policy. In: Modern studies at the Main Geophysical Observatory to its 150th anniversary. v. 1 (Ed. Berly, M. E. and Meleshko, V. P.). Hydrometeorological Publishers, St. Petersburg, pp. 99-126.

Gifford, F. A., 1959, Statistical plume model. In: Atmospheric diffusion and air pollution (Ed. F.N. Frenkiel, P.A.Sheppard), Academic Press, pp. 143-164.

Kiselev, V. B., Gorelova, V. V., 1979, The study of the atmospheric diffusion equation by means of perturbation theory for functionals. *Trudy GGO (Proc. Main Geoph. Observatory*) **436**, pp. 37–42 (in Russian).

Lidbetter, M. R., Lingren, G., Rootzen, H., 1986, Extremes and related properties of random sequences and processes, Springer Verlag.

Nieuwstadt, F. T. M., Van Dop, H. (Eds.), 1982, Atmospheric turbulence and air pollution modelling, D. Reidel Publishing.

Pugachev, V. S., 1979, Probability theory and mathematical statistics, Nauka Publishers (in Russian).

Solaile, J., 1972, Etude experimentalle de la couche limite turbulente en ecoulement stratifie instable, Universite Claude Bernard, Lyon.

HIGHER ORDER NON-CONFORMING FEM UP-WINDING

Krassimir Georgiev and Svetozar Margenov
Institute for Parallel Processing, Bulgarian Academy of Sciences, Acad. G. Bonchev str. Bl. 25-A, 1113 Sofia, Bulgaria

Abstract: The implementation of rotated bilinear non-conforming finite elements in the advection-diffusion part of an Eulerian model for long-range transport of air pollutants is considered. The study is aimed at a possible implementation of such elements in the operational version of the Danish Eulerian Model (DEM). One-dimensional linear finite elements are currently used during the space discretization of the advection-diffusion part of DEM. The application of more advanced numerical techniques in this part of the model is desirable.
Two variants of the rotated bilinear non-conforming finite element method (FEM) are analysed. They differ in the interpolation operator used to determine the local FEM basis of the related square reference element. In both cases, the nodal unknowns are associated with the midpoints of the edges.
The proposed monotone time-stepping scheme generalises the up-winding techniques originally developed for the case of finite difference method (FDM). It is of second order with respect to both space and time discretization parameters. The presented numerical results illustrate the behaviour of the considered new up-wind scheme. The well known in the community *rotational test* is used as a benchmark problem.

Key words: Air pollution modelling, partial differential equations, finite difference method (FDM), finite element method (FEM), non-conforming FEM, up-wind schemes.

I. Farago et al. (eds.), Advances in Air Pollution Modeling for Environmental Security, 209–218.

1. INTRODUCTION

Let us denote with q the number of the pollutants, and let $\mathbf{\Omega}$ be the domain under consideration. The Eulerian approach leads to the following system of partial differential equations describing the air pollution model:

$$\frac{\partial c_s}{\partial t} = -u\frac{\partial c_s}{\partial x} - v\frac{\partial c_s}{\partial y} + \frac{\partial}{\partial x}\left(K_x\frac{\partial c_s}{\partial x}\right) + \frac{\partial}{\partial y}\left(K_y\frac{\partial c_s}{\partial y}\right) \tag{1}$$
$$+ E_s + Q_s\left(c_1, c_2, \ldots, c_q,\right) - (k_{1s} + k_{2s})c_s, \qquad s = 1,2,\ldots,q,$$

where $c_s = c_s(x, y, t)$ are the concentrations of the pollutants; (u, v) are the wind velocities; K_x and K_y are the diffusion coefficients; the emission sources are described by E_s; k_{1s} and k_{2s} are the deposition coefficients, and finally, the chemical reactions are described by $Q_s = Q_s(c_1, c_2, \ldots, c_s)$. To get the system (1), it is assumed that the wind velocity field is *divergence free*, i.e., $u_x + v_y = 0$. As usual in the atmospheric models, an operator splitting is used (see, e.g., Zlatev, 1995). The main advantage of this approach is that when each process is treated separately, the most efficient numerical integration technique can be chosen. Here, we focus our attention on the advection-diffusion submodel described by q decoupled partial differential equations in the form

$$\frac{\partial c}{\partial t} = -u\frac{\partial c}{\partial x} - v\frac{\partial c}{\partial y} + \frac{\partial}{\partial x}\left(K_x\frac{\partial c}{\partial x}\right) + \frac{\partial}{\partial y}\left(K_y\frac{\partial c}{\partial y}\right). \tag{2}$$

To solve numerically this problem we consider FDM and FEM as discretization methods in space combined with a two–layer scheme in time. The resulting time stepping procedure can be written in the form

$$\mathbf{T}\hat{\mathbf{C}} = \mathbf{F}(\mathbf{C}), \tag{3}$$

where $\hat{\mathbf{C}}$ and $\mathbf{C}$ stand for the nodal vectors of the concentrations corresponding to the current and previous time layers, and $\mathbf{T}$ is the transfer matrix of the time stepping procedure. Our study is aimed at the construction of a robust discretization method, based on rotated bilinear non–conforming FEM in combination with up–wind approximation of the advection terms. The last is determined in terms of monotone properties of the transfer matrix.

2. HIGHER ORDER UP-WINDING IN FEM

Consider the partial differential operator

$$Lc \equiv \frac{\partial c(x,y,t)}{\partial t} + u(x,y,t)\frac{\partial c(x,y,t)}{\partial x} + v(x,y,t)\frac{\partial c(x,y,t)}{\partial y}. \qquad (4)$$

Let ω_h be a uniform mesh in Ω with mesh-sizes h_x and h_y, and time step τ. The following second order finite difference scheme, approximating the operator (4) on a five-point stencil is known as *central differences* scheme (CD), see Figure 1 (a).

CD-scheme:

$$L_h c_h \equiv \frac{(\hat{c}_{i,j} - c_{i,j})}{\tau} + \frac{1}{4h_x}\hat{u}_{i,j}\left(\hat{c}_{i+1,j} - \hat{c}_{i-1,j}\right) + \frac{1}{4h_y}\hat{v}_{i,j}\left(\hat{c}_{i,j+1} - \hat{c}_{i,j-1}\right)$$
$$+ \frac{1}{4h_x}u_{i,j}\left(c_{i+1,j} - c_{i-1,j}\right) + \frac{1}{4h_y}v_{i,j}\left(c_{i,j+1} - c_{i,j-1}\right).$$

Here $c_{i,j} = c(x_i, y_j, t^n)$, $\hat{c}_{i,j} = \hat{c}(x_i, y_j, t^{n+1})$. This scheme has second order accuracy in both space and time, i.e., $\left|Lc - L_h c\right| = O\left(h^2 + \tau^2\right)$. Unfortunately, the CD-scheme is not monotone. This leads to oscillations of the numerical solution producing negative concentrations as well.

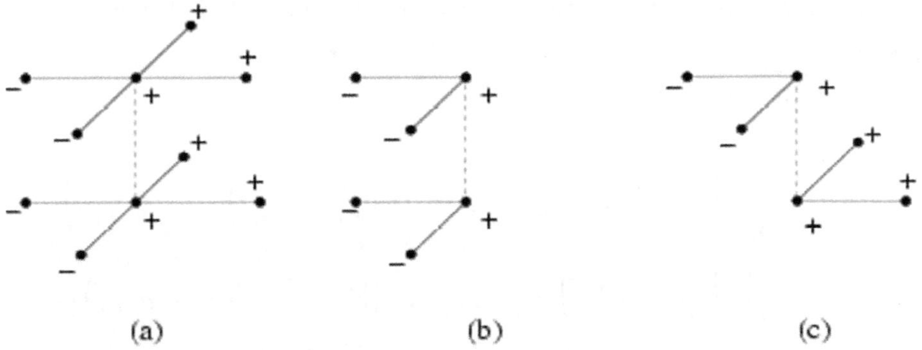

Figure 1: Central differences (a); fist order up-wind (b); and second order up-wind (c).

The first order up-wind (UW1) scheme is derived to avoid the oscillations. Figure 1(b) illustrates the FDM stencil corresponding to the case (*u—0*, *v—0*). The discretization can be presented in the following general form:

UW1-scheme:

$$L_h c \equiv \frac{\left(\hat{c}_{i,j} - c_{i,j}\right)}{\tau} +$$

$$\frac{1}{4h_x}\left[\left(\left|\hat{u}_{i,j}\right| + \hat{u}_{i,j}\right)\left(\hat{c}_{i,j} - \hat{c}_{i-1,j}\right) + \left(\left|\hat{u}_{i,j}\right| - \hat{u}_{i,j}\right)\left(\hat{c}_{i,j} - \hat{c}_{i+1,j}\right)\right] +$$

$$\frac{1}{4h_y}\left[\left(\left|\hat{v}_{i,j}\right| + \hat{v}_{i,j}\right)\left(\hat{c}_{i,j} - \hat{c}_{i,j-1}\right) + \left(\left|\hat{v}_{i,j}\right| - \hat{v}_{i,j}\right)\left(\hat{c}_{i,j} - \hat{c}_{i,j+1}\right)\right] +$$

$$\frac{1}{4h_x}\left[\left(\left|u_{i,j}\right| + u_{i,j}\right)\left(c_{i,j} - c_{i-1,j}\right) + \left(\left|u_{i,j}\right| - u_{i,j}\right)\left(c_{i,j} - c_{i+1,j}\right)\right] +$$

$$\frac{1}{4h_y}\left[\left(\left|v_{i,j}\right| + v_{i,j}\right)\left(c_{i,j} - c_{i,j-1}\right) + \left(\left|v_{i,j}\right| - v_{i,j}\right)\left(c_{i,j} - c_{i,j+1}\right)\right]$$

The scheme is unconditionally monotone. This means that the transfer matrix **T** is strictly diagonally dominated by non-positive off-diagonal entries. Unfortunately, the accuracy in space is of first order only, i.e.,

$$\left|Lc - L_h c\right| = O\left(h + \tau^2\right).$$

The presented below second order up-wind scheme (UW2) is based on a simple modification of UW1 (see, e.g., Pintchukov, 1991), which can be easily interpreted as a mesh implementation of the method of characteristics.

UW2-scheme:

$$L_h c \equiv \frac{\left(\hat{c}_{i,j} - c_{i,j}\right)}{\tau} +$$

$$\frac{1}{4h_x}\left[\left(\left|\hat{u}_{i,j}\right| + \hat{u}_{i,j}\right)\left(\hat{c}_{i,j} - \hat{c}_{i-1,j}\right) + \left(\left|\hat{u}_{i,j}\right| - \hat{u}_{i,j}\right)\left(\hat{c}_{i,j} - \hat{c}_{i+1,j}\right)\right] +$$

$$\frac{1}{4h_y}\left[\left(\left|\hat{v}_{i,j}\right| + \hat{v}_{i,j}\right)\left(\hat{c}_{i,j} - \hat{c}_{i,j-1}\right) + \left(\left|\hat{v}_{i,j}\right| - \hat{v}_{i,j}\right)\left(\hat{c}_{i,j} - \hat{c}_{i,j+1}\right)\right] -$$

$$\frac{1}{4h_y}\left[\left(\left|v_{i,j}\right| - v_{i,j}\right)\left(c_{i,j} - c_{i,j-1}\right) + \left(\left|v_{i,j}\right| + v_{i,j}\right)\left(c_{i,j} - c_{i,j+1}\right)\right] -$$

$$\frac{1}{4h_y}\left[\left(\left|v_{i,j}\right| - v_{i,j}\right)\left(c_{i,j} - c_{i,j-1}\right) + \left(\left|v_{i,j}\right| + v_{i,j}\right)\left(c_{i,j} - c_{i,j+1}\right)\right]$$

The scheme is of second order, i.e., the following error estimate holds: $\left|Lc - L_h c\right| = O\left(h^2 + \tau^2\right)$. The matrix **C** is again unconditionally monotone.

The important property of both considered up-wind schemes is, that if τ is small enough, then the right hand side in (3) is non-negative, and therefore the concentrations stay always non-negative. The related stability condition reads as $\tau \approx \frac{h_m^2}{M}$, where $h_m = \min(h_x, h_y)$ and $M = \max(u, v)$.

3. ROTATED BILINEAR NON-CONFORMING FEM

Among the most frequently used space discretization methods are the finite volume method, the Galerkin finite element method (conforming or non-conforming) or the mixed finite element method. Each of these methods has its advantages and disadvantages when the problem is used in a particular application. For example, for applications related to highly heterogeneous media the finite volume and mixed finite element methods have proven to be accurate and locally mass conservative. While applying the mixed FEM, Lagrange multipliers are introduced. Arnold and Brezzi (1985) have demonstrated that the Schur complement system for the Lagrange multipliers is equivalent to a discretization of the original problem by Galerkin method using linear non-conforming finite elements. Further, such a relationship between the mixed and non-conforming finite element methods has been studied and simplified for various finite element spaces.

The rotated bilinear elements are introduced by Rannacher and Turek (1992) as a simplest non-conforming approximation tool for the Stokes equations. Our study is further motivated by the stability properties of these elements for a wide class of parameter dependent problems. The regular sparsity structure of the FEM matrices is another advantage which can be also important for the development of efficient parallel implementation of the algorithms (see Bencheva and Margenov, 2003).

Here, we propose a new non-conforming FEM scheme for advection-diffusion problems. Some earlier results of this kind were recently published in Antonov et al., 2003. Two variants of rotated bilinear non-conforming finite elements (RbiL FEM) are considered, which differ only in the interpolation operator used to define the local basis functions $\{\varphi_i;\ i=1,\dots,4\}$ for the reference square element, see Figure 2.

Variant MP:

$$\varphi_i(x_j) = \delta_{ij}, \quad i, j = 1, \dots, 4$$

Variant MV:

$$\frac{1}{|\Gamma_j|} \int_{\Gamma_j} \varphi_i(x) dx = \delta_{ij}, \quad i, j = 1, \dots, 4$$

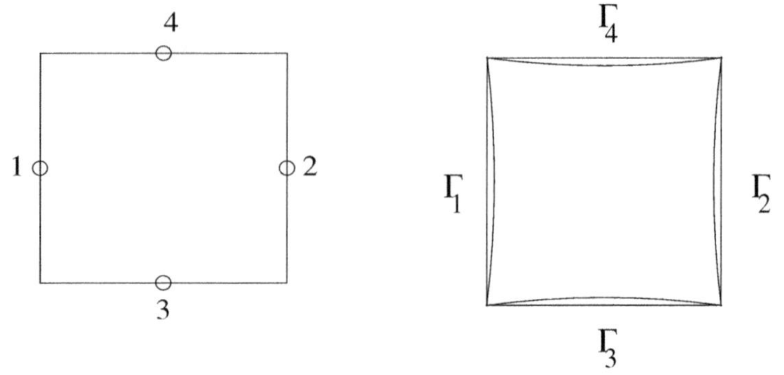

Figure 2: Reference elements for rotated bilinear non-conforming FEM: midpoint variant MP (left) and integral mid-value variant MV (right).

The standard FEM formulation of the pure advection problem can be written in the form

$$\left[\mathbf{A}_{00} + \frac{\tau}{2}\left(\mathbf{A}_{10}^{(u)} + \mathbf{A}_{01}^{(v)}\right)\right]\hat{\mathbf{C}} = \left[\mathbf{A}_{00} - \frac{\tau}{2}\left(\mathbf{A}_{10}^{(u)} + \mathbf{A}_{01}^{(v)}\right)\right]\mathbf{C} \qquad (5)$$

where $\mathbf{A}_{00}$ stands for the mass matrix and $\mathbf{A}_{10}$, $\mathbf{A}_{01}$ are the matrices representing the advection terms in x and y direction, respectively. Then, the global matrices read as

$$\mathbf{A}_{00} = \sum_{e\in T} \mathbf{A}_{00;e}, \qquad \mathbf{A}_{10}^{(u)} = \sum_{e\in T} u(e)\mathbf{A}_{10;e}, \qquad \mathbf{A}_{01}^{(v)} = \sum_{e\in T} v(e)\mathbf{A}_{01;e},$$

where $\sum$ stands for the FEM assembling procedure, applied to the current element matrices. The accuracy of this scheme is $O(h^2+\tau^2)$, but unfortunately it is not monotone.

4. HIGHER ORDER UP-WINDING IN NC FEM

The second order up-wind scheme for the case of RbiL FEM is a straightforward FEM extension of the already considered FDM construction. It is based on the explicit formulae for the element matrices, which in the variant MP are given below.

$$\mathbf{A}_{0\,;e} = \frac{h^2}{180}\begin{bmatrix} 113 & -7 & 37 & 37 \\ -7 & 113 & 37 & 37 \\ 37 & 37 & 113 & -7 \\ 37 & 37 & -7 & 113 \end{bmatrix},$$

$$\mathbf{A}_{1\,;e} = \frac{h}{6}\begin{bmatrix} 5 & 1 & 3 & 3 \\ -1 & -5 & -3 & -3 \\ -2 & 2 & 0 & 0 \\ -2 & 2 & 0 & 0 \end{bmatrix}, \quad \mathbf{A}_{0\,;e} = \frac{h}{6}\begin{bmatrix} 0 & 0 & 2 & -2 \\ 0 & 0 & 2 & -2 \\ -3 & -3 & -5 & -1 \\ 3 & 3 & 1 & 5 \end{bmatrix}$$

It is important that for RBiL FEM, the off-diagonal entries of the global matrices have a contribution from one FE only. This helps for a simpler generalisation of the FDM second order up-wind scheme.

RbiL FEM UW2

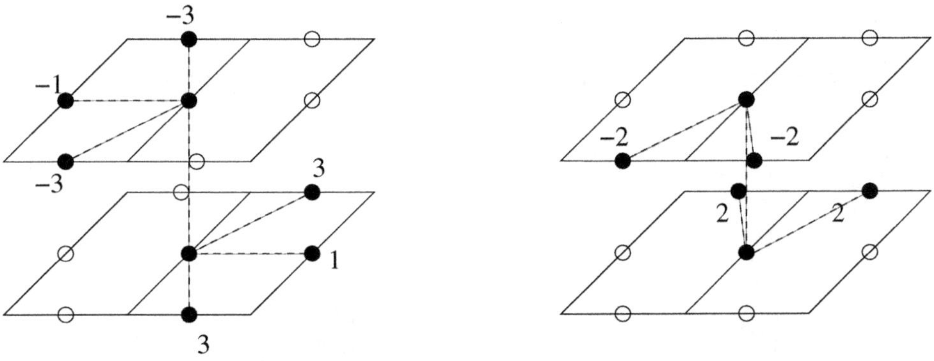

(a) (b)

Figure 3: (a) *x*-component of the advection; (b) *y*-component of the advection.

The RBiL FEM up-winding in the particular case of $u(e)>0$, $v(e)>0$ is illustrated by Figure 3. This scheme can again be viewed as a discrete interpretation of the method of characteristics. The scheme is of second order in space and in time, i.e., the accuracy is $O(h^2 + /^2)$. It also follows from the construction, that the transfer matrix **C** from (3) is always monotone. The time stepping stability condition that ensures the non-negativity of the computed concentrations follows in the same way and has the same form as in the case of the FDM up-winding procedure.

5. ROTATIONAL TEST

The *rotational test* is the most popular tool among the researchers in the fields of meteorology and environmental modelling for testing the numerical algorithms used in the large air pollution models. From a mathematical point of view, it is a pure advection problem, which easily represents a lot of the typical difficulties in the numerical treatment of such problems. The governing partial differential equation is as follows:

$$\frac{\partial c}{\partial t} = -(1-y)\frac{\partial c}{\partial x} - (x-1)\frac{\partial c}{\partial y}, \quad (x,y) \in (0,2)^2 ,$$

$$c_0(x,y) = \begin{cases} 100\left(1-\dfrac{r^*}{r}\right) & if \quad r^* < r \\ 0 & if \quad r^* < r \end{cases}$$

$$x_0 = 0.5, \quad y_0 = 1, \quad r = 0.25, \quad r^* = \sqrt{(x-x_0)^2 + (y-y_0)^2}.$$

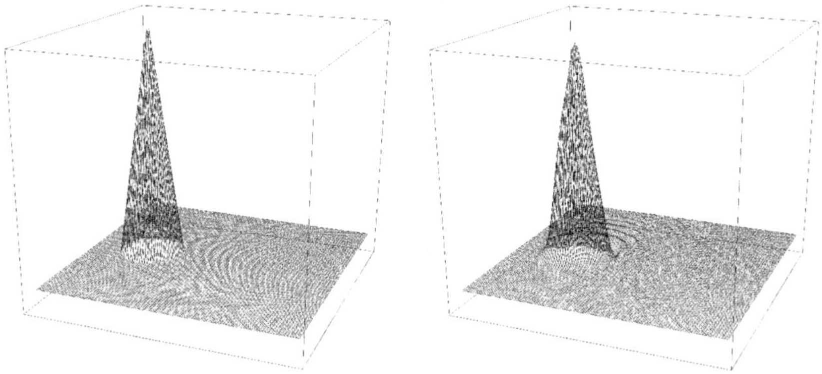

Figure 4: Cone at the beginning: h_{max} = *100* (left); Cone after one rotation for MV NC FEM: h_{max} = *94.751* (right).

Let us remind that the solution $c(x,y,t_{rot})$ after one full rotation is equal to $c_0(x,y)$. Our analysis and the derived conclusions are based on the comparison between the exact and the numerical solution. The initial cone and the numerical result obtained after one full rotation of 1600 time steps are shown below, $h = 2/63$. Our observation is that the qualitative and the quantitative behaviour of the approximate solution well fit with the theoretical forecast. Both, MP and MV variants of RBiL FEM up-wind schemes were examined, showing some stable advantages of the last one. Let us note that some other authors have also reported advantages of the

6. CONCLUDING REMARKS

We consider a new second order up-wind scheme in the framework of the rotated bilinear non-conforming FEM.

- The rotated bilinear finite elements are originally introduced as a simple approach for the numerical solution of the Stokes problem. During the years, they were recognised as a robust discretization tool for a wide class of parameter dependent problems in strongly heterogeneous media.
- The major contribution of this article is the proposed simple technique, which allows us to combine the locally conservative in-space approximation of the RBiL FEM with a higher order up-wind treatment of the advection terms.
- One additional advantage of the RBiL FEM is due to the special sparsity structure of the global matrices which provides a potential for efficient parallel implementation in the case of large-scale coupled models.

The derived new up-wind scheme is primarily addressed but not restricted to applications in air pollution models. The numerical treatment of convection-diffusion-reaction problems in porous media is another environmental problem where the proposed approach could be successfully applied.

Acknowledgements: This research was supported by the NATO Scientific Programme Grant CRG 960505, the European Commission under the Program for supporting the Centres of Excellence (contract ICA1-CT-2000-70016) and the Bulgarian NSF Grant IO-01/03.

REFERENCES

Antonov, A., Georgiev, K., Komsalova, E. and Zlatev, Z., 2003, Bilinear nonconforming finite elements in an Eulerian air pollution model: Rotational tests. *LNCS,* 2542, pp. 379-386.

Arnold, D. N. and Brezzi, F., 1985, Mixed and nonconforming finite element methods: implementation, postprocessing and error estimates. *RAIRO, Model. Math. Anal. Numer.* **19**, pp. 7-32.

Bencheva, G. and Margenov, S., 2003, Parallel incomplete factorization preconditioning of rotated linear FEM systems. *J. Comp. Appl. Mech.* V.4(2), pp. 105-117.

Pintchukov, S., 1991, On construction of monotone predictor-corrector schemes of arbitrary order of approximation. *Math. Modelling* **3**(9), pp. 95-103.

Rannacher, R. and Turek S., 1992, Simple nonconforming quadrilateral Stokes elements. *Numer. Meth. for PDE's* **8**, pp. 97-111.

Zlatev, Z., 1995, Computational treatment of large air pollution models. Kluwer Academic Publishers, Dordrecht-Boston-London.

EMISSION CONTROL IN SINGLE SPECIES AIR POLLUTION PROBLEMS

Krassimir Georgiev [(1)], Svetozar Margenov [(1)] and Vladimir M. Veliov [(2)]

[(1)] *Institute for Parallel Processing, Bulgarian Academy of Sciences, Sofia, Bulgaria;* [(2)] *Institute of Mathematics and Informatics, Bulgarian Academy of Sciences, Acad. G. Bonchev str., Bl. 8, 1113 Sofia, Bulgaria* and *Vienna University of Technology, Argentinierstr. 5/119, 1040 Vienna, Austria*

Abstract: To control the emission of dangerous pollutants in order to ensure safety of human beings and ecosystems within prescribed limits becomes a common practice in the developed countries. The control policies are implemented as emission tax, limits on the surface traffic intensity, penalties, etc., and are usually targeted to a long-term effect. This paper presents and investigates an optimal control model of a short run (operative) control of the emissions of air pollutants in a bounded area, possibly exposed also to external pollution. The main scenario assumes that the safety bounds would be violated as a result of an abnormal additional pollution and/or unfavourable weather conditions (wind) if temporary restrictions on the emissions are not imposed. The aim of control is either to prevent violation of the safety bounds at a minimal cost (if possible at all), or to find an optimal trade-off between the cost of emission reduction and the damage caused by the pollution. Mathematically, the model involves a transport PDE with varying wind, (partly controllable) emission and non-linear deposition, and an objective function incorporating the cost of control and the cost of damage caused by the pollution. The theoretical background and a solution approach to the arising optimal control problem are presented in this paper. An analytic solution is obtained in a simplified scenario, which is useful for numerical tests.

Key words: Air pollution modeling, emission reduction, control theory, optimal control, control policies, partial differential equations.

I. Farago et al. (eds.), Advances in Air Pollution Modeling for Environmental Security, 219–228.

1. THE EMISSION CONTROL MODEL

Let $\Omega \subset \mathbb{R}^2$ be a convex compact domain (see Figure 1), $t \geq 0$ the time, and $x = (x_1, x_2)$ the spatial coordinates. The domain Ω is interpreted as that containing the area where the concentration of pollutant, $y(t, x)$, matters. Moreover, Ω is assumed to contain all sources of emission that may influence considerably the concentration in Ω and which can be controlled. We consider the following scenario.

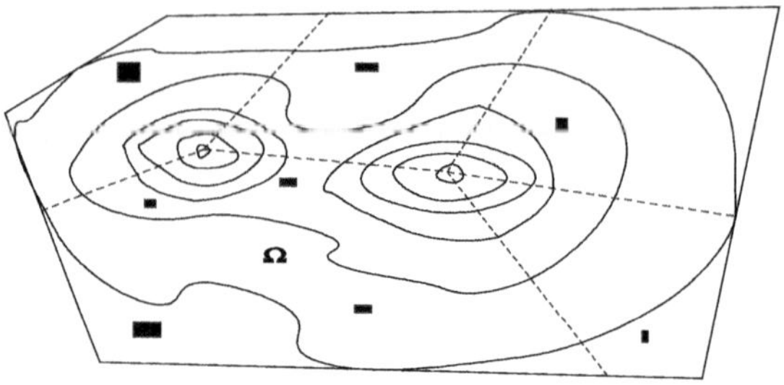

Figure 1: The computational domain with a possible distribution of some controllable emission sources (the black spots and the dashed lines) and the isolines of the population density.

Assume that at time $t = 0$ a forecast is available, which predicts what the wind speed, $v(t, x)$, and the emission intensity, $f(t, x)$ will be for $x \in \Omega$, and the concentration, $\hat{y}(t,x)$, for $x \in \partial\Omega$ ($\partial\Omega$ is the boundary of Ω) in the time interval $[0, T]$. This forecast could be based on a weather prediction, emission prediction, and computation of $\hat{y}(t,x)$ in a model that involves a larger domain than Ω. If it happens (presumably as an extraordinary event) that the predicted values of the concentration are not acceptable, then the control model described below is involved.

1.1 The dynamics

Let us assume that a part of the predicted emissions, $f(t, x)$, can be reduced to a (minimal possible) value $f_0(t, x) \geq 0$. That is,

$$f(t,x) = f_0(t,x) + (1 - u(t,x))\ f_1(t,x) \tag{1}$$

where f_0 is the uncontrollable emission, $f_1 = f - f_0$ is the controllable part of the emission, $u(t, x) \in [0, 1]$ is a control variable. Let us mention that if some point-sources can be controlled then one can take

$$u(t,x) = u_0(t,x) + \sum u_i(t,x)\delta_{x_i}(x),$$

where $x_i \in \Omega$ are the point sources. Denoting the deposition intensity with $g(t, x, y)$, the dynamics of the control model (without diffusion) is described by the equation

$$y_t + \langle v(t,x), \nabla y \rangle = f_0(t,x) + (1-u(t,x)) f_1(t,x) - g(t,x,y), \qquad (2)$$

where $(t,x) \in [0,T] \times \Omega$,

$$y(0,x) = \hat{y}_0(x), \quad x \in \Omega, \qquad (3)$$

$$y(t,x) = \hat{y}(t,x), \quad x \in \Gamma_-(t) \subset \partial\Omega, \qquad (4)$$

where

$$\Gamma_-(t) = \left\{ (t,x) \in \partial\Omega : \underset{l \in N_\Omega(x)}{\text{am x}} \langle l, v(t,x) \rangle < 0 \right\},$$

and $N_\Omega(x)$ is the external normal cone to the (convex compact) set Ω at x (see, e.g., Marchuk, 1982).

1.2 The control cost

Short-term reduction of the emission of production facilities is usually accompanied with a decrease in production, therefore with losses. Traffic restrictions that could also be used as a control instrument, also cause losses. These losses represent the cost of the control.

The control cost function is assumed in the general form

$$J(u) = \int_0^T \int_\Omega B(t,x,u(t,x))\, \mathrm{d}x\, \mathrm{d}t. \qquad (5)$$

If a distributed control is difficult to implement (e.g. in the case of a restriction to mobile sources) one can take a non-distributed control $u(t)$, and then

$$J(u) = \int_0^T B(t,u(t))\, dt.$$

A reasonable example is

$$B = \beta(t,x)u + \gamma(t,x)u^r, \qquad \gamma(t,x) \geq 0, r > 1 \qquad (6)$$

or

$$B = \beta(t)u + \gamma(t)u^r, \qquad \gamma(t) \geq 0, r > 1.$$

For the industrial units these dependences should be known. In the case of personal cars the coefficient β is related to the marginal efficiency of the investments in the public transportation.

1.3 The aim of control

Let $c(x)$ be the maximal admissible concentration at point $x \in \Omega$, that is the *viability constraint*

$$y(t,x) \le c(x) \quad \text{for all} \quad x \in \Omega \tag{7}$$

should hold on $[0,T]$ (see Aubin, 1991). Two different interpretations of the above constraint could be given. The "hard" one requires satisfaction of (7), no matter how high the control costs are, if this is possible at all. Of course, among all controls that ensure (7), the cheapest one is sought.

The soft interpretation allows (7) to be unsatisfied, but penalizes its violation. The most natural way of penalization is to evaluate the *cost of damage* caused by the violation, and to add it to the control cost.

More generally, with every concentration function $y(\cdot,\cdot)$ we associate the monetary equivalent of the corresponding damage

$$D(t) = \int_{\Omega} H(t,x,y(t,x))\,\mathrm{d}x. \tag{8}$$

In the case where the cost of the damage is related to violation of a soft viability constraint, H could have the form

$$H(x,z) = \begin{cases} 0 & \text{if} \quad z \le c(x) \\ \alpha(x)\varphi(z) & \text{if} \quad z > c(x), \end{cases}$$

where $\alpha(x)$ depends on the density of the population at x, and $\varphi(z)$ is an increasing function.

1.4 The overall model

Bringing the above considerations together we come up with the following optimal control problem:

$$\text{minimize} \int_0^T \int_{\Omega} L(t,x,y(t,x),u(t,x))\,\mathrm{d}x\,\mathrm{d}t, \tag{9}$$

subject to (2), (3), (4), and the control constraint

$$u \in [0,1]. \tag{10}$$

Here $L = B + H$ in case of the soft interpretation of (7), and $L = B$ in case of the hard interpretation. In the latter case (7) should be explicitly added in the problem formulation.

The following will be assumed: all functions involved in the model are continuous and continuously differentiable with respect to x and u; B is strongly convex in u, uniformly in x.

Remark 1: In the case of changing wind it may happen that some air leaves and later re-enters Ω. If a control had been applied before this air has left, then the concentration of the pollutant in it could be smaller at the re-entry than the corresponding $\hat{y}$ calculated in advance for the uncontrolled system. Thus in the case of possible re-entry the optimization stands "on the safe side".

Remark 2: The problem considered here is meaningful only if T is sufficiently large, so that the "extraordinary" situation that causes use of control "dies out" by the end of the time horizon. In the example below we shall point out cases where this is not fulfilled. In such a situation other control formulations should be involved.

2. OPTIMALITY CONDITIONS

Under the assumptions formulated at the end of the previous section one can prove the existence and uniqueness of an optimal solution to problem (2), (3), (4), (9), (10).

The objective function can be considered as a functional, $u \to J(u)$ over the space of control functions, for which it is convenient to take $L_\infty([0,T]\times\Omega)$. Indeed, for every admissible control function, u, the equations (2), (3), (4) determine a unique solution, y, which plugged into (9) together with u gives the value $J(u)$. A key part of obtaining a solution procedure and necessary optimality conditions for the problem is to find the derivative of the functional J.

Let us denote

$$\Gamma_+(t) = \partial\Omega \setminus \Gamma_-(t).$$

For a reference control-trajectory pair (u,y) we define the adjoint equation

$$\lambda_t + \operatorname{div}(\lambda v) = -\lambda(t,x)\, g_y(t,x,y(t,x)) + L_y(t,x,y(t,x),u(t,x)), \quad (11)$$

with an end-time condition

$$\lambda(T,x) = 0,$$

and a boundary condition

$$\lambda(t,x) = 0, \quad \text{for} \quad t \in [0,T], \quad x \in \Gamma_+(t).$$

Then one can prove that

$$\frac{\mathrm{d}J}{\mathrm{d}u(\cdot)} = L_u(t,x,y(t,x),u(t,x)) + \lambda(t,x) f_1(t,x). \quad (12)$$

As a consequence, if (u, y) is optimal, then the maximum principle (in local form) holds:

$$-L_u(t,x,y(t,x),u(t,x)) - \lambda(t,x)\, f_1(t,x) \in N_{[0,1]}(u(t,x)),$$

where $N_U(u)$ is the external normal cone to the convex set U at $u \in U$, that is

$$N_{[0,1]}(u) = \begin{cases} (-\infty, 0] & \text{if } u = -1, \\ 0 & \text{if } u \in (0,1), \\ [0, +\infty) & \text{if } u = 1. \end{cases}$$

Using formulae (11) and (12) one can implement a standard steepest descend method for approximation of the optimal solution.

3. TEST EXAMPLE

The test example below concerns the more difficult case of a hard viability constraint. We obtain an analytic solution of the problem, which can be used to test the numerical solvers. Let us take the following data specifications:

- $t \in [0,T]$, $\Omega = [0, \omega] \in \mathrm{R}^1$,
- $v(t, x) = 1$ (the value 1 is not a restriction, since one can replace any $v > 0$ with $v = 1$, replacing also ω with ω / v),
- $f_0(t,x) = f_0$,
- $f_1(t,x) = f_1$,
- $g(t,x,y) = ay$,
- $y_0(x) = 0$.

Here $\Gamma_-(t) = \{0\}$ and we assume that the prediction $\hat{y}(t,0)$ has the form

$$\hat{y}(t,0) = \begin{cases} \hat{y}(t) & \text{for } t \in [0, \delta], \\ 0 & \text{for } t \in (\delta, T]. \end{cases}$$

The system returns to the "nominal" regime before the end of the optimization horizon if $\delta + \omega \le T$. In fact, we may assume that $\delta + \omega = T$.

We consider the objective function

$$\int_0^T \int_\Omega \left(u(t,x) + \frac{\alpha}{2}(u(t,x))^2 \right) \mathrm{d}x \, \mathrm{d}t,$$

and the viability constraint (7) with $c(t, x) = c$.

Obviously the above problem splits into a family of auxiliary optimal control problems for ordinary equations, parameterized by the initial condition:

$$\min \int_0^\omega \left(w(t) + \frac{\alpha}{2}(w(t))^2 \right) \mathrm{d}t,$$

$$z \quad \dot{z} = a + f_0 + (1-w)f_1, \quad z(0) = z_0, \quad t \in [0,\omega], \tag{13}$$

with state constraint $z(t) \le c$. If $\hat{w}_{z_0}(\cdot)$ is an optimal control for the last problem, then

$$\hat{u}(t,x) = \begin{cases} 0 & \text{if } t \le x \text{ or } t > \delta \\ \hat{w}_{\hat{y}(t,x)}(x) & \text{if } t \in [x,\delta]. \end{cases}$$

is an optimal control for the original one.

First of all we introduce the condition

$$\max_{t\in[0,\omega]} \left[e^{-at} \max_{\xi\in[0,\delta]} y_0(\xi) + \int_0^t e^{-a(t-s)} f_0 \, ds \right] \le c, \tag{14}$$

which is necessary and sufficient for the solvability of the original problem. More explicitly (since f_0 and f_1 are constants) this reads as

$$\max_{t\in[0,\omega]} e^{-at} \left(\max_{\xi\in[0,\delta]} y_0(\xi) - \frac{f_0}{a} \right) + \frac{f_0}{a} \le c, \tag{15}$$

which can be checked by evaluation either at $t = 0$ or at $t = \omega$, depending on the sign in the parentheses.

Now we focus on the auxiliary problem with z_0 belonging to the range of $y_0(\cdot)$. The analysis is a routine but we include for completeness, and even more, in order to help the intuition. Condition (15) implies that for every such z_0 the auxiliary problem has a solution.

Clearly, the interesting case is

$$\max_{t\in[0,\omega]} e^{-at} \left(z_0 - \frac{f_0 + f_1}{a} \right) + \frac{f_0 + f_1}{a} > c, \tag{16}$$

since otherwise the unique optimal solution is trivial: $v(t) = 0$. Therefore, further we assume also (16). We will discuss separately the cases $\alpha = 0$ and $\alpha > 0$.

1. $\alpha = 0$.

Let us compare two admissible trajectories, $z_1(\cdot)$ and $z_2(\cdot)$, corresponding to control functions $w_1(\cdot)$ and $w_2(\cdot)$, respectively, both starting from a point p at $t = \beta$ and reaching $z=q$ at time $t = \gamma > \beta$, and for which $z_1(t) \le z_2(t)$ on $[\beta,\gamma]$. Then

$$\int_\beta^\gamma w_1(s)\,\mathrm{d}s = \int_\beta^\gamma \frac{1}{f_1}(-az_1(s) + f_0 + f_1 - \dot{z}_1(s))\,\mathrm{d}s \tag{17}$$

$$= \frac{1}{f_1}\left[z\int_\beta^\gamma (-a\ _1(s) + f_0 + f_1)\,\mathrm{d}s - (q-p)\right]$$

$$\geq z\frac{1}{f_1}\left[\int_\beta^\gamma (-a\ _2(s) + f_0 + f_1)\,\mathrm{d}s - (q-p)\right] = \int_\beta^\gamma w_2(s)\,\mathrm{d}s. \tag{18}$$

Therefore, $w_2(\cdot)$ is "cheaper" than $w_1(\cdot)$. This, of course, makes sense, since the deposition is higher for higher concentration.

We distinguish the following two sub-cases:

(A) the value $z = c$ is an admissible equilibrium of (15):

$$0 \leq w^* = \tfrac{1}{f_1}(f_0 + f_1 - ac) \leq 1.$$

Note that (14) and (16) imply that $w^* > 0$, therefore in this case w^* is an admissible control value.

(B) $w^* > 1$.

First we consider first the case **(B)**. In this case the condition $z = c$ is not achieved by the optimal trajectory $\hat{z}(\cdot)$ in $[0,T)$. Then the optimal control $\hat{u}(\cdot)$ satisfies the classical Pontryagin maximum principle: there is $\overline{\lambda}$ such that $\hat{w}_{z_0}(t)$ minimizes

$$w(1 + f_1 e^{-a(\theta - t)}\overline{\lambda}), \quad v \in [0,1],$$

that is, $w_{z_0}(t) = 0$, if $1 + f_1 e^{-a(\theta-t)}\overline{\lambda} \geq 0$, and $w_{z_0}(t) = 1$ otherwise. If $\overline{\lambda} \geq 0$, then $\hat{w}(t) = 0$ on $[0,\theta]$ (this is the trivial case where (16) does not hold). If $\overline{\lambda} < 0$, then $\hat{w}_{z_0}(t)$ is either constant, or has one switching point from 0 to 1. According to (17), (18), this switching point, denoted by $\tau(z_0)$, is determined by the equation

$$e^{-aw} z_0 + \frac{f_0 + f_1}{a}(e^{-a(w-\tau)} - e^{-aw}) + \frac{f_0}{a}(1 - e^{-a(w-\tau)}) = c,$$

which is,

$$e^{a\tau}\left(\frac{f_1}{a}e^{-a\omega}\right) + e^{-aw} z_0 - \frac{f_0 + f_1}{a}e^{-aw} + \frac{f_0}{a} = c.$$

Under the conditions (14) and (16), in case **(B)** the optimal control is

$$\hat{w}_{z_0}(t) = \begin{cases} 0 & \text{for} \quad t \in [0,\ \tau(z_0)), \\ 1 & \text{for} \quad t \in [\tau(z_0),\ \omega). \end{cases}$$

In case **(A)** one can use a similar argument to prove that

$$w_{z_0}(t) = \begin{cases} 0 & \text{for } t \in [0,\theta), \\ w^* & \text{for } t \in [\theta,\omega), \end{cases}$$

where θ is determined by the equation

$$e^{-a\theta} z_0 + \frac{f_0 + f_1}{a}(1 - e^{-a\theta}) = c,$$

which is

$$e^{-a\theta}\left(z_0 - \frac{f_0 + f_1}{a}\right) + \frac{f_0 + f_1}{a} = c.$$

Summarizing, one can formulate the optimal policy as follows: keep the control equal to zero as long as it is possible, so that afterwards the system can be kept viable till the pollutant leaves the area of interest. In the terminology introduced in Quincampoix and Veliov, 1998, in the spirit of the general viability theory (see Aubin, 1991), this means the following: keep $w(t) = 0$ as long as the boundary of the viability kernel of the auxiliary system extended to the state-time space, with target $t = \omega$, is reached (if at all); after that – stay on the boundary of the kernel until the target $t = \omega$ is reached.

2. $\alpha > 0$.

This case is the more realistic one due to the usual increasing cost to scale. We assume that the solvability condition (15) is fulfilled. If (16) does not hold, then, as before, $w_{z_0}(t) = 0$ is optimal, therefore we assume also (16).

If on some interval (τ,θ) the state constraint is not active, then the maximum principle implies that the optimal control has in this interval the form

$$w_{z_0}(t) = \begin{cases} 0 & \text{if } \left(f_1 e^{-a(\theta-t)}\lambda - 1\right)/\alpha < 0, \\ \dfrac{f_1 e^{-a(\theta-t)}\lambda - 1}{\alpha} & \text{if } \left(f_1 e^{-a(\theta-t)}\lambda - 1\right)/\alpha \in [0,1], \\ 1 & \text{if } \left(f_1 e^{-a(\theta-t)}\lambda - 1\right)/\alpha > 1, \end{cases}$$

where λ is a nonnegative number. Having this in mind we define for a parameter $\rho \geq 0$ the function

$$\varphi(\rho;t) = \frac{1}{\alpha}\left(\rho e^{at} - 1\right),$$

and also

$$w(\rho;t) = \begin{cases} 0 & \text{if } \varphi(\rho;t) < 0, \\ \varphi(\rho;t) & \text{if } \varphi(\rho;t) \in [0,1], \\ 1 & \text{if } \varphi(\rho;t) > 1. \end{cases}$$

We consider again the cases **(A)** and **(B)** as above.

In the case **(B)** obviously $\hat{z}(t) < c$ on $[0,\omega)$, and one can prove by an obvious control variation that $\hat{z}(\omega) = c$. Then the optimal control is

$$\hat{w}_{z_0}(t) = w(\rho;t),$$

where $\rho \geq 0$ is determined as the unique solution to the equation $\hat{z}(\omega) = c$. Case **(A)** is more complicated and different sub-cases should be considered and compared in order to obtain a complete solution. Therefore, we describe only the structure of the optimal control, namely,

$$w_{z_0}(t) = \begin{cases} w(\rho;t) & \text{for } t \in [0,\theta), \\ w^* & \text{for } t \in [\theta,\omega). \end{cases} \tag{19}$$

Here there are two unspecified parameters, ρ and θ, for which we have one equation:

$$\hat{z}(\theta) = c\,. \tag{20}$$

The condition of minimum of the objective function as a function of these two parameters should be added to this equation to determine both ρ and θ. However, (19), (20) are, in principle, enough for verification of the numerical results.

Acknowledgement: This research of the first two authors was partially supported by the NATO Scientific Program Grant (CRG 960505), the European Commission under the Program for supporting the Centres of Excellence (contract ICA1-CT-2000-70016) and the Bulgarian Ministry of Education and Science under Grant I-901/99. The third author was partly supported by the Austrian Science Foundation under contract 14060-OEK. The authors thank the Austrian Science and Research Liaison Office (Sofia, Bulgaria) for cooperation and assistance.

REFERENCES

Aubin, J.-P., 1991, Viability Theory, Birkhauser, Boston, Basel, Berlin.

Marchuk, G. I., 1982, Mathematical models for environmental problems, Nauka, Moscow, (in Russian, English translation is available by North-Holland, Amsterdam, 1986).

Quincampoix, M. and Veliov, V. M., 1998, Viability with a target: theory and applications. *Applications of Mathematics in Engeneering*, Cheshankov, B. I. and Todorov, M. D. (Eds.), Proc. XXIII Summer School, Sozopol'97, Herron Press, Sofia, pp. 47-54.

A NEW OPERATOR SPLITTING METHOD AND ITS NUMERICAL INVESTIGATION

Boglárka Gnandt
Eötvös Loránd University, Department of Meteorology, Pázmány Péter sétány 1/A, 1117 Budapest, Hungary

Abstract: Several splitting methods are used in the fields of applied mathematics. In this paper we construct a new operator splitting method, the so-called additive splitting, which is a first-order scheme and is parallelizable on the operator level. We perform numerical comparisons between the additive splitting and some other existing splitting schemes using different types of numerical methods. On a certain class of problems the new method gives better results than the traditional splitting methods of the same order, and proves to be competitive with the more expensive, higher-order splitting techniques. We see good possibilities to apply the new method in real-life air pollution models.

Key words: Operator splitting, splitting error, additive splitting, air pollution modelling.

1. INTRODUCTION

In large-scale air pollution modelling and some other fields of applied mathematics huge systems of partial differential equations have to be treated numerically. The number of chemical species involved in a modern air pollution model sometimes reaches 200, while the number of spatial grid points even many millions. This means that the system of ordinary differential equations obtained after spatial discretization is extremely big. Moreover, the model equations contain terms that have different physical meanings and so different mathematical properties (e.g. linear, nonlinear, stiff and non-stiff). Therefore, it is impossible to find such a universal numerical method which would perform well when applied directly to the original system. The

I. Farago et al. (eds.), Advances in Air Pollution Modeling for Environmental Security, 229–241.

application of operator splitting allows us to treat the different physical terms separately.

Operator splitting is a kind of problem decomposition: we divide the spatial differential operator of the global system into a few simpler operators and solve the corresponding problems one after the other, by connecting them through their initial conditions.

The simpler systems which are obtained in this manner, and are sometimes called sub-systems, might have some special properties that can be exploited in the numerical solution. The sub-systems are usually easier to treat numerically than the whole system.

Splitting can be performed in several ways. We expect the method to be accurate as well as efficient enough. The latter property depends on the number of computations and the possibility of performing the computations in parallel. Taking into account the latter requirement, we made attempts to construct a new splitting scheme which does not require a lot of computational work, and is parallelizable on the operator level.

The paper is organized as follows. In Section 2 examples of splitting methods are presented and analysed briefly. In Section 3 we introduce our new splitting method, the so-called additive splitting and investigate some of its properties. In Section 4 the results of our numerical experiments are presented and illustrated with figures.

2. DIFFERENT SPLITTING METHODS

In this section some frequently used splitting methods are presented.

The system of partial differential equations included in mathematical models is transformed by spatial discretization into a system of ordinary differential equations. Our task is to determine the solution of the Cauchy problem obtained in this manner.

Let us consider the simple initial value problem

$$\begin{cases} \dfrac{du(t)}{dt} = Au(t) + Bu(t)\,, \quad t \in (0,T], \\ u(0) = u_0, \end{cases} \tag{1}$$

where $u : \mathrm{R} \to \mathrm{R}^N$ is the unknown function, $u_0 \in \mathrm{R}^N$ is a given vector, and A and B are given matrices of type $\mathrm{R}^{N\mathrm{x}N}$, $t \in [0,T]$.

Let $\tau > 0$ denote the splitting time step.

Let $u(\tau)$ and $u_{sp}(\tau)$ denote the exact solution of (1) and the splitting solution of the split problem, respectively at time τ. Then their difference is called local splitting error, i.e.,

$$Err_{sp}(\tau) = u(\tau) - u_{sp}(\tau)\,. \tag{2}$$

When $Err_{sp}(\tau) = O(\tau^{p+1})$, then the splitting method is called p-th order.

In the following we summarize some widely used splitting methods. (For more details, see Hundsdorfer and Verwer, 2003, and Zlatev, 1995.)

2.1 Sequential splitting

The scheme of this method is the following: as a first step, we solve the system with operator A using the initial condition of the original problem, and then, applying the obtained solution at time τ as an initial condition, we solve the system with operator B. The solution obtained in this way is considered as the splitting solution in τ. This procedure is performed cyclically in the following way:

$$\begin{cases} \dfrac{du_k^{(1)}(t)}{dt} = Au_k^{(1)}(t), \quad t \in ((k-1)\tau, k\tau], \\ u_k^{(1)}((k-1)\tau) = u_{k-1}^{(2)}((k-1)\tau), \end{cases} \tag{3}$$

$$\begin{cases} \dfrac{du_k^{(2)}(t)}{dt} = Bu_k^{(2)}(t), \quad t \in ((k-1)\tau, k\tau], \\ u_k^{(2)}((k-1)\tau) = u_k^{(1)}(k\tau), \end{cases} \tag{4}$$

for k = 1, 2,…, m, where $u_0^{(2)}(0) = u_0$. The splitting solution in kτ is defined as

$$u_{sp}(k\tau) = u_k^{(2)}(k\tau), \qquad k = 1, 2, \ldots, m. \tag{5}$$

It can be shown that the sequential splitting is a first-order method (see e.g. Havasi et al., 2001).

2.2 Strang-Marchuk splitting

Using this method, at each time step we begin and end the computation with operator A (we apply it over a distance τ/2 twice) and put B to the middle (we apply it over a distance τ once) as follows:

$$\begin{cases} \dfrac{du_k^{(1)}(t)}{dt} = Au_k^{(1)}(t), \quad t \in ((k-1)\tau, (k-\tfrac{1}{2})\tau], \\ u_k^{(1)}((k-1)\tau) = u_{k-1}^{(3)}((k-1)\tau), \end{cases} \tag{6}$$

$$\begin{cases} \dfrac{du_k^{(2)}(t)}{dt} = Bu_k^{(2)}(t), \quad t \in ((k-1)\tau, k\tau], \\ u_k^{(2)}((k-1)\tau) = u_k^{(1)}((k-\tfrac{1}{2})\tau), \end{cases} \tag{7}$$

$$\begin{cases} \dfrac{du_k^{(3)}(t)}{dt} = Au_k^{(3)}(t), & t \in ((k-\tfrac{1}{2})\tau, k\tau], \\ u_k^{(3)}((k-\tfrac{1}{2})\tau) = u_k^{(2)}(k\tau), \end{cases} \tag{8}$$

for k = 1, 2,..., m, where $u_0^{(3)}(0) = u_0$. The splitting solution in kτ is

$$u_{sp}(k\tau) = u_k^{(3)}(k\tau), \qquad k = 1, 2, \ldots, m. \tag{9}$$

This method is of second order.

2.3 Weighted splitting

This method can be obtained by symmetrizing the sequential splitting in the following way: in each time step we apply sequential splitting both in the order $A \to B$ as follows:

$$\begin{cases} \dfrac{du_k^{(1)}(t)}{dt} = Au_k^{(1)}(t), & t \in ((k-1)\tau, k\tau], \\ u_k^{(1)}((k-1)\tau) = u_{sp}((k-1)\tau), \end{cases} \tag{10}$$

$$\begin{cases} \dfrac{du_k^{(2)}(t)}{dt} = Bu_k^{(2)}(t), & t \in ((k-1)\tau, k\tau], \\ u_k^{(2)}((k-1)\tau) = u_k^{(1)}(k\tau), \end{cases} \tag{11}$$

and $B \to A$ as follows:

$$\begin{cases} \dfrac{dv_k^{(1)}(t)}{dt} = Bv_k^{(1)}(t), & t \in ((k-1)\tau, k\tau], \\ v_k^{(1)}((k-1)\tau) = u_{sp}((k-1)\tau), \end{cases} \tag{12}$$

$$\begin{cases} \dfrac{dv_k^{(2)}(t)}{dt} = Av_k^{(2)}(t), & t \in ((k-1)\tau, k\tau], \\ v_k^{(2)}((k-1)\tau) = v_k^{(1)}(k\tau). \end{cases} \tag{13}$$

The splitting solution in kτ is defined as

$$u_{sp}(k\tau) = \theta u_k^{(2)}(k\tau) + (1-\theta) v_k^{(2)}(k\tau), \qquad \text{for k} = 1, 2, \ldots, m, \tag{14}$$

where $u_{sp}(0) = u_0$, and $\theta \in [0,1]$ is some fixed weight parameter.

The weighted splitting is a second-order method if $\Theta = 0.5$, otherwise it has first order. An important property of the weighted splitting is that it can be parallelized in a natural way (on the operator level). The theoretical investigation of the method can be found in Csomós et al. (2003), and an application of the method on a one-column transport-chemistry model is described in Botchev et al. (2003).

2.4 Comparison of the methods

As a summation, we collect the basic properties of the splitting methods presented above.

The sequential splitting has low accuracy and it is not parallelizable on the operator level.

The Strang-Marchuk splitting is of higher accuracy, but is not parallelizable on the operator level, either.

The symmetrically weighted splitting has also a higher accuracy and it can be performed in parallel, but it requires a lot of computational work.

In applications, instead of very high accuracy we rather need the possibility of parallelization and little computational work for every processor.

Our aim was to construct a new method that combines the advantage of the capability of parallelization with the simplicity of the sequential splitting.

3. THE ADDITIVE SPLITTING

In this section, we define our new splitting method, which will be called additive splitting in the sequel.

The algorithm of the additive splitting is defined as follows:

$$\begin{cases} \dfrac{du_k^{(1)}(t)}{dt} = Au_k^{(1)}(t), \qquad t \in ((k-1)\tau, k\tau), \\ u_k^{(1)}((k-1)\tau) = u_{sp}((k-1)\tau), \end{cases} \tag{15}$$

$$\begin{cases} \dfrac{du_k^{(2)}(t)}{dt} = Bu_k^{(2)}(t), \qquad t \in ((k-1)\tau, k\tau), \\ u_k^{(2)}((k-1)\tau) = u_{sp}((k-1)\tau), \end{cases} \tag{16}$$

$$u_{sp}(k\tau) = u_k^{(1)}(k\tau) + u_k^{(2)}(k\tau) - u_{sp}((k-1)\tau), \qquad \text{for } k = 1, 2, \ldots, m, \tag{17}$$

where $u_{sp}(0) = u_0$.

Clearly, the steps (15) and (16) can be performed in parallel because they use the same initial value. Furthermore, we apply each sub-operator once, just as in the case of the sequential splitting.

In the sequel we determine the accuracy of the additive splitting.

The exponential of a matrix can be defined by its Taylor expansion, so for the exact solution at t = τ we have

$$\begin{aligned} u(\tau) &= \exp((A+B)\tau)u_0 = \\ &= (I + (A+B)\tau + \tfrac{1}{2}(A+B)^2\tau^2)u_0 + O(\tau^3), \end{aligned} \tag{18}$$

and the additive splitting solution in τ can be given as

$$u_{sp}(\tau) = \exp(A\tau)u_0 + \exp(B\tau)u_0 - u_0 =$$
$$= (I + A\tau + \tfrac{1}{2}A^2\tau^2)u_0 + (I + B\tau + \tfrac{1}{2}B^2\tau^2)u_0 - u_0 + O(\tau^3) = \quad (19)$$
$$= (I + (A+B)\tau + \tfrac{1}{2}(A^2 + B^2)\tau^2)u_0 + O(\tau^3).$$

Using the above formulae the local splitting error has the following form:

$$Err_{sp}(\tau) = u(\tau) - u_{sp}(\tau) = \tfrac{1}{2}((AB + BA)\tau^2)u_0 + O(\tau^3). \quad (20)$$

Therefore, the additive splitting is a first-order splitting method. It should be noted that the splitting error is independent of the commutativity of the sub-operators, while in the case of the other splitting methods the commutativity of the sub-operators implies zero splitting error (Dimov et al., 2001). On the other hand, we note that when A and B are orthogonal matrices ($AB = BA = 0$), then the accuracy of the method is higher.

In practice, in order to solve the sub-problems obtained by splitting, we always use some numerical method with discretization step size Δt. During our experiments we observed that the additive splitting has an interesting property related to the explicit Euler method. If we apply the explicit Euler method to the sub-problems of the additive splitting and the time step of the numerical integration is equal to the splitting time step, then the splitting solution can be calculated in the following way:

$$u_k^{(sp)} = u_k^{(1)} + u_k^{(2)} - u_{k-1}^{(sp)} = (I + A\tau)u_{k-1}^{(sp)} + (I + B\tau)u_{k-1}^{(sp)} - u_{k-1}^{(sp)} =$$
$$= (I + (A+B)\tau)u_{k-1}^{(sp)}, \quad (21)$$

for k = 1, 2,..., m, where $u_0^{(sp)} = u_0$. One can see that the result is equivalent to solving the original problem by the use of the explicit Euler method in itself.

It should be noted that the above assertion is not true for the case where the time integration step is smaller than the splitting time step.

As it follows from the principle of splitting, the numerical solution of the sub-problems obtained by splitting is less costly than that of the original problem. In order to characterize the efficiency of the methods, we define the speed of the computation as follows. We assume that applying any of the sub-operators A and B once requires one unit of CPU time. In the case of the parallelizable splittings, depending on the algorithm and the number of processors, more than one sub-problems can be solved in unit time.

Table 1 presents the main properties of the different splitting methods.

Splitting	Accuracy	Perform parallel on operator level	Speed (unit of CPU time)
Sequential	low	no	2
Strang-Marchuk	high	no	3
Weighted	high	yes	2
Additive	low	yes	1

Table 1: Properties of the splitting methods.

4. COMPUTATIONAL EXPERIMENTS

In this section some plots showing the results of our numerical experiments are presented. In order to investigate the additive splitting in comparison with the other methods we prepared a program in MATLAB.

In the experiments we used linear systems of ordinary differential equations with constant coefficients.

As usual, these systems are solved by the use of numerical integration methods. The step size is defined by dividing the splitting time step τ into $n \geq 1$ parts, i. e., $\Delta t = \frac{\tau}{n}$.

4.1 Additive splitting and the explicit Euler method

To check the relation (21) experimentally, we solved the test problem numerically both by using the explicit Euler method in itself and applying the different splitting techniques together with the explicit Euler method. Figure 1 shows the global errors obtained by the different splitting methods for the case of $n = 1$, i.e., when the property (21) is valid.

We notice that the splitting methods give better results than the numerical methods without applying splitting, even though using a splitting method represents a further error source. In Figure 2 one can see that if the numerical time step is smaller than the splitting time step, then (21) does not hold anymore.

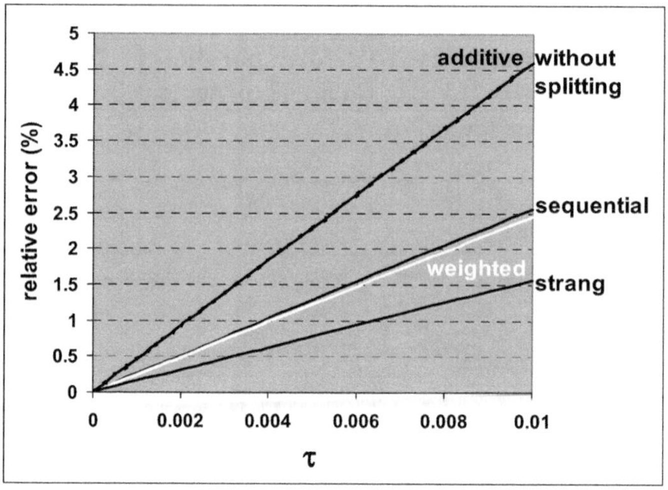

Figure 1: Global errors – using the explicit Euler method, n = 1.

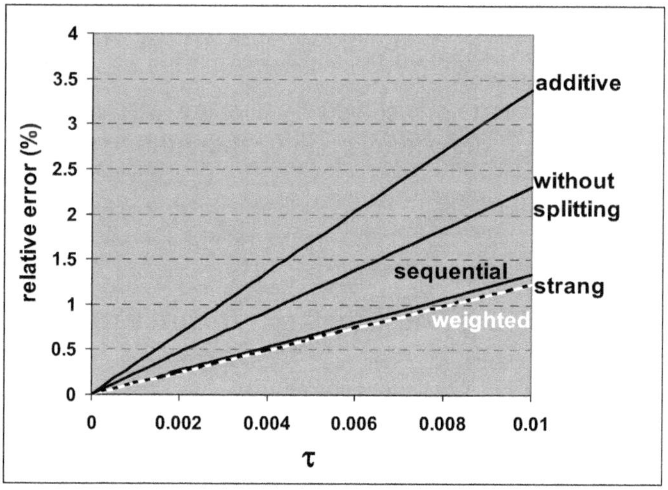

Figure 2: Global errors – using the explicit Euler method, n = 2.

4.2 Additive splitting with different numerical methods

Figure 3 shows the results obtained with the additive splitting by using different numerical methods for solving the sub-problems.

Even if the middle-point method has second order, the additive splitting gives best results with the first-order implicit Euler method; this shows that the application of a higher order numerical method does not necessarily imply higher accuracy. We also observed that when the time step of the numerical integration is reduced, then the numerical solutions obtained both with the explicit and the implicit Euler method approach the solution obtained by the use of the middle-point method.

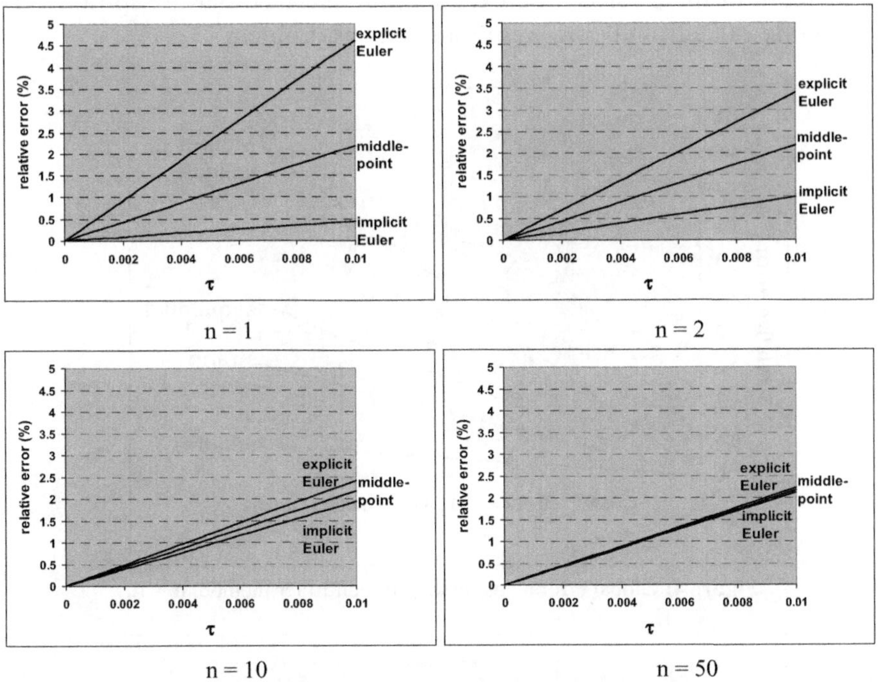

Figure 3: Global errors – additive splitting method with explicit Euler method, middle-point method and implicit Euler method.

4.3 The best results using the additive splitting

We solved several test problems with the implicit Euler method by using different splitting methods. Figure 4 shows that the additive splitting gives the best results if the time integration step is equal to the splitting time step. If however we decrease the integration step-size (n increases), the results are improving for all of the splitting methods, except for the additive splitting, where the errors are increasing. When $n \geq 4$, the error is highest in the case of the additive splitting (Figure 5).

This feature can be seen from the following formula, which expresses the global error of the additive splitting applied together with the implicit Euler method:

$$Err_{sp}(T) = \frac{T}{2}\tau\left[\frac{1}{n}(A^2 + B^2) - (AB + BA)\right]u_0 + O(\tau^2). \tag{22}$$

The limit of this expression as n tends to infinity is

$$\lim_{n\to\infty}\left\|Err_{sp}\right\| = \left\|(AB+BA)u_0\right\|\frac{T}{2}\tau\,. \tag{23}$$

The increase of the error can be explained by the fact that the splitting error tends to this value from below in the studied test problem.

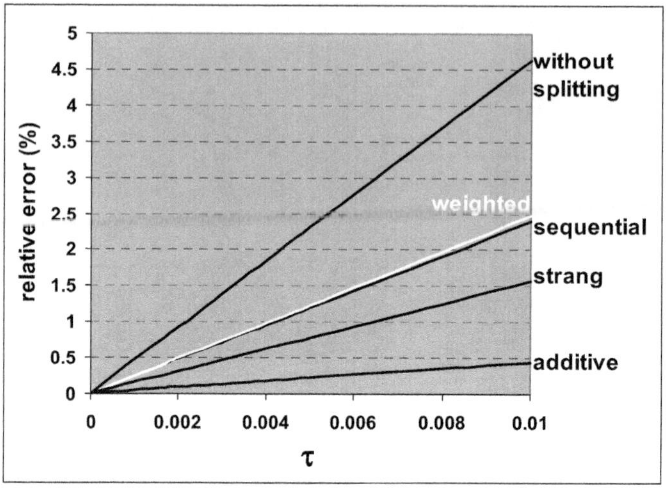

Figure 4: Global errors – using the implicit Euler method, n = 1.

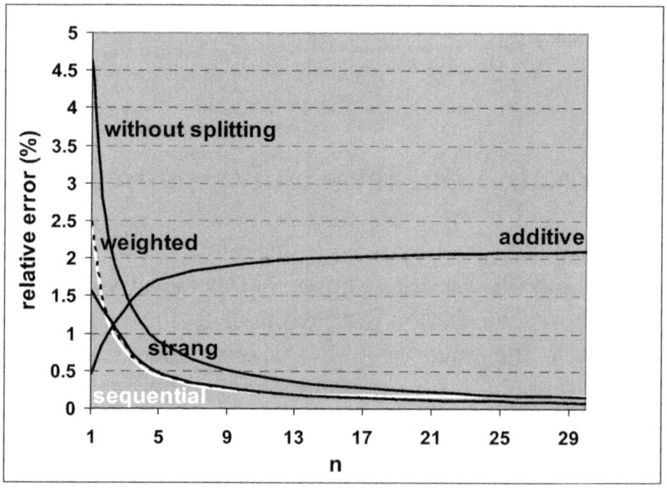

Figure 5: Global errors – using the implicit Euler method, τ = const.

4.4 Investigation of the orders

In Section 3 we showed that the additive splitting is a first-order method (see (20)). We did some computations to check this property experimentally.

Figure 6 shows the local errors obtained with the different splitting schemes by using numerical methods of the same order for each.

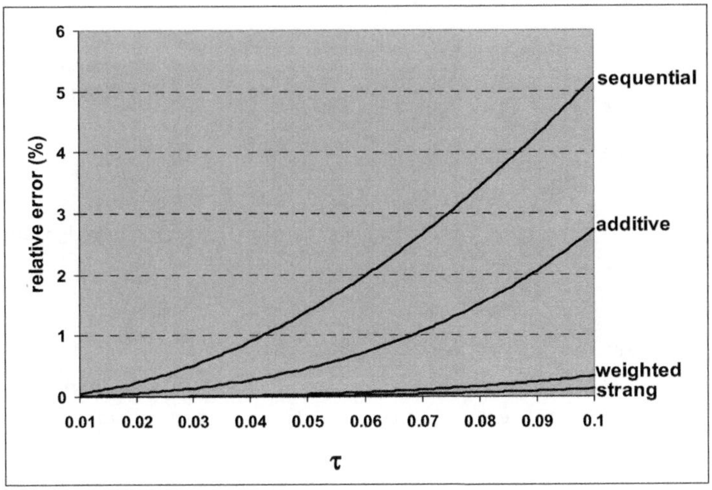

Figure 6: Local errors – splitting methods with numerical methods of the same order.

One can see that the shape of the curve corresponding to the additive splitting is similar to that representing the also first-order sequential splitting. On the other hand, in point of the accuracy we can notice that the additive splitting gives better results than the sequential method, with local errors approximately halfway between the sequential splitting and the second order methods.

4.5 Commuting matrices

In the last numerical experiment we investigated the effect of the commutativity on the numerical results.

Figure 7 shows the global errors obtained with the different splitting methods by using commuting matrices and the same numerical method in the test problem.

At the end of Section 3 we mentioned that the commutativity of the sub-operators has no effect on the error of the additive splitting, while in the case of the other splitting methods the commutativity of the sub-operators implies zero splitting error. Moreover, Figure 7 shows that we can obtain the best results by using the additive splitting even in the case of commuting matrices.

We also noticed that the errors of the other splittings were supposed to disappear in this case. The experience that they did not vanish can be explained by the fact that the sub-problems are not solved exactly, but using some numerical method, the error of which is in complex interaction with the splitting errors.

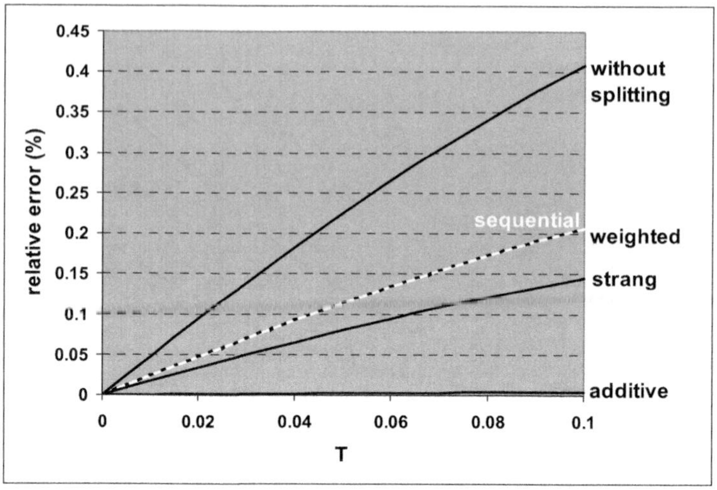

Figure 7: Global error changing in time – using the implicit Euler method, commuting matrices.

5. CONCLUDING REMARKS

The main aim of this paper was to investigate the method of additive splitting - theoretically and experimentally - in comparison with some traditional splitting methods.

Our conclusions and experiences are the following:

- If the time integration step is equal to the splitting time step, the additive splitting with the explicit Euler method is equivalent to the explicit Euler method in itself.
- Although the additive splitting is a first-order splitting method, in some cases it gives better results than the also first-order sequential splitting.
- In the case of using the implicit Euler method, the additive splitting gives more accurate solutions than the other splitting procedures, thus proves to be competitive with the more expensive, higher-order splitting techniques.
- The solution obtained by using the additive splitting is independent of the commutativity of the operators.
- The major advantage of the new splitting method is that it can be parallelized on the operator level.

Finally, we remark that we see good possibilities to apply the additive splitting in real-life air pollution problems.

6. REFERENCES

Botchev, M., Faragó, I., Havasi, Á., 2003, Testing weighted splitting schemes on a one-column transport-chemistry model. *Int. J. Env. Pol.* **22**, Nos. 1/2, to appear.

Csomós, P., Faragó, I., Havasi, Á., 2003, Weighted sequential splittings and their analysis. *Comput. Math. Appl.,* to appear.

Dimov, I., Faragó, I., Havasi, Á., Zlatev, Z., 2001, L–commutativity of the operators in splitting methods for air pollution models. *Annales Univ. Sci. Sec. Math.* **44**, pp. 127–148.

Havasi, Á., Bartholy, J., Faragó, I., 2001, Splitting method and its application in air pollution modeling. *Időjárás* **105**, pp. 39–58.

Hundsdorfer, W., Verwer, J. G., 2003, Numerical solution of time-dependent advection-diffusion-reaction equations. Springer Series in Computational Mathematics.

Zlatev, Z., 1995, Computer treatment of large air pollution models. Kluwer.

ADVANCES IN URBAN DISPERSION MODELLING

Sven-Erik Gryning[(1)] and Ekaterina Batchvarova[(2)]
[(1)]Risø National Laboratory, DK-4000 Roskilde, Denmark; [(2)]National Institute of Meteorology and Hydrology, 66 Blvd. Tzarigradsko Chaussee, BG 1784 Sofia, Bulgaria

Abstract: The ability to predict the measured crosswind spread of the plume as well as the centerline (maximum) concentrations in the urban environment based on tracer experiments was found generally to be within a factor of 2.

Key words: Atmospheric dispersion, urban environment, applied modeling, lateral plume spread, maximum concentrations, tracer experiments.

1. INTRODUCTION

The number of major urban scale tracer experiments with adequate meteorological measurements is small. The Saint Louis Dispersion Study (McElroy, 1969) was carried out in 1963-65, the data formed the initial basis for the description of atmospheric dispersion in regulatory modelling. The Copenhagen experiment (Gryning and Lyck, 1984) was carried out in 1978-79 to study dispersion from elevated sources over urban areas. Recently three major urban scale tracer experiments have been performed. The Urban 2000 Field Experiment (Allwine et al., 2002; Hanna, 2003) was carried out in Salt Lake City during October 2000 to investigate transport and dispersion around a single downtown building, through the downtown area and into the greater Salt Lake City area. The Barrio Logan experimental study (Venkatram et al., 2004) in California was performed to add to the understanding of dispersion in an urban area characterized by buildings with heights less than 10 m. The BUBBLE tracer experiment (Rotach et al., 2004) represents the dispersion characteristics from a roof level source over a typical European city. These experiments embrace the combined effects of the street canyons on the scale

I. Farago et al. (eds.), Advances in Air Pollution Modeling for Environmental Security, 243–252.

of 1 to 10 km.

2. TRACER EXPERIMENTS

Measurements drawn from the Copenhagen and BUBBLE full-scale tracer experiments are used to investigate the ability to model the dispersion process in an urban area.

2.1 Copenhagen experiment

Atmospheric dispersion experiments were carried out in the Copenhagen area to investigate the dispersion process of a tracer released from an elevated source in the urban/residential area (Gryning and Lyck, 1984). The tracer sulphur hexafluoride, an inert gas tracer, was released from a tower at a height of 115 m and collected near ground level in crosswind arcs 2 to 6 km from the source, see Figure 1. Data from the experiment are available in Gryning and Lyck (2002).

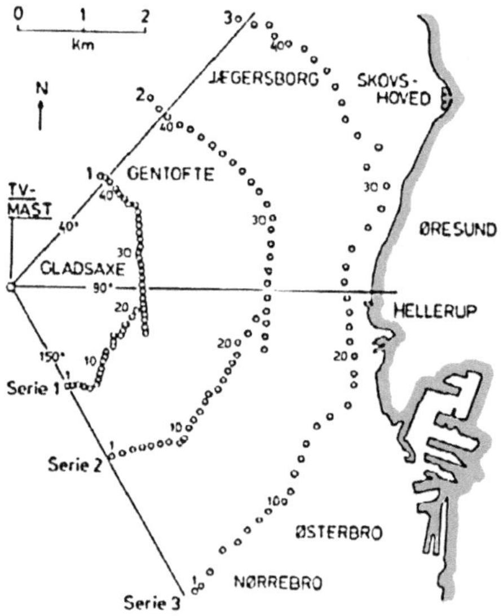

Figure 1: Tracer sampling-unit setup for the experiments in Copenhagen. Typically 20 locations in each arc situated in the actual plume direction for the individual experiments were used.

The tracer sampling time was 1 hour. The site in both the upwind and downwind directions was mainly residential. The meteorological measurements included turbulence at the height of the tracer release, profiles of temperature and wind along the mast and the standard routine radiosoundings launched 4 km northeast of the tracer release point. All tracer experiments were performed during daytime in neutral to slightly convective atmospheric conditions.

2.2 BUBBLE tracer experiment

The BUBBLE tracer experiment represents the dispersion characteristics over a typical European city (Rotach et al., 2004). The tracer experiment was designed as an urban low source tracer experiment (Figure 2), to distinguish from the ground source tracer releases of the experiments in Salt Lake City and Barrio Logan. Low level does not refer to street level, but means that the release was performed a few meters above the roof level - typically 20 m above the street level. Thus the experiment is not meant to elucidate small-scale dispersion process within the single canyons, but will embrace the combined effects of the street canyons on the scale of the experiment. Four tracer experiments were performed, all during very unstable meteorological conditions. The tracer sampling was performed at roof level in arcs up to 1.6 km from the tracer release. A considerable amount of information on the meteorology over the urban area is available. Of special interest for this study are observations in the center of the tracer sampling area of the turbulence profile throughout the street canyon and above. The data for the tracer experiments are available in Gryning et al. (2004).

Figure 2: Panorama view of the BUBBLE experimental area, taken from a point in the middle of the area.

3. PLUME DISPERSION IN URBAN AREAS

Based on measurements from the Copenhagen and BUBBLE tracer experiments aspects of the lateral spread of tracer plumes, σ_y, and the maximum concentration $C_{\max}$ as a function of distance in urban areas will be investigated. The Copenhagen experiment represents an elevated source and near neutral and convective meteorological conditions with relatively high wind velocities, the BUBBLE experiment represents near roof level tracer releases in an urban area during more convective conditions and lower wind speeds.

The study is based on well-known and commonly used semi-empirical estimates that as starting point take Taylor's famous formula for plume dispersion (Taylor, 1921):

$$\sigma_y = \sigma_v t\, f_y(t/T_y) \qquad \sigma_z = \sigma_w t\, f_z(t/T_z) \tag{1}$$

where t is travel time of the plume and f_y and f_z are functions of the dimensionless travel time, where T_y and T_z are Lagrangian time scales for the lateral and vertical dispersion processes. The approximations

$$f_y = \left(1 + \sqrt{t/2T_y}\right)^{-1} \qquad f_z = \left(1 + \sqrt{t/2T_z}\right)^{-1} \tag{2}$$

are often recommended for applied dispersion modelling (Gryning et al., 1987). For ground-level sources $T_y = 200$ s is recommended and $T_y = 600$ s for elevated sources and when the vertical extent of the plume is larger than 10% of the depth of the mixing layer. For unstable atmospheric conditions $T_z = 300$ s.

When measurements of σ_w and σ_v are not available, these parameters can be obtained from parameterisations. Following the discussions in Batchvarova and Gryning (2005a) and Batchvarova and Gryning (2005b) we apply the parameterisations (Gryning et al., 1987):

$$\sigma_w^2 = u_*^2\left[1.5\left(\frac{z}{z_i}\right)^{2/3}\left(\frac{w_*}{u_*}\right)^2 \exp\left(-2\left(\frac{z}{z_i}\right)\right) + \left(1.7 - \left(\frac{z}{z_i}\right)\right)\right], \tag{3}$$

where the convective velocity scale is $w_* = \left((g/T)\overline{w'T'}\, z_i\right)^{1/3}$, with g for the acceleration due to gravity and T for temperature, and

$$\sigma_v^2 = 0.35 w_*^2 + (2 - z/z_i)u_*^2 \quad . \tag{4}$$

The expressions (3) and (4) have been tested in the report of the COST Action 710 (Cenedese et al., 1998) and found to work well for a variety of data sets.

3.1 Lateral dispersion

The simulation of the lateral spread is performed in two ways. In the first one we use the observed σ_v values. For the Copenhagen experiment observations at the tracer release height (115 m) are used, for BUBBLE the values at the highest level of observation (31.7 m) are used. Figure 3 shows the measurements and model simulations of σ_y from the Copenhagen experiment, using measured values of σ_v (left panel) and parameterized values (right panel). In both cases the agreement with the measurements are within a factor of two.

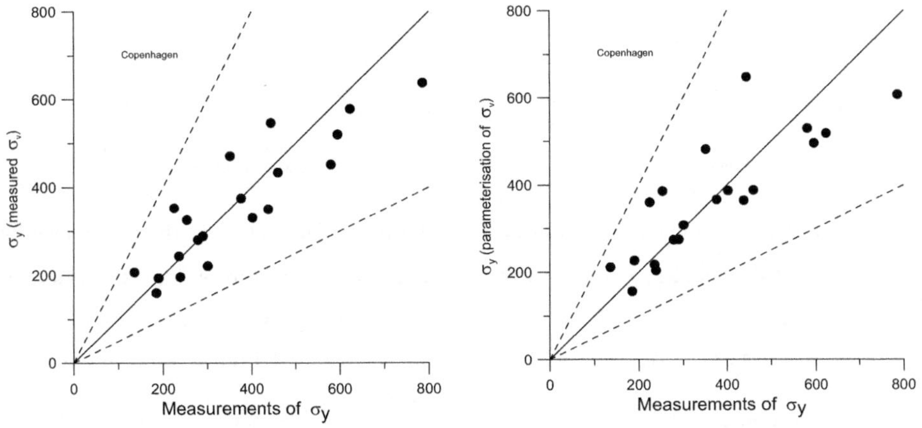

Figure 3: Measured and modelled values of σ_y for the Copenhagen experiment. The left panel shows simulations based on measurements of σ_v at 115 m. The right panel shows simulations using parameterized values of σ_v, also at 115 m. The lines show the 1:1 relationship and its factor of two ranges.

A similar analysis for the BUBBLE experiments is illustrated in Figure 4. Because the tracer release is from a low level source but the plume dispersion in the vertical is expected to be rapid, simulations were performed for both $T_y = 200$ s, which is the recommended value for ground level sources, and

$T_y = 600$ s as suggested for plumes larger than 10% of the mixed layer height. Both assumed values of T_y in combination with the observed σ_v are seen to provide a fair estimate of the lateral spread with the best estimate given by the use of $T_y = 600$ s. This suggests that the high values of both the friction velocity and convective velocity typically for urban areas disperses the plume very effectively in the vertical which makes it behave more like an elevated source than a ground level source.

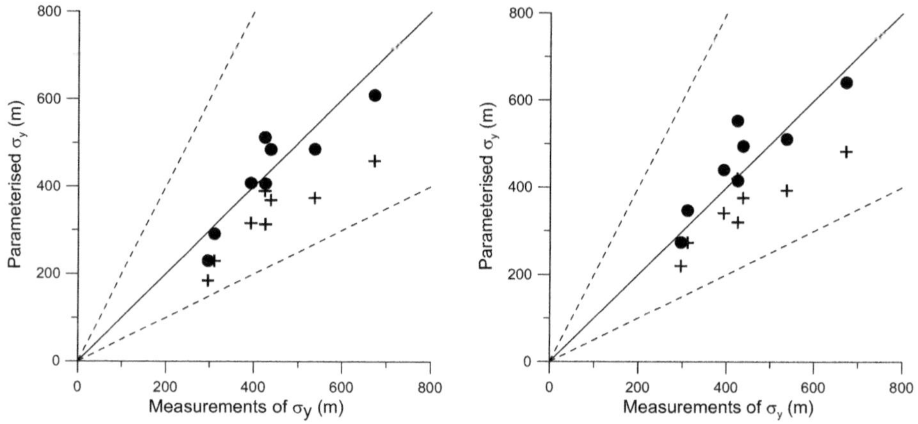

Figure 4: Measured and modelled values of σ_y for the BUBBLE tracer experiment done for $T_y = 600$ s (filled circles) and $T_y = 200$ s (crosses). Left panel shows simulations based on measurements of σ_v at 31.7 m from Sperrstrasse. The right panel is obtained using parameterized values of σ_v, also at 31.7 m. The lines show the 1:1 relationship and its factor of two ranges.

In an urban environment applied expressions for the lateral spread, originally developed and tested over flat, homogeneous terrain, resulted in agreement for both the Copenhagen and BUBBLE experiments better than a factor of two, which is similar to the uncertainty reported for flat terrain.

3.2 Maximum concentrations

Here we apply a very simple modelling approach to the very complex dispersion process of the urban area.

For the Gaussian plume model the ground-level centreline concentration $C_{max}(x)$ at downwind distance x can be expressed as:

$$C_{max}(x) = \frac{Q}{\pi \sigma_y \sigma_z u} \exp\left(-\frac{h^2}{2\sigma_z^2} \right) \tag{5}$$

where h is the tracer release height, and Q the release rate. Figure 5 illustrates the results from a comparison between model simulations and measured tracer concentrations from the Copenhagen experiment. The measured arc-wise maximum concentration has been compared to the modelled centreline concentration. The value of σ_y and σ_z was derived from expressions (1) and (2) using $T_y = 600$ and $T_z = 300$ s. The left panel in Figure 5 illustrates the comparison when the measured values of σ_v and σ_w were applied in the expressions for σ_y and σ_z. The right panel refers to the case when the parameterized values of σ_v and σ_w were applied. The agreement can be seen to be within a factor of two.

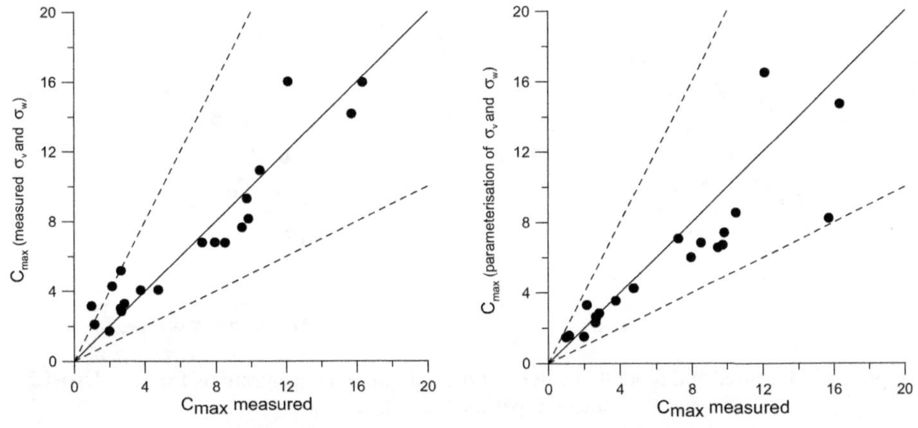

Figure 5: Measured and modelled normalized values of the maximum concentrations for the Copenhagen experiment. The left panel shows simulations based on hourly measurements of σ_v at 115 m. The right panel is obtained by using parameterized values of σ_v, also at 115 m. The lines show the 1:1 relationship and its factor of two ranges.

A somewhat similar comparison is shown for the BUBBLE tracer experiment on 26 June, Figure 6, being the only of the four tracer experiments with a well developed tracer plume that is covered by the tracer sampling network. Because it is a low level release, an even simpler modeling approach as compared to the Copenhagen experiment is used. Inserting the near field expressions for σ_y and σ_z:

$$\sigma_y = \sigma_v x/u \qquad\qquad \sigma_z = \sigma_w x/u \tag{6}$$

in the Gaussian plume formula, the ground level concentration at the centreline (maximum) C_{max} at a given downwind distance x can be expressed as:

$$C_{max}(x) = \frac{Qu}{\pi \sigma_v \sigma_w x^2} \tag{7}$$

where Q is the tracer release rate. Input to the model are measurements of σ_v, σ_w and wind speed u at a representative height for the dispersion of the plume.

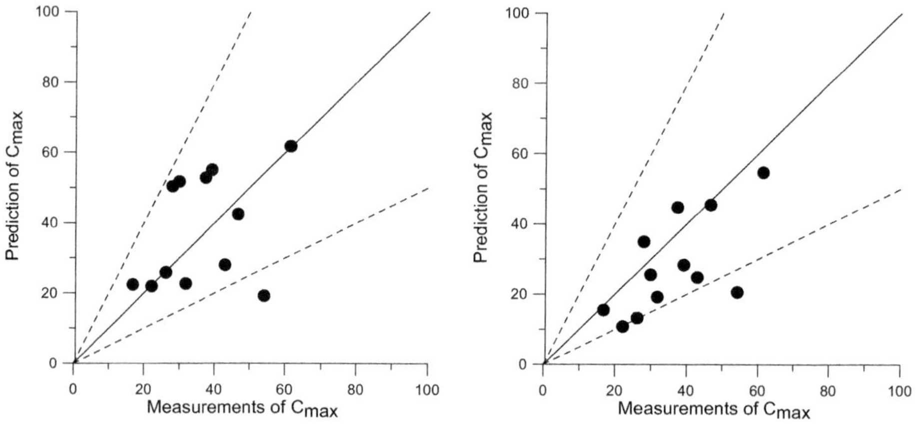

Figure 6: Measured and modelled values of the maximum concentrations for the BUBBLE tracer experiment on 26 June.

Figure 6 shows the comparison of the maximum observed half-hourly tracer concentration every half hour in each of the two tracer sampling arcs (12 values in total) during the tracer experiment on 26 June, and the model prediction. The maximum tracer concentrations are drawn from two arcs about 700 and 1200 m from the tracer source. Input to the model is half hourly observations of σ_v, σ_w and the wind speed at a height of 31.7 m. The left panel shows simulations based on measurements of σ_v at 31.7 m from Sperrstrasse. The right panel is obtained by using parameterized values of σ_v, also at 31.7 m. The lines show the 1:1 relationship and its factor of two ranges. Concentration units are in ng/m^{-3} with a tracer release rate of 0.0503

gs^{-1}. It can be seen that 11 out of 12 half hourly concentration fall within the range of a factor of 2.

4. CONCLUDING REMARKS

We have applied simple models for the lateral and vertical atmospheric dispersion in an urban environment and found an agreement of about a factor of two between model results and measurements. This result can be considered very promising when considering the complex structure of the urban boundary layer. Using particle model for the dispersion on June 26 BUBBLE tracer experiment (Rotach et al., 2004) does not perform better than a factor of 2 on the hourly scale. It seems that any attempt to model the dispersion in the urban atmospheric boundary layer better than a factor of 2 on the hourly scale is quite a hopeless task.

Apparently there is a limitation in the capability of models (simple and comprehensive) to represent realistically the multi-layer structure of the urban boundary layer on the hourly scale. The reason is hidden within the inherent and unavoidable lack of representativeness of actual conditions both for models and measurements.

Acknowledgement: The authors are thankful to the BUBBLE community, and especially to Dr. Mathias Rotach (Meteo Swiss), Dr. Roland Vogt and Andreas Christen (University of Basel) for the discussions on the tracer experiment results. This study was supported by a NATO-CLG (EST-CLG-979863). The work of EB was also supported by the Swiss National Science Foundation (grant 7IP 065650.01) and the BULAIR (EVK2-CT-2002-80024) project.

REFERENCES

Allwine, K. J., Shinn, J. H., Streit, G. E., Clawson, K. L., Brown, M., 2002, Overview of Urban 2000, *Bull. Am. Meteorol. Soc.* **83**, pp. 521-536.

Batchvarova, E., Gryning, S.-E., 2005a Advances in urban meteorology modelling. NATO advanced research workshop: Advances in air pollution modelling for environmental security, Borovetz (BG), 8-12 May 2004. *This volume.*

Batchvarova, E.; Gryning, S.-E., 2005b, Progress in Urban Dispersion Studies. Accepted for publication in *Theoretical and Applied Meteorology*.

Cenedese, A., Cosemans, G., Erbrink, H., Stubi, R., 1998, Vertical profiles of wind, temperature and turbulence. In: Harmonisation of the pre-processing of meteorological data for atmospheric dispersion models. COST action 710 Final report. Office for Official Publications of the European Communities.

Gryning, S.-E., Lyck, E., 1984, Atmospheric Dispersion from Elevated Sources in an Urban Area: Comparison between Tracer Experiments and Model Calculations. *J. Climate Appl. Meteorol.* **23**, pp. 651-660.

Gryning, S.-E., Holtslag, A. A. M., Irwin, J. S., Sivertsen, B., 1987, Applied Dispersion Modelling Based on Meteorological Scaling Parameters. *Atmos. Environ.* **21**, pp. 79-89.

Gryning, S.-E., Lyck, E., 2002, The Copenhagen tracer experiments: Reporting of measurements. Risø-R-1054(rev.1)(EN) 74 p. (only available on the Internet: www.risoe.dk/rispubl/VEA/ris-r-1054_rev1.htm).

Gryning, S.-E., Batchvarova, E., Rotach, M. W., Christen, A., and Vogt, R., 2004, Roof-level SF_6 tracer experiments in the city of Basel - *Report published in the institute series of IAC-ETH.*

Hanna, S. R., 2003, Tracer Cloud Transport In Salt Lake City and Los Angeles, ICUC5, Lodz, Poland, Sept 2003, CD-ROM.

McElroy, J. L., 1969, A comparative study of urban and rural dispersion. *Journal of Applied Meteorol.* **8**, pp. 19-31.

Rotach, M. W., Gryning, S.-E., Batchvarova, E., Christen A., Vogt, R., 2004, Pollutant dispersion close to an urban surface - BUBBLE Tracer experiment. Accepted for publication in *Journal of Meteorology and Atmospheric Physics*.

Taylor, G. I., 1921, Diffusion by continuous movements. *Proc. London Math. Soc.* **20**, pp. 196-202.

Venkatram, A., Isakov, V., Pankratz, D., Heumann J., Yuan, J., 2004, The analysis of data from an urban dispersion experiment. *Atmos. Env.* **38**, pp. 3647-3659.

INTERNET-BASED MANAGEMENT OF ENVIRONMENTAL SIMULATION TASKS

Kostas Karatzas
Aristotle Uuniversity, Dept. of Mechanical Engineering, Box 483, 54124 Thessaloniki, Greece

Abstract: Urban air quality information originates either from observations or from mathematical tools – models and estimations. While the former correspond to the current status of air quality, and may be directly interpreted in terms of human health risk and eco-system degradation potential or effect, the latter provide forecasting information in advance, thus offering decision makers the opportunity to take preventive measures that would "smooth" or alter the results of a forecasted "episode" or even "crisis". Both "information categories" have a strong spatial/temporal dimension, and are an ideal application domain for World Wide Web (Web)-based information dissemination methods. In the present paper, the use of web-based technologies for the management of environmental simulation tasks within the air quality domain is discussed.

Key words: Environmental simulation, internet, client-server, air quality modelling, wizards.

1. INTRODUCTION

Environment related authorities have to deal with a twofold problem: on the one hand, they have to provide environmental information to policy makers and the public while on the other they need to setup and maintain a mechanism (including instrumentation, administrative and bureaucracy structures) that allows them to monitor and manage the quality of the urban environment. Taking just the air quality domain as an example, this problem is characterized by:

- multiple sources of information, including on-line monitoring systems;
- a dynamic and spatially distributed nature involving multiple temporal and spatial scales for the complex dispersion and transformation processes that "translate" emissions into ambient air quality conditions;

I. Farago et al. (eds.), Advances in Air Pollution Modeling for Environmental Security, 253–262.

- distributed (and mobile) emission sources with pronounced temporal patterns that include industry, households, and traffic sector and may be modelled as a network (dynamic) equilibrium process;
- accidental releases that may not be categorised within the existing "emission profile" of an urban area; these releases may typically include industrial accidents, accidents related to the transportation of dangerous goods, urban scale "disasters" (e.g. fire in a shopping mall), releases of dangerous gases or biological compounds by mistake (e.g. laboratory faults), criminal-terrorist activities, etc.;
- direct regulatory and indirect economic control on emission sources;
- multiple objectives and criteria at different spatial and temporal scales for the different actors and the regulatory framework.

Similar characteristics can more or less be observed in other environmental domains such as noise or water quality. To provide relevant information to the management and decision making process, information technology and analytical tools such as simulation models may be combined into powerful information and decision support systems.

2. THE REGULATORY FRAMEWORK

2.1 Urban air quality management

The life cycle of environmental quality information within an urban domain is represented in Figure 1. Thus, while the environment is the data "generator" per se, the legislative framework specifies procedures, methods and terms under which environmental monitoring and modelling should take place for regulatory purposes (resulting in data "production"), and defines urban environment management actions by which environmental quality related goals should be reached in an area of interest. The latter may be considered as a post-processing and "information generation" phase.

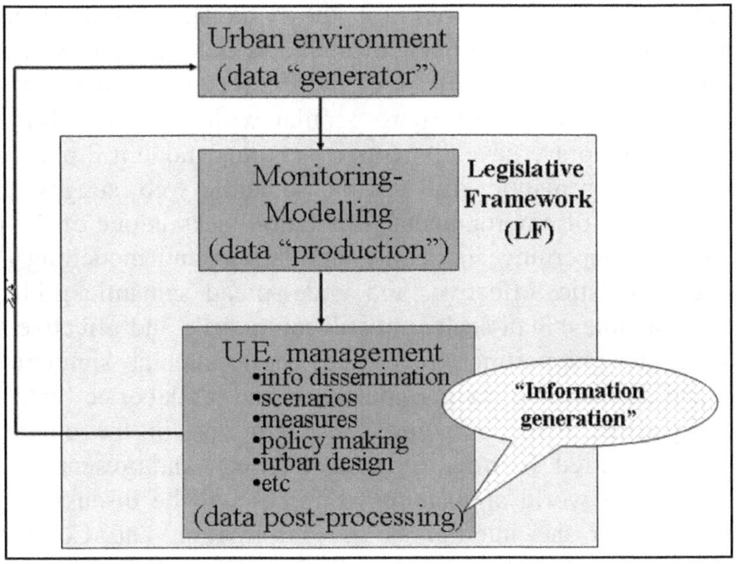

Figure 1: Environmental Information Life Cycle.

Focusing on air quality, is should be noted that the latter is one of the more advanced environmental fields regarding the EU legal framework developed. The first EU legislation concerning air quality information exchange and availability was Dir. 82/459, which was replaced by Dir. 97/101 which established a reciprocal exchange of information and data from networks and individual stations measuring ambient air pollution within the Member States. The 96/62 directive on ambient air quality assessment and management, also called "framework directive for air quality", which was adopted by the European Council in September 1996, and the Daughter Directives issued thereinafter, stress the need of model application as a supplementary assessment method for reporting of monitoring data and for managing and policy making support (European Commission, 2002; van den Hout, 2002).

Accordingly, there is a need for the European countries to build upon the scientific expertise and experiences of each other and to harmonise model development in some respects (URL 1). This is within the core of a number of scientific initiatives in this area, like the EUROTRAC-2 subproject SATURN (URL 2) and the European Initiative on "Harmonisation within Atmospheric Dispersion Modelling for Regulatory Purposes" (URL 3). Yet, one of the main problems still existing is that the use of an air quality (AQ) model is related to the use of a "sophisticated" tool which may need the support of experts in order to be set up (including data preparation and pre-processing) and applied (including post-processing and analysis). This is usually the case where various, distributed in space, application areas are employed, and the users are not familiar with model input and have no knowledge in interpreting model output. In addition, some of these models require considerable computational resources to be made available, which may not be the case for a single application site, regardless of the diminishing cost of PC clusters, which are becoming more and more popular within the modelling society. Moreover, contemporary developments concerning modern Problem Solving Environments, the semantic grid and the semantic web, suggest that web-based management of environmental simulation tasks is one of the principal ways to follow concerning air quality simulations and modelling (URL 4). Yet, before a holistic, effective, and wide-spread semantic grid and web platform is available, small-scale, but still multi-scale and effective solutions are required, for supporting web-based environmental simulation. Such solutions should be able to communicate and collaborate with existing modelling systems, while allowing for integration in the emerging grid community, if required. In order to demonstrate past and present technologies available, two real world application examples will be discussed hereafter. Both make use of the internet as the Information and Communication Technology backbone, but represent different approaches, thus providing a good example of web-based technology developments.

3. A TCP/IP – HTTP APPROACH IN MANAGING ENVIRONMENTAL SIMULATIONS

The first application presented concerning web-based management of air quality simulation is related to ECOSIM, an EU supported project under the umbrella of Environmental Telematics of the 4th FP (1996-98, URL 5). The main objective of this project was to design, implement and test an integrated urban environmental management information system using distributed information resources in order to provide easy-to-use but scientifically sound information to a broad range of users. To integrate a number of information resources in an efficient way, a client-server architecture based on TCP/IP and http was proposed and materialised by the project co-ordinator, Environmental Software and Services GmbH (ESS, URL 6; system architecture description and details available via Fedra, 1997). The main components were model servers, where the different simulation models were executed (for different environmental domains of interest, like air quality, coastal and ground water quality), and data base servers including the on-line connection to monitoring networks. The main server (provided by ESS) coordinated the various information resources and provided the elements of the user interface: graphical display, GIS, and an embedded expert system that supports users with the definition of scenarios for analysis. The ECOSIM Demonstrator was physically located in Athens, at the premises of the Greek Ministry of Environment, Physical Planning and Public Works, while the air quality model system server was located in AUT's premises in Thessaloniki (Lab. of Heat Transfer and Environmental Engineering). The server was making use of *http* protocol and *cgi* script applications. The Demonstrator initiated the model (through X11 GUI) at the model server side, through a *cgi* written in C. Information was passed from the user (Demonstrator/client) to the model (server) via POST, and a connection was then created and was left open for the receipt of the model results (Figure 2). Results concerning the system pilot application are described in detail elsewhere (Fedra et. al., 1999, Karatzas et. al., 2003).

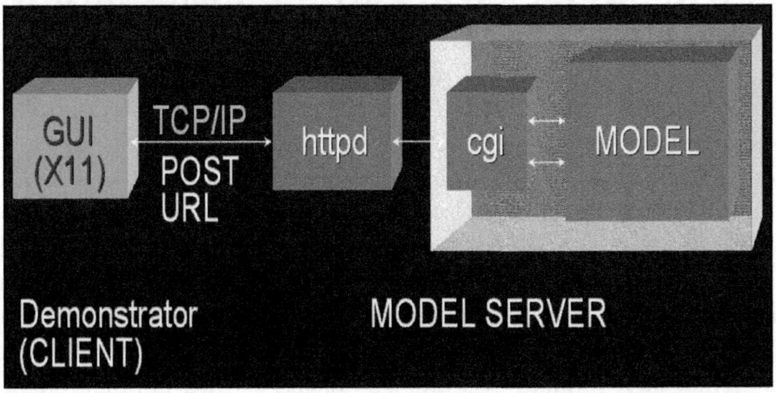

Figure 2: The generic client-server architecture integrating remotely located models to the ECOSIM system.

The ECOSIM system server was integrated with the Mesoscale Model MEMO (Moussiopoulos et. al., 1993, Moussiopoulos et. al., 1997) and the photochemical dispersion model MUSE (Sahm and Moussiopoulos, 1999), two "typical", state-of-the-art models that may be considered to be representative concerning input data needs, preparation requirements and execution time: both models require initial and boundary condition information (including meteorological conditions and emissions), plus information representative for the area of application like orography and land use, and a set of controlling parameters (more information and details may be found via the Model Documentation System of the European Environment Agency, URL 7). The main users of the system were employees of the Department of Air Quality of the Greek Ministry of Environment, having a considerable knowledge in the scientific and monitoring part of the air quality problems in Athens, but no access to or previous experience in AQ modelling tools. For this reason, the capability provided to apply state-of-the-art modelling tools for the Greater Athens Area was highly appreciated. From the comments received, it was suggested that users with a clear view concerning the influence of meteorological conditions and emissions in the area of interest, were also highly interested in related applications and decision support systems. On the other hand, users that had an everyday AQ management "routine" and short-term forecasting as their main concern, were mostly interested in accompanying applications that would automatically create short reports including predefined textual information and graphs, compiled "on the fly". These comments were taken into consideration from the project co-ordinator, for further developments of the product.

4. A WIZARD-BASED APPROACH IN MANAGING ENVIRONMENTAL SIMULATIONS

Air quality management calls for the usage of appropriate modelling tools for evaluating a set of parameters that support efficient environmental decision making. Yet, it is evident that such modelling tools are software programmes that require expertise in input data selection and preparation as well as running and post-processing of the results. A wizard is basically a series of screens or dialogue boxes that users follow through the completion of a task. Generally, each wizard screen asks users to enter information, either by making selections, or filling in fields (Bollaert, 2001). In the case of environmental-AQ simulations, the web-based wizard application helps the user in going through the whole AQ simulation process by providing a workflow scenario to be followed, accompanied by logical checks and support functions. To this end, wizards should be considered application services which are designed in such a way that they:

- Help the user to use a state-of-the-art AQ modelling tool in an easy way
- "Save" time by remembering user's previous actions and choices/decisions
- Explain every step needed to continue until the final objective is accomplished

In order to make use of an AQ simulation model, the *Model User's Interface* (MUI) has been developed, which is an interface application that allows for remote workflow management of environmental simulation tasks. Usually, a high amount of CPU and memory resources is required by the system running the AQ model; thus a client - server approach has been used for the implementation of the service, with the client being the user interface collecting parameters and data input and sending them to the server waiting for connections on a dedicated machine on the local network/internet. The implementation of the MUI is based on Java Web Start technologies for the client and Tomcat4 servlet container for the server. Thus, the platform used for development is J2SDK1.4 (http://java.sun.com). The client is a Java2 Swing application that makes use of Java Web Start technology to enable remote application invocation. The server is currently built on Java Servlet2.3 technology and the development environment was a GNU/Linux system with Tomcat4.1 servlet container and PostgreSQL7.3 database.

The user may invoke a graphical interface running on his/her workstation to prepare the case study. Then, the system provides a log-in prompt requiring user name and password to be accessed. These are issued by the system operator, are of personal basis, being checked for validity via a secure procedure, and may be unique per organisation or single person user. This is made by connecting to a special script that checks the personal data supplied in the login dialog box against an authentication database. Then, the user has access to his/her simulation tasks, and their status. Currently, three status categories are identified: a simulation case may either be waiting in the queue to be "served" by the calculations server (or cluster or distributed calculation environment), may be in process (where the percentage of completion may be added to the information that the user receives), or may already be completed. The user can then select one out of three functions: drop a case study and completely remove it from the list, download the resulting data from a finished case study (run), or create a new case study. It should be noted that the database managing the simulation tasks may also provide calculation priorities per user (thus different user categories may have different task priorities), and also timestamps the results, so that they will be deleted or archived in the case that the user does not download them within a certain time frame (parameterized my the service provider).

In case that the user wishes to define and submit a new case, model input and related parameters have to be provided. During this phase, a number of logical tests may be applied to model input in order to minimise mistakes. After the user commits a case for calculation, the client connects to the model server (which can physically be anywhere on the Internet) and pass the data required. The server application authenticates the call, stores the data until model execution, manages the requests and stores the results locally for a finite time of period, or until the user downloads them, and finally returns the status of the service request. The client interface can provide functions like checking the state and progress of the model execution at any time. The architecture of the application, which was developed and tested in the frame of the e-content EU supported project ENV-e-CITY (URL 8), is introduced in Figure 3, while a set of screen dumps demonstrating the application at work is shown in Figure 4. The MUI was tested with the aid of the OFIS model (Sahm et. al., 2000, Moussiopoulos and Sahm, 2001) within the frame of the ENV-e-CITY project, and was found to greatly improve the access of non-experts to the use of advanced AQ models, and to minimise model application errors while simplifying the use of AQ models for routine applications.

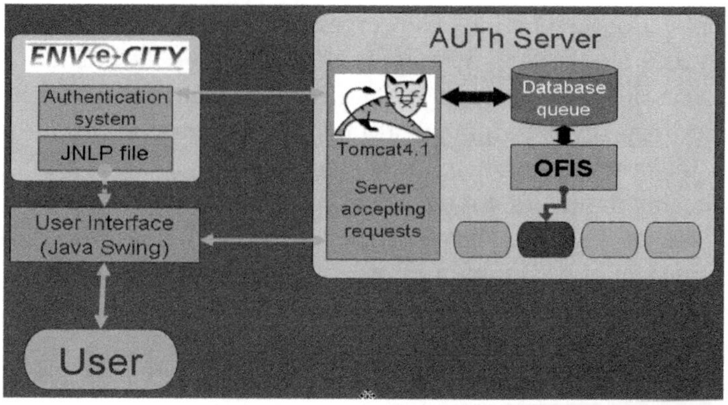

Figure 3: Basic architecture of the Model User's Interface service.

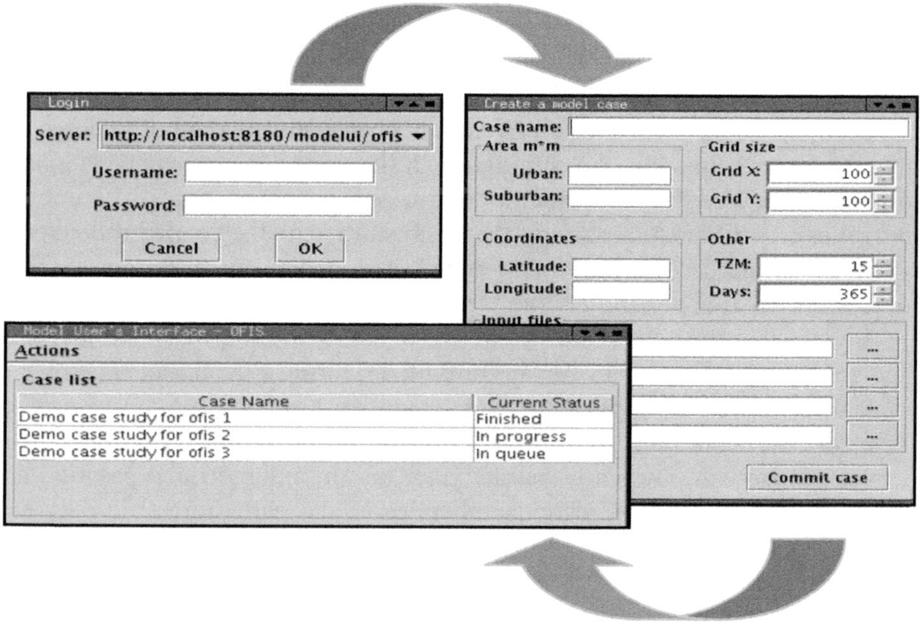

Figure 4: An example based on draft interfaces for the Model User's Interface service, demonstrating sequence of actions.

It should be noted that the MUI is a fully parameterised, user-tailored and localised application (now supporting multilingual environments), easily adaptable to any type of AQ model.

5. DISCUSSION AND CONCLUSIONS

Two AQ modelling related applications are presented that demonstrate the effectiveness and the advantages of internet-based environmental simulation tasks. Especially when one focuses on technologies like Java, which have gained the trust of software developers and provide with wide interoperability, and platform independent functions, it is evident that such applications may well be modified, tailored and applied for various AQ modelling tools and environments, thus creating added value services on top of traditional scientific tools. In addition, such technologies allow support and promote security on data management, transfer and dissemination (an issue of high importance in the AQ modelling domain). As AQ model applications call for multi-domain, secure, interdisciplinary frameworks for tool development and implementation, environmental simulation tasks call for internet based, flexible, secure, adaptable, low budget, scalable solutions. If these capabilities are combined with ontology concerning air quality management and modelling, then the latter may become a reference domain for the application of emerging technologies like semantic grid and semantic web to scientific computing.

Acknowledgements: The author gratefully acknowledges the European Commission for supporting projects ECOSIM and ENV-e-CITY, Prof. Nicolas Moussiopoulos who co-ordinated the participation of the Laboratory of Heat Transfer and Environmental Engineering of the Aristotle University of Thessaloniki to these projects, and provided guidance and support concerning the use of models MEMO, MUSE and OFIS, and Messrs Bassoukos, Papaioannou and Naneris for providing technical support in the development and set-up of the applications.

REFERENCES

Bollaert, J., 2001, Crafting a wizard: Fifteen dos and don'ts for designing wizards that make complex tasks easier for your users. IBM Developer Works, http://www-106.ibm.com/developerworks/library/us-wizard/?article=wir.

European Commission, 2002, Guidance on assessment under the EU air quality directives. http://www.europa.eu.int/comm/environment/air/guidanceunderairquality.pdf.

Fedra, K., 1997, ECOSIM project Architecture Description and Assessment Methodology. http://www.ess.co.at/ECOSIM/Deliverables/D0404.html.

Fedra, K., Karatzas, K. and Moussiopoulos, N., 1999, Integrated urban environmental management: monitoring, simulation, decision support. *Proceedings of the 3rd International*

Exhibition and Conference on Environmental Technology (HELECO 99) (A. Andreadakis et al., eds), Thessaloniki, Greece, 3-6 June, Vol. 2, pp. 219-225.

Karatzas, K., Fedra, K., Moussiopoulos, N., Lascaratos, A., Perivoliotis, L., Loizidou, M., and Fatta, D., 2003, Environmental Management of a coastal urban area: the ECOSIM project. *Proceedings of the 8th Int. Conference on Environmental Science and Technology*, (T.D. Lekkas, ed), Vol. B., pp. 393-400.

Moussiopoulos, N., Flassak, Th., Sahm, P., and Berlowitz, D., 1993, Simulations of the wind field in Athens with the nonhydrostatic mesoscale model MEMO. *Environmental Software* **8**, pp. 29-42.

Moussiopoulos, N., Sahm, P., Kunz, R., Vögele, T., Schneider, Ch., and Kessler, Ch., 1997, High resolution simulations of the wind flow and the ozone formation during the Heilbronn ozone Experiment. *Atmos. Environ.* **31**, pp. 3177-3186.

Moussiopoulos, N., and Sahm, P., 2001, The OFIS model: an efficient tool for assessing ozone exposure and evaluating air pollution abatement strategies. *International Journal of Environment and Pollution* **14**, pp. 597-606.

Sahm, P., and Moussiopoulos, N., 1999, The OFIS model: A new approach in urban scale photochemical modelling. EUROTRAC Newsletter 21/99, EUROTRAC-ISS, Garmisch-Partenkirchen, pp. 22-28.

Sahm, P., Moussiopoulos, N. and Janssen, J., 2000, The ozone fine structure model: Model concept and options. *Environmental Monitoring and Assessment* **65**, pp. 305-312.

URL 1, (May 31, 2004); http://www.meteo.bg/EURASAP/42/past1.html

URL 2, (May 31, 2004); http://aix.meng.auth.gr/saturn/

URL 3, (May 31, 2004); http://www.harmo.org/

URL 4 , (May 31, 2004); http://www.geodise.org/

URL 5, (May 31, 2004); http://www.ess.co.at/ECOSIM

URL 6, (May 31, 2004); http://www.ess.co.at/

URL 7, (May 31, 2004); http://air-climate.eionet.eu.int/databases/MDS/index_html

URL 8, (May 31, 2004); http://www.env-e-city.com , http://www.env-e-city.org

van den Hout, 2002, Air Quality Assessment Requirements Under The New EU Air Quality Directives, in *Urban air quality management systems* (Moussiopoulos, N., and Karatzas, K., eds), SATURN-EURASAP Workshop Proceedings, EUROTRAC-2 International Scientific Secretariat, GSF – National research Centre for Environment and Health, Munich, Germany, September 2002.

AIR POLLUTION ASSESSMENT INSIDE AND AROUND IRON ORE QUARRIES

Mykola Kharytonov[(1)], Alexandr Zberovsky, Anatoly Drizhenko[(2)], and Andriy Babiy[(3)]
[(1)]State Agrarian University, Voroshilov st. 25, 49027, Dnepropetrovsk, Ukraine; [(2)]National Mining Academy, K. Marx Av. 19, 49027, Dnepropetrovsk, Ukraine; [(3)]Institute for Nature Management Problems and Ecology, National Academy of Sciences of Ukraine, Moskovskaya st. 6, 49000, Dnepropetrovsk, Ukraine

Abstract: The mathematical description of natural ventilation processes in quarries requires the updating of huge data bases. In order to describe air circulation in a quarry, the known mathematical model was adopted, which takes into account the process of thermal convection from natural as well as from artificial heat sources. The model includes the conventional Navier-Stokes, heat and mass transfer equations. In the hydrodynamic equation the buoyancy force was taken into account as well. The system of differential equations for describing the problem of free thermal convection in the quarry was treated numerically. The developed programs were used to obtain current estimations of the ventilation circuits in quarries in order to determine the pollution levels, needed for the forthcoming practical recommendations. We also examined the distance at which iron, nickel and cobalt spread from the quarry, and the effect of heavy metal pollution on the activity of certain soil enzymes.

Key words: Air quality, weather, modeling, assessment, numerical forecast, quarry, dust-gas cloud transport.

I. Farago et al. (eds.), Advances in Air Pollution Modeling for Environmental Security, 263–274.

1. INTRODUCTION

The microclimate of quarries was investigated in the former USSR by the State Committee of Hydrometeorology in the sixties and seventies of the past century. The majority of the studies in Ukraine were prepared in the period 1972-1979 for the quarries of Krivbass with a maximal depth of 200 m. Recently, the depth of the Ukranian quarries has been continuously growing and will exceed 400 m in the near future. However, it was established that the earlier data series cannot be used for quarries more than 350 m in depth (Zberovsky, 1997). It has become clear that the mathematical description of natural ventilation processes in quarries and the prediction of air pollution levels inside and outside them requires the updating of a huge amount of microclimatic data.

The amount and type of emissions in different parts of the Dnepropetrovsk province, a significant industrial region of Ukraine, depend on the kind of industrial enterprises located in the region. In particular, iron ore mining-metallurgy results in high levels of industrial dust, oxides of sulphur, carbon and nitrogen. In the last decade, the annual average emission of industrial dust was 1.25 million tons for the iron ore mining region of the Dnepropetrovsk province (Babiy et al., 2003). The iron ore mining in the quarries involves blasting, which also has a considerable environmental effect.

That is why air pollution assessment inside and outside iron ore quarries was made.

2. MATERIALS AND METHODS

The investigations were conducted in the Krivoyrozhsky zone with high levels of air and soil pollution due to iron ore mining. Krivoyrozhsky zone or Krivbass includes Shirokovsky, Krivorozhsky and Pyatikhatsky districts with a total area of about 2660 km^2. There are five iron ore mining enterprises within Shirokovsky and Krivorozhsky districts. There are 200 open-pit mines here as well. The area of each quarry reaches 3 km^2.

A 2D mathematical model of air circulation in a quarry offered by Potashnik (1980) was applied by taking into account the process of thermal convection from natural and artificial heat sources (Zberovsky, 1997). That is, a two-dimensional non-stationary model of mixed convection for a turbulent liquid was used in a "trench" with curvilinear section. With the help of advanced model we solved the problem of free convection in the quarry under calm conditions.

The following parameters were used in the calculations:

- longitudinal size of the quarry on the top: L = 1000 m,
- quarry depth: 460 m,

- the temperature difference between the quarry and the surrounding air: $\Delta T_{\infty} = 6^{\circ} C$.

Chemical assessment of soil pollution as an effect of blasting was made independently from our attempt to model the air circulation in the quarry.

Soil and quarry dust samples were taken at the distances of 0.5-1.5, 3-5, 5-7 and 10-12 km from the edge of the quarry. The activities of enzymes in the samples were measured after one week at 28 $^{\circ}C$. The soil samples were prepared for chemical analysis by using $1N\,HCl$. The heavy metal content of the samples was determined by the use of a flame atomic-absorption spectrophotometer. Agrochemical and biochemical analyses of the soil samples were carried out by means of the standard methods.

3. RESULTS AND DISCUSSION

Meteorological elements greatly influence the technical, economic and social parameters of quarries. For example, low temperatures reduce the productivity and increase the frequency of occupational diseases among the workers. Calm or weak wind leads to an increase of the pollution level in the quarry; icy grounds and the fog reduce the productivity and can cause quarry accidents (Kharytonov et al., 2003). The interactions between the gas-dynamic parameters in the quarry and the microclimatic characteristics (e.g. geothermal and radiating parameters) as well as some ecological and social characteristics are highly complex. Therefore, the system "quarry – environment – human" can be considered as a diffuse system with complicated straight and feedback relations. That is why it is difficult to separate the different phenomena taking place in this system. At the same time, the nature, physical essence and scales of these phenomena are completely different. Three approaches can be applied to overcome these difficulties: physical, heuristic and combined (Kharytonov et al., 2003). The mathematical description of the processes of natural ventilation in quarries requires the updating of databases containing a huge number of parameters. The description of the natural ventilation includes the definition of the airflow speed in each point of the quarry space. It is a complex (multidimensional) problem of flow kinematics for a real liquid. In this connection, the free thermal convection problem inside the quarry was solved in the framework of a mathematical model by means of numerical integration for the system of differential equations.

3.1. Estimation of thermal effects in the quarry by using a two-criterion mathematical model of natural air re-circulation

From a general point of view, the logical sense of the basic physical model equations is as follows. We consider two-dimensional, non-stationary mixed convection of a turbulent liquid in a “trench” with curvilinear bottom border section. For the description of the turbulent heat and mass transfer, effective constant factors were used. In this manner we are led to the usual Navier-Stokes, heat conduction and diffusion equations.

The expansion of the model equations through the inclusion of the equations of heat transfer, the introduction of buoyancy forces in the hydrodynamic equations and a source term in the heat transfer equation allows us to take into account the thermal effects on the microclimate of the quarry. At the same time, it is reasonable to consider the free and forced convection processes both separately and in interaction. Background values are set as analytical relationships fitted to experimental data or the information received from reports concerning the behavior of meteorological elements in large and middle scale processes above the area of interest. In general, background values for speed constants, temperature and other elements were calculated to study background processes.

The system of perturbation equations in the tensor form (using Einstein’s convention on summation with respect to repeated indices) has been obtained by transition from the complete Navier-Stokes and heat conduction equations to the Boussinesq approach as below.

$$\frac{\partial u_i}{\partial x_i} = 0 ;$$

$$\frac{\partial u_i}{\partial t} + u_j \frac{\partial u_i}{\partial x_j} = -\frac{\partial p'}{\partial x_i} + \beta_T g T' \delta_{i3} + \frac{\partial}{\partial xj}\left(k \frac{\partial u_i}{\partial x_j} \right);$$

$$\frac{\partial T'}{\partial t} + u_j \frac{\partial T'}{\partial x_j} = \omega \left[\gamma_\phi (Z) - \gamma_a \right] + \frac{\partial}{\partial x_j}\left(k_T \frac{\partial T'}{\partial x_j} \right); \qquad (1)$$

$$\frac{\partial c}{\partial t} + u_j \frac{\partial c}{\partial x_j} = \frac{\partial}{\partial x_j}\left(k_d \frac{\partial c}{\partial x_j} \right).$$

where u_i, ϖ are the velocity vector projections on the horizontal and vertical axes (x_i, z), p' is the atmospheric pressure deviation from the hydrostatic one $p_\Phi(z)$ caused by the background density stratification, and T' is the temperature perturbation; c is the volume concentration; k, k_T, k_d are the coefficients of turbulent viscosity, temperature conductivity and diffusion, respectively; β_T is the coefficient of thermal expansion, $\beta_T = \frac{1}{T_\Phi}$, $\gamma_\Phi(Z)$ is the adiabatic gradient of background temperature: $\gamma_\Phi(z) = \frac{dT_\Phi(z)}{dz}$; g is the gravitational acceleration, and c_p the heat capacity of air under constant pressure. Indices 1 and 3 correspond to coordinates and vector projections in the x- and z- directions, respectively; the diagonal components of the Kronecker tensor are equal to 1, the others are zero.

It was necessary to transform the above mentioned equations to dimensionless form in order to obtain more information from the system and to decrease the number of the determining parameters.

The final form of the equations is the following:

$$\frac{\partial u}{\partial x}+\frac{\partial w}{\partial z}=0;$$

$$\frac{\partial u}{\partial t}+u\frac{\partial u}{\partial x}+w\frac{\partial u}{\partial z}=-\frac{\partial p'}{\partial x}+\frac{1}{Re_m}\left[\frac{\partial}{\partial x}\left(k\frac{\partial u}{\partial x}\right)+\frac{\partial}{\partial z}\left(k\frac{\partial u}{\partial z}\right)\right];$$

$$\frac{\partial w}{\partial t}+u\frac{\partial w}{\partial x}+w\frac{\partial w}{\partial z}=-\frac{\partial p'}{\partial z}+\frac{1}{Re_m}\left[\frac{\partial}{\partial x}\left(k\frac{\partial w}{\partial x}\right)+\frac{\partial}{\partial z}\left(k\frac{\partial w}{\partial z}\right)\right]+\frac{Gr_m}{(\mathrm{Re}_m)^2}T'; \quad (2)$$

$$\frac{\partial T'}{\partial t}+u\frac{\partial T'}{\partial x}+w\frac{\partial T'}{\partial z}=\frac{1}{\mathrm{Pr}_T\,\mathrm{Re}_m}\left[\frac{\partial}{\partial x}\left(k\frac{\partial T'}{\partial x}\right)+\frac{\partial}{\partial z}\left(k\frac{\partial T'}{\partial z}\right)\right]+N(z)w;$$

$$\frac{\partial c}{\partial t}+u\frac{\partial c}{\partial x}+w\frac{\partial c}{\partial z}=\frac{1}{Pr_d\,Re_m}\left[\frac{\partial}{\partial x}\left(k\frac{\partial c}{\partial x}\right)+\frac{\partial}{\partial z}\left(k\frac{\partial c}{\partial z}\right)\right].$$

The similarity criteria that appear in the mixed convection equations are the following:

- Turbulent numbers of Reynolds and Grasshof, determining the value of the fraction $Gr_m/(Re_m)^2$ which describes the ratio between Archimedes forces and convective terms;
- the thermal and diffusive turbulent Prandtl numbers $Pr_T=k/k_T$ and $Pr_d=k/k_d$;
- Criterion $N(z)=\frac{L}{\Delta T_\infty}[\gamma_\Phi(z)-\gamma_Q]$
- or Eqert' s criterion, describing the heating of air as a result of compression. The characteristic velocity value for the free convection can be by making equal the ratio $Gr_m/(\mathrm{Re}_m)^2$ to 1, from which we obtain $u_\infty=\sqrt{g\beta_T\Delta T_\infty L}$.

The Grasshof number Gr_m is a key parameter of the free movement, which defines a mode of fluid motion arising due to temperature differences.

Let us move on to the analysis of some concrete results. Criteria Re_m, Gr_m, Pr_T, which determine the flow physics, on the scales under consideration had the following values: $Gr_m/(Re_m)^2 = 1$; $Re_m = (Gr_m)^{\frac{1}{2}} \approx 150 \sim 200$; $\Pr_T = 0.7$.

Parameter $N(z) = \frac{L}{\Delta T_\infty}[\gamma_\Phi(z) - \gamma_Q]$ for the problem of single-layer linear background instability was accepted equal to unity which corresponded to $\gamma_\Phi(z) = +0{,}02^\circ C/m$. The boundary condition for the dimensionless temperature perturbation was defined as follows:

$$T'(x,z,t)\big|_{z=h(x)} = \begin{cases} 1 & \text{th for e right side} \\ 0 & \text{for the rest of the boundary} \end{cases}$$

The initial velocity field in the quarry was defined according to Zberovsky, 1997 and Kharytonov et al., 2003. The right wall of the quarry was assumed to be $6^\circ C$ warmer than the air and the other walls. The background temperature gradient was equal to $+0.02^\circ C$. These boundary conditions have allowed us to model the radiative heating of the quarry wall in the morning and evening hours. The nature of the convective flow in the quarry can be explained by two circumstances. Firstly, above the heated horizontal parts of the land surface due to the continuous receipt of solar radiation rising air columns are formed. Secondly, if the heated surface is also inclined, the so-called "slope effect" takes place because of the Archimedes force. In our case, the joint action of the "slope effect" and the thermal heterogeneity of the surface are complicated with the curvilinear profile. The offered model has allowed us to establish the stationary regime of the convective airflows. Visually it represents a set of two whirlwinds rotating in opposite directions. However, it is necessary to develop further the final algorithm, taking into account the above two factors. The considered technique for estimating convective airflows in a quarry allows us to evaluate operatively and precisely the convective flow speed at any point of the quarry space, both for calm conditions and under the influence of wind. This is important for the prediction of dust-gas cloud development in the quarry and around it. The developed programs were used to obtain current estimations of the ventilation circuits in quarries in order to determine the pollution levels, needed for the forthcoming practical recommendations. Calculating the distribution of harmful atmospheric substances in the quarry has allowed determining the contours and volumes of zones with concentrations above 1 MPC (maximum permissible concentration).

3.2. The origin, formation and spreading of dust-gas clouds (DGC) and farmland pollution caused by heavy metals

The maximum volume of explosive substance (ES) released into the air due to one blasting in a quarry in the Krivbass iron ore mining region is about 1000 tons. 63-80% of the dust-gas cloud particles that dropped out around the iron ore quarry had a diameter of $< 1.4\ \mu m$ as usual. Two types of dust-gas clouds were determined: primary and secondary. The cloud formation process lasts 30-45 s after the blasting. Then intensive dropping of larger dust particles from cloud takes place within 60-120 s. It was established experimentally that gases and small particles spread at long distances (up to $10000-12000$ m) depending on the wind velocity (Mikhaylov, 1981; Kharytonov et al, 2002b).

The approaches to determine the time for quarry to be ventilated after blasting and the distances of dust-gas cloud spreading allow us to choose the sanitary-protected zone correctly. It was established that cloud spreading is beginning just after blasting and lasts more than 5-30 min. The volume of dust - gas cloud is within $10-20\,m\ln m^3$. The DGC height can reach 150 m.

Possible ways to decrease the dust-gas cloud spreading after numerous explosions in the quarries include organized and technological actions, watering the explosion zone, the application of foam, and artificial quarry ventilation (Zberovsky, 1997; Kharytonov et al., 2003). A generally applicable tool for the evaluation of polluted areas, based on the total toxicity of pollutants, soil resistance, efficiency of remediation costs as well as on legal aspects of land-use, in order to establish the priorities for soil remediation practices has also been developed (Kharytonov et al, 2002; Kharytonov et al, 2002 b; Ruth and Kharytonov, 2003; Kharytonov et al, 2003).

The soil pollution caused by heavy metal spreading after iron ore blasting in the quarry along northern wind rose direction is shown in Figure 1.

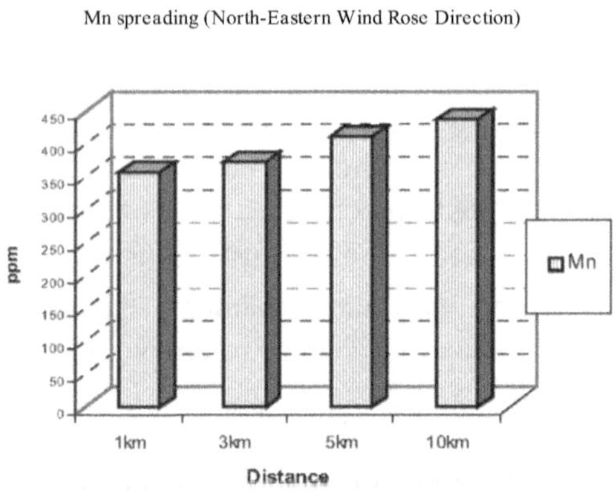

Figure 1/a.

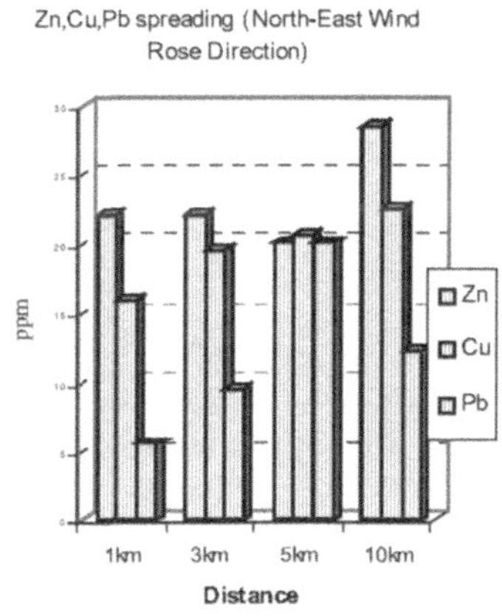

Figure 1/b: Heavy metal spreading along northern wind rose direction.

It was established that the distance of heavy metals spreading with an iron ore dust-gas cloud along northern wind rose direction is more than 10 km. Additional information on the iron soil pollution assessment in the Shirokovsky district of the Dnepropetrovsk region is presented in Figure 2.

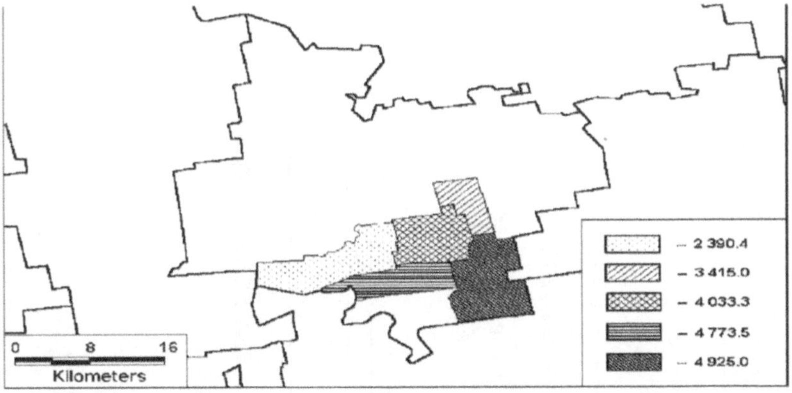

Figure 2: Iron distribution in farmlands within an ore mining region, mg/kg.

In addition to this, the spreading with dust clouds of heavy metals, related to iron, within Shirokovsky district was evaluated, too (Figure 3).

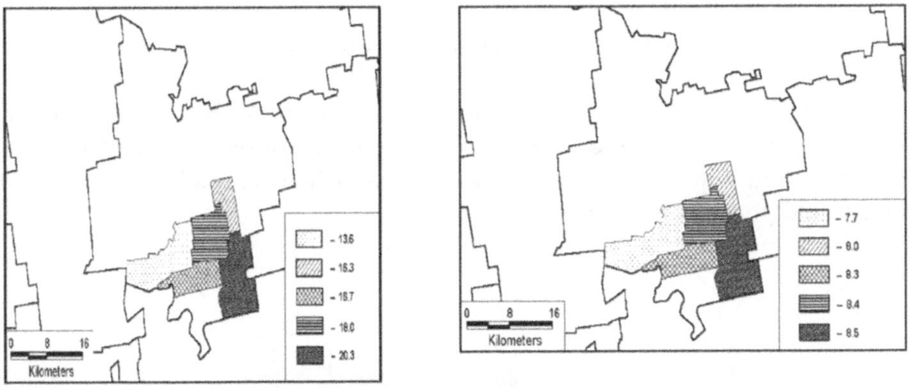

Nickel **Cobalt**

Figure 3: Heavy metal spreading after iron ore blasting, mg/kg.

Iron, nickel and cobalt spreading cause the same character of soil pollution. Two farms with different levels of soil pollution caused by airborne heavy metals were selected (Figure 4).

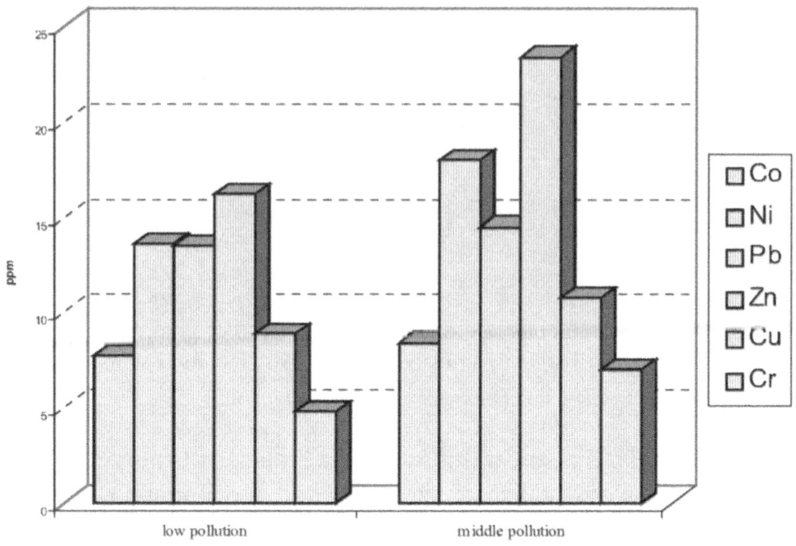

Figure 4: Heavy metals load assessment for two farms.

It was also established that long-term pollution of soil by quarry dust led to violation of its biochemical status. The negative consequences of the black soil pollution with the quarry dust were established for such enzymes as the urease and phosphatase (Figure 5).

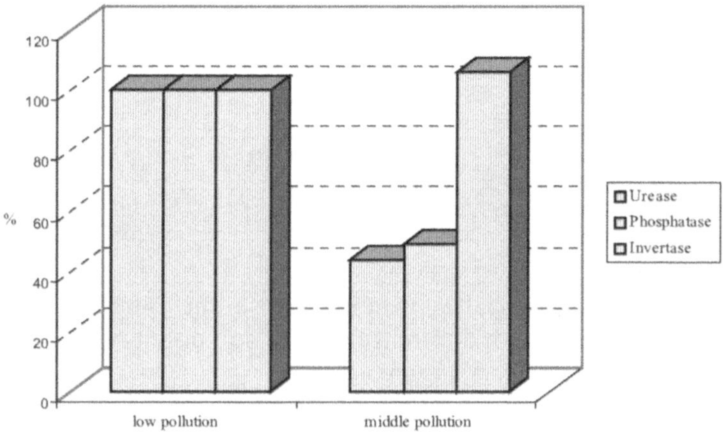

Figure 5: Soil enzymes testing for two levels of heavy metal pollution.

It was found that the urease and phosphotase activity was reduced by a factor of 2.5-3 at a distance 4-5 km from the quarry. Obviously, this results from the fact that the total heavy metal aerosol spreading has a negative impact on the soil activity. Model experiments showed differences in the rehabilitation effects of soils depending on dose, soil buffering, microorganisms, etc. It was shown also that humic fertilizers and vermicompost are more applicable as remediation amendments than ordinary fertilizers (Kharytonov et al, 2002).

4. CONCLUDING REMARKS

- Prospects for a two-dimensional mathematical model of natural ventilation processes in quarries were considered.
- Iron, nickel and cobalt spreading with dust-gas iron ore cloud cause the same character of soil pollution.
- The distance of heavy metal spreading with iron ore dust-gas cloud along northern wind rose direction is more than 10 km.
- The heavy metal pollution reduced the activity of soil enzymes (urease & phosphatase) by a factor of 2.5-3 at a distance of 4-5 km from the quarry.

REFERENCES

Babiy, A. P., Kharytonov, M. M., Gritsan, N. P., 2003, Connection between emissions and concentrations of atmospheric pollutants. Melas, D. and Syrakov, D. (eds.), Air Pollution Processes in Regional Scale, NATO Science Series, IV: Earth and environmental sciences. Kluwer Academic Publishers. Printed in the Netherlands, pp. 11-19.

Kharytonov, M., Gritsan, N., and Anisimova, L., 2002b, Environmental problems connected with air pollution in the industrial regions of Ukraine. In: Global atmospheric change and its impact on regional air quality. Barnes, I. (Ed). NATO Science Series, IV: Earth and environmental sciences. Kluwer Academic Publishers 16, pp. 215-222.

Kharytonov, M., Korbanyuk, R., Bulgakova, M., Protective effects of the humic substances against total pollutants toxicity. NATO - ARW in Zvenigorod (Russia), September 23rd-29th, 2002 "Use of humates to remediate polluted environments». Abstracts Proceedings, pp. 21-22.

Kharytonov, M., Zberovsky, A., Drizhenko, A., Babiy, A., 2003, Blasting impact assessment of the dust-gas clouds spreading in the iron ore mining region of Ukraine, Data fusion for Situation Monitoring, Incident Alert and Response Management, NATO ASI, August 18-29 2003, Armenia, CD-ROM.

Mikhaylov, V. A., Beresnevich, P. V., Borisov, V. G., Loboda, A. I., 1981, Protection from dust in the mining quarries , Moscow, Nedra, 262 p. (in Russian).

Potashnik, E. L., 1980, Thermal effects estimation in the bi - dimensional mathematical models for the quarry natural ventilation processes. Ventilation of shafts and mines, 7, The Interuniversity collection, Leningrad, pp. 56-63 (in Russian).

Ruth, E., Kharytonov, M., 2003, Integrated approach for Assessment of Polluted Areas. In: Social and Environmental Impacts in the North: Methods in evaluation of socio-economic and environmental consequences of mining and energy production in the Arctic and Sub-Arctic. Rasmussen, R. O. and Koroleva, N. E. (eds.). NATO Science Series, IV: Earth and environmental sciences., Kluwer Academic Publishers, 31, pp. 57-66.

Zberovsky, A. V.,1997, Air protection in the quarry – environment - human system. Dnepropetrovsk, p. 136 (in Russian).

DATA ASSIMILATION OF RADIONUCLIDES AT SMALL AND REGIONAL SCALE

Some preliminary results

Monika Krysta[(1)], Marc Bocquet[(1)], Olivier Isnard[(2)], Jean-Pierre Issartel[(1)], and Bruno Sportisse[(1)]

[(1)]CEREA (Centre d'Enseignement et de Recherche en Environnement Atmosphérique), Joint Research Laboratory ENPC/EDF R&D (Ecole Nationale des Ponts et des Chaussées / Electricité de France), Cité Descartes, Champs sur Marne, 77455 Marne la Vallée Cedex, France; [(2)]IRSN (Institut de Radioprotection et de Sûreté Nucléaire), 92262 Fontenay-aux-Roses Cedex, France

Abstract: We show in this paper possibilities offered by inverse modelling to improve the description of atmospheric dispersion of radionuclides in an emergency case. We describe how to couple models to observations in order to refine some input data such as source terms or certain physical parameters. We present our preliminary results obtained in synthetic experiments and indicate the directions of further studies leading to the capacity of adapting the parameter values on the basis of real observations. Two cases are investigated: the local scale (0-10 km) with the use of operational analytic models and the regional/continental scale with the use of 3D Eulerian models.

Key words: Atmospheric dispersion, radionuclides, data assimilation, inverse modelling.

1. INTRODUCTION

The purpose of this paper is to give an account of the exploration of opportunities offered by inverse modelling to achieve a better description of an accidental radioactive release. Even though no effort is spared to reduce the likelihood of a radioactive leak into the atmosphere, the possibility of such an accident cannot be excluded. It is absolutely vital for the monitoring agencies to be able to produce rapidly a reliable evaluation of the environmental consequences so that the decisions aiming at the protection of the population could be taken.

I. Farago et al. (eds.), Advances in Air Pollution Modeling for Environmental Security, 275–284.

The relevance of certain problems linked to a reliable description of accident consequences vary according to the size of the domain affected by the release. Small-scale atmospheric dispersion models (0-10 km) meant for operational use are usually analytic, Gaussian-like. These simplified models use estimates of the input parameters, such as meteorological conditions, depletion processes or emission terms, which might be incorrect. Inverse modelling compensates for uncertainties of such a description with information carried by the observations acquired after the leak has already taken place. Since the dependence of model outputs on input parameters is either linear or non-linear, we use inverse modelling based on the variational approach. This approach is an example of data assimilation that is often used in geosciences (Tarantola, 1987). It provides us with the refined input parameters obtained by minimizing the discrepancy between model outputs and observations (Robertson and Perrson, 1993; Quélo et al., 2004).

The questions may be different for regional scale. In a typical situation the observations showing an unusual level of radioactive contamination are the first indication of an accident. In these circumstances the problem to be solved is establishing the position of the emission source. An inverse modelling method must focus on the location problem. The one that we use is coupled to the regional three-dimensional Eulerian dispersion model (POLAIR3D), and is strongly related to the linearity of the problem (Issartel and Baverel, 2003; Issartel, 2003).

The objective of this paper is to briefly review the methodological questions arising in both small and regional scale. In the first section we introduce a small-scale model and give some preliminary results of inverse modelling based on twin experiments. In the second section we show a case study for regional scale illustrated with some linear algorithms for inverse modelling of the sources.

2. SMALL SCALE

2.1 Direct model

For small-scale considerations we limit ourselves to the vicinity of a nuclear power station, typically a distance of 10 kilometres. We mimic a time-dependent continuous source with a sequence of pollutant emissions. Each pollutant emission is considered to be instantaneous and results in an activity described by the advection-diffusion equation:

$$\frac{\partial act}{\partial t} + \nabla \cdot (act \cdot \vec{u}) = \nabla \cdot (D_{md} \cdot \nabla act) + S,$$

where an instantaneous activity $act = ACT + act'$ is defined as a sum of the mean activity and the activity fluctuation act'. $\vec{u} = \vec{U} + \vec{u}'$ is an instantaneous wind velocity also consisting of a mean $\vec{U}$ and fluctuating $\vec{u}'$ component. D_{md} is the molecular diffusivity tensor and S the source term. As a consequence, the mean activity of a pollutant at a given point $\vec{x}$ at time t can be obtained through the superposition principle by summing up the activities contributions from all pollutant emissions.

In a widely used approach of the Gaussian Puff Model (GPM) an instantaneous pollutant emission is identified with a Gaussian puff. Gaussian puff is an analytic solution to the advection-diffusion equation obtained under some simplifying hypotheses (Hanna, 1982). It gives us the following formula for the mean pollutant activity:

$$ACT(\vec{x},t) = \sum_{i=1}^{N_p} Q_i F_{di}(z,t) \frac{1}{\sqrt{(2\pi)^3}\,\sigma_{yi}^2(t)\sigma_{zi}(t)} \times$$
$$e^{-\frac{(x-x_i(t))^2}{2\sigma_{yi}^2(t)}} e^{-\frac{(y-y_i(t))^2}{2\sigma_{yi}^2(t)}} \times \left[e^{-\frac{(z-h)^2}{2\sigma_{zi}^2(t)}} + e^{-\frac{(z+h)^2}{2\sigma_{zi}^2(t)}} \right],$$

where h is the height of the source, N_p is the number of puffs, Q_i is the activity carried away by the ith puff, $\sigma_{yi}(t)$ and $\sigma_{zi}(t)$ are the standard deviations of puffs geometrical dimensions given by some parameterizations. All the depletion processes such as radioactive disintegration F_{rad} or deposition (dry deposition velocity v_d, scavenging coefficient Λ) are included in:

$$F_{di}(z,t) = \exp\left(-v_d \int_{t_i}^{t} \frac{\left[e^{-\frac{(z_{ref}-h)^2}{2\sigma_{zi}^2(\tau)}} + e^{-\frac{(z_{ref}+h)^2}{2\sigma_{zi}^2(\tau)}} \right]}{2\sigma_{zi}(\tau)} d\tau \right) \exp\left(-\int_{t_i}^{t} \Lambda(\tau) d\tau \right) F_{rad},$$

where z_{ref} is the reference height and t_i is the emission time of the ith puff. The puff movement is governed by time-dependent meteorological conditions $\vec{U}$:

$$\frac{d\vec{x}_i}{dt} = \vec{U}, \quad \vec{x}_i = (x_i, y_i, z_i).$$

2.2 Inverse modelling

Let $\vec{p}$ be a vector of dispersion model input parameters. Typically it comprises emission characteristics or quantities related to physical processes in the atmosphere (such as point Q_p or diffusive source Q_d emission rate, dry deposition velocity or wet scavenging). In our case its first estimate $\vec{p}^b$ comes either from other models or from measurements and may differ substantially from the true value $\vec{p}^t$. The aim of inverse modelling is to modify input parameter values $\vec{p}^*$ so that they approached the true ones and led to activities, $ACT(\vec{p}^*)$, which would be closer to reality than activities $ACT(\vec{p}^b)$ obtained with the first estimate.

A cost function is defined in order to measure the discrepancy between model outputs and other sources of information such as observations $\overrightarrow{obs}$. In order that comparison could take place, $\vec{p}$ must be transformed by the observation operator $H(\vec{p})$. In our framework it is the direct model that plays the role of the observation operator, $H_k(\vec{p}) = \overrightarrow{ACT}_k(\vec{p})$ for $k = 1,\ldots,N_{obs}$, where N_{obs} is the number of observations. Thus, the cost function is given by:

$$J(\vec{p}) = \frac{1}{2}\left\|\vec{p} - \vec{p}^b\right\|_{B^{-1}}^2 + \frac{1}{2}\left\|\overrightarrow{ACT}(\vec{p}) - \overrightarrow{obs}\right\|_{R^{-1}}^2,$$

where $\overrightarrow{ACT}(\vec{p})$ is a vector formed by mean activities and its kth component is the activity at the kth observation point. The weights associated with the two terms are given by error covariance matrices B (background error covariance matrix) and R (observation error covariance matrix).

The modified parameter values $\vec{p}^*$ are obtained by minimizing the cost function with an appropriate algorithm (L-BFGS-B). The cost function value required by the algorithm is computed at each iteration step by a direct model while the cost function gradient is obtained with the help of an automatic differentiator in an adjoint mode (Faure and Papegay, 1998). The advantage of the adjoint mode consists in the fact that all the gradient components are obtained in a single execution of a program.

2.3 Some results

Before applying the above method to the atmospheric dispersion model we are forced to cope with a few difficulties. In order to take advantage of the inverse modelling procedure we must be able to know the statistics of background and observation errors. We make a simplifying hypothesis of independence of errors one from another. As a result, the observation and background error covariance matrices reduce to the diagonal form having respectively the observation $(\sigma_k^o)^2$ or background $(\sigma_i^b)^2$ error variance on diagonal. Denoting by N_p the number of optimized parameters we can write the cost function as:

$$J(\vec{p}) = J^b(\vec{p}) + J^o(\vec{p}) = \frac{1}{2}\sum_{i=1}^{N_p}\frac{(\vec{p}_i - \vec{p}_i^b)^2}{(\sigma_i^b)^2} + \frac{1}{2}\sum_{k=1}^{N_{obs}}\frac{(ACT_k(\vec{p}) - obs_k)^2}{(\sigma_k^o)^2}.$$

Inverse modelling has been performed with the help of synthetic observations. In this case the true entry parameter values $\vec{p}^t$ as well as the variances of error statistics are known and the error distribution is assumed to be Gaussian. The observations are model outputs perturbed with a Gaussian noise according to the formula $obs_k = ACT_k(\vec{p}^t) + \xi_k$, where $\xi_\kappa \sim N(0, \sigma_\kappa^0)^2)$ and $(\sigma_k^o)^2$ is the kth observation error variance.

To illustrate the method we have chosen a linear problem and differentiated the cost function with respect to the point source emission rate Q_p. The derivative calculated by the adjoint module $\nabla^{ADJ} J(Q_p)$ has been compared to the derivative approximated by the finite difference method:

$$\nabla^{DF} J(Q_p, \delta Q_p) = \frac{J(Q_p) - J(Q_p + \delta Q_p)}{\delta Q_p},$$

where δQ_p is a small perturbation. The behaviour of the ratio of $\nabla^{ADJ} J(Q_p)$ to $\nabla^{DF} J(Q_p)$ as a function of δQ_p being known (close to unity for a certain range of δQ_p, far from unity outside this range), the comparison is a reliable test of accuracy of the adjoint module.

The results of the optimization process appear in Table 1. It shows different frameworks that refer to the cost function with or without the background term making use of either ideal or noisy observations.

In case of a cost function reduced to the observation term and ideal observations the optimized value equals to the true one, but this is not the case any more for noisy observations.

Then we add the background term to the cost function and choose the initial value to play the role of the first estimate $\vec{p}^b$. The result depends on the ratio between the weights associated with the background and observation terms. If the background error is very small, the optimized value is necessarily close to the first estimate. As the relative importance of the observations grows, the optimized value is pulled from the first estimate towards the true value. The true value may only be attained in case of ideal observations. In case of noisy observations the growing background error also eliminates the influence of the first estimate. Nevertheless, the optimized value is not equal to the true one, but to the one obtained for the cost function without the background term that employed the same set of noisy observations.

	Observations	Point source emission rate [Bq s^{-1}]			
		Q_p^t	Q_p^b	$(\sigma_{Q_p}^b)^2$	Q_p^*
$J^o(Q_p)$	Ideal	1.0	0.5	-	1.000
	Noisy	1.0	0.5	-	0.997
$J^b(Q_p)+J^o(Q_p)$	Ideal	1.0	0.5	10^{-4}	0.625
		1.0	0.5	10^{0}	0.999
		1.0	0.5	10^{4}	1.000
	Noisy	1.0	0.5	10^{-4}	0.624
		1.0	0.5	10^{0}	0.997
		1.0	0.5	10^{4}	0.997

Table 1: Inverse modelling results for the point source emission rate based on synthetic observations.

3. REGIONAL/CONTINENTAL SCALE

In this part of our paper we describe the use of adjoint techniques applied to the inversion of radionuclides dispersion at continental scale. Variational data assimilation methods would perfectly suit this purpose. However the linearity of the isotopes dispersion allows for a simpler formalism of inversion. In particular, the measurements obtained in the field can be linearly (thus simply) related to the source to invert through the adjoint solutions of the problem, called *retroplumes* in this context (Marchuk, 1995 and references therein; Issartel and Baverel, 2003).

3.1 Transport and reaction equations

The physical system we are investigating is described at each point $(\vec{x},t)$ by a set of concentrations $c_k(\vec{x},t)$, where $1 \le k \le n$ is the index of the isotopes. Measured activities per volume are combinations of those concentrations. If $\Omega = D\times\left[t_i,t_f\right]$ is the space-time domain, then the vector of concentration fields is an element of a functional space denoted as $E(\Omega)\otimes E_I$ where E_I is the n-dimensional isotopic space. In the absence of advection/diffusion, the system evolves according to the laws of an isotopic chain:

$$\frac{dc}{dt} = Mc,$$

where c is a vector of E_I. M is the *kinetic* matrix composed of real negative numbers which are the decay rates $\{\lambda_k\}_{1\le k\le n}$ of all the members of the isotopic family. In general it is a real $n \times n$ matrix but in the cases investigated here we assume that it can be put into triangular form.

The concentration vector c of the radioelements dispersed throughout the atmosphere follows:

$$\frac{\partial c}{\partial t} + \nabla \cdot (c\vec{u}) - \nabla \cdot (D_{td} \cdot \nabla c) - Mc = S.$$

It is assumed that the air density is uniform and constant. Note that wet scavenging, convection, dry deposition could also be included in this model (Brandt et al., 2000). Their simplest parameterizations do not break the linearity of the transport equation. If M is triangular and if the rates $\{\lambda_k\}_{1\le k\le n}$ are all distinct, then the matrix can be diagonalized. If none of the operators distinguishes between species, then the transport equation can be brought into diagonal form. The inversion problem would then be simplified.

3.2 Measurement and Lagrange relation

The measurement equation is:

$$\mu_i = \int_{\Omega} d\vec{x}dt \left\langle c(\vec{x},t), \pi_i(\vec{x},t) \right\rangle,$$

where $\langle .,. \rangle$ is the canonical scalar product in E_I. π_i is the sampling function related to μ_i. μ_i could be an activity or a concentration, depending on the choice of the vector field π_i. It is the formal description of the radioelements sampling performed during the measurement procedure. The measurements are indexed by $1 \le i \le p$, where p is the total number of measurements.

Let us introduce the scalar product in $E(\Omega) \otimes E_I$:

$$(\varphi, \psi) = \int_{\Omega} d\vec{x}dt \left\langle \varphi(\vec{x},t), \psi(\vec{x},t) \right\rangle.$$

This allows us to write down the measurement equation as:

$$\mu_i = (c, \pi_i).$$

Because of the linearity of the problem, a linear operator L relates the source term S to the concentrations c, $c = L \cdot S$. As a consequence:

$$\mu_i = (c, \pi_i) = (L \cdot S, \pi_i) = (S, L^* \cdot \pi_i).$$

It is not difficult to check that the action of the adjoint of L, denoted by L^*, on π_i is the retroplume vector c_i^*, which satisfies the backward equation:

$$-\frac{\partial c_i^*}{\partial t} - \vec{u} \cdot \nabla c_i^* - \nabla \cdot \left(D_{td} \cdot \nabla c_i^* \right) - M^T c_i^* = \pi_i,$$

where M^T is the transpose of the kinetic matrix (note that the turbulent diffusion tensor D_{td} is supposed to be self-adjoint). The Lagrange duality relation between c and c_i^* becomes:

$$\mu_i = \left(c, \pi_i \right) = (S, c_i^*).$$

It is an explicit relation between the source and the observations.

3.3 Case of an accidental release

Suppose we are interested in a three-element chain including a grandmother, a mother and a daughter designated by the labels g, m and d, respectively. The above notation corresponds to:

$$S(\vec{x},t) = \begin{bmatrix} S_g(\vec{x},t) \\ S_m(\vec{x},t) \\ S_d(\vec{x},t) \end{bmatrix}, \quad \pi_i = \begin{bmatrix} \pi_{i,g} \\ \pi_{i,m} \\ \pi_{i,d} \end{bmatrix}, \quad c_i^* = \begin{bmatrix} c_{i,g}^* \\ c_{i,m}^* \\ c_{i,d}^* \end{bmatrix}.$$

The measurements could be written accordingly as:

$$\mu_i = \int_{\Omega} d\vec{x} dt (S_g c_{i,g}^* + S_m c_{i,m}^* + S_d c_{i,d}^*).$$

A classical problem related to the management of industrial risk is spot source localisation. The source is defined by its position $\left(\vec{x}_0, t_0 \right)$ and by the amounts of grandmother, mother and daughter released from $\vec{x}_0$ at t_0, Q_g, Q_m and Q_d, respectively. Now the measurements can be written as:

$$\mu_i = Q_g c_{i,g}^* (\vec{x}_0, t_0) + Q_m c_{i,m}^* (\vec{x}_0, t_0) + Q_d c_{i,d}^* (\vec{x}_0, t_0).$$

The identification of the source, in particular its localisation, amounts to a six-parameter problem for the ground position $\vec{x}_0$, the time t_0 and the quantities Q_g, Q_m and Q_d. Ideally, the problem could be completely solved if six independent measurements were available. In reality it is impossible for two reasons; firstly, the measurements necessarily contain errors; secondly, the adjoint retroplumes $c^*_{i,\gamma}$ (where $\gamma = g, m$ or d) must be obtained from a model which is never perfect either.

3.4 Localisation of the release site

This ideal diagnostics would not be very useful in real situations. Even if an accidental source generally corresponds to the fixed position $\vec{x}_0$ at ground level of some industrial facility, the duration of the release may often be as long as one day. This situation may be investigated with an extension of the simplex algorithm for the dispersion of a single passive tracer (Issartel and Baverel, 2003). A source at the position $\vec{x}_0$ is represented by the rates of release $\dot{Q}_g \geq 0$, $\dot{Q}_m \geq 0$, $\dot{Q}_d \geq 0$. Each measurement now becomes:

$$\mu_i = \int_\Omega dt(\dot{Q}_g c^*_{i,g}(\vec{x}_0,t_0) + \dot{Q}_m c^*_{i,m}(\vec{x}_0,t_0) + \dot{Q}_d c^*_{i,d}(\vec{x}_0,t_0)).$$

Suppose the influence of measurements error is reflected in division and multiplication of a measurement by a factor of 2. Then, the following set of constraints may be formed:

$$\dot{Q}_g \geq 0, \quad \dot{Q}_m \geq 0, \quad \dot{Q}_d \geq 0,$$

$$\frac{\mu_i}{2} \leq \int_\Omega dt(\dot{Q}_g c^*_{i,g}(\vec{x}_0,t_0) + \dot{Q}_m c^*_{i,m}(\vec{x}_0,t_0) + \dot{Q}_d c^*_{i,d}(\vec{x}_0,t_0)) \leq 2\mu_i.$$

$$Q_g + Q_m + Q_d = \int_\Omega dt(\dot{Q}_g + \dot{Q}_m + \dot{Q}_d)$$

Under these constraints it is possible to determine a minimum value $Q^{\min}_{tot}(\vec{x}_0)$ of the total release of elements of the radioactive chain:

In fact $Q^{\min}_{tot}(\vec{x}_0)$ should be calculated for all positions $\vec{x}_0$ in order to determine which positions may be actually considered as possible sources. If the number of measurement constraints is sufficient, $Q^{\min}_{tot}(\vec{x}_0)$ should be excessively large or undetermined for most positions. This diagnostics provides a good discrimination for the areas where the accident may be investigated.

4. FUTURE PROSPECTS

In the short-range case, further inverse modelling development will be twofold. On one hand other uncertain parameters, such as dry deposition velocity (strongly related to the aerosol nature of radionuclides), will have to be inverted. On the other hand we must prepare the system so that the real observations coming from a wind tunnel experiment performed on a scale model of a nuclear power station could be embedded in the analysis.

A crucial issue to be investigated is the compatibility of the model outputs and observations. The model outputs provide us with the mean activities only while the observations at our disposal may be either mean or instantaneous activities. Thus, in the framework of instantaneous observations we will be forced to handle representativeness error, which is a constituent of observation error and depicts the incongruity between mean and instantaneous activities.

At regional scale, the coupling with a small-scale model is a key feature of an operational system so that a point source could be described in a less diffusive way. The resulting data assimilation system will therefore be associated with two kinds of observational data (either on a coarse regional mesh or on a fine small-scale one) and the representativeness problem will also have to be addressed.

Acknowledgements: Part of this work is carried on within MIRA project (Inverse Modelling of Atmospheric Release) undertaken by IRSN in collaboration with ENPC and ECL (Ecole Centrale de Lyon).

REFERENCES

Brandt, J., Christensen, J. H., Frohn, L. M., Zlatev, Z., 2000, Numerical modelling of transport, dispersion, and deposition validation against ETEX-1, ETEX-2 and Chernobyl. Environmental Modelling & Software 15, issues 6-7.

Faure, Ch., Papegay, Y., 1998, Odyssée User's Guide Version 1.7. Technical Rapport, INRIA.

Hanna, S. R., Briggs, G. A., Hosker, R. P., Jr., 1982, Handbook on atmospheric transport. Technical Information Center, U.S. Department of Energy.

Issartel, J.-P., Baverel, J., 2003, Inverse transport for the verification of the Comprehensive Nuclear Test Ban Treaty. Atmospheric Chemistry and Physics 3, p. 475.

Issartel, J.-P., 2003, Rebuilding sources of linear tracers after atmospheric concentration measurements. Atmospheric Chemistry and Physics 3, p. 2111.

Marchuk, G. I., 1995, Adjoint Equations and Analysis of Complex Systems. Mathematics and its Applications, Hazewinkel (Ed), Kluwer Academic Publishers.

Quélo, D., Sportisse, B., Isnard, O., 2004, Data assimilation of linear tracers at small scale: a feasibility study. In revision for Journal of Environmental Radiology.

Robertson, L. and Perrson, C., 1993, Attempts to apply four dimensional data assimilation of radiological data using the adjoint technique. Radiation Protection Dosimetry 50, nos 2-4, p. 333.

Tarantola, A., 1987, Inverse Problem Theory, Methods of Data Fitting and Model Parameter Estimation. Elsevier.

THE IMPACT OF SEA BREEZE ON AIR QUALITY IN ATHENS AREA

Dimitrios Melas[(1)] , Ioannis Kioustioukis[(1)] and Mihalis Lazaridis[(2)]

[(1)] *Laboratory of Atmospheric Physics, Aristotle University of Thessaloniki, email:* melas@auth.gr [(2)] *Department of Environmental Engineering, Technical University of Crete*

Abstract: In the present paper, numerical simulation results are analysed in order to quantitatively assess the air pollution potential in Athens area during two days of diverse air quality and meteorological conditions, namely the 12th and 15th September 1994. During the 12th September, Athens was under the influence of Etesians while during the 15th the synoptic forcing was weak with the subsequent development of sea breezes. The results were produced using a 3-D, higher-order turbulence closure model, which has been previously validated against an extensive data set. The results revealed that the ventilation during the Etesian day is approximately half an order of magnitude larger than the respective one during the sea breeze day. This is attributed to large differences both in wind speeds and in boundary-layer depths. In addition, the results indicate that, during sea breeze days, Athens basin is prone to horizontal recirculation of pollutants.

Key words: Sea breeze circulation, Athens, air quality.

1. INTRODUCTION

Meteorological conditions are known to have a major impact on air quality, especially in urban areas where the high density of pollutant emissions constitutes a permanent threat. In many cities, unfavourable meteorological conditions are responsible for the majority of air pollution episodes (Ziomas et al., 1995). It is, thus, of primary importance to understand the meteorological processes that result in adverse conditions for dispersal of air pollutants. Air quality in urban agglomerations located in the coastal zone is strongly influenced by meteorological phenomena like sea/land breezes and Thermal Internal Boundary Layers (TIBL's). These phenomena have a large impact on the spatial and temporal distribution of the winds and the turbulence and determine to a large extent the dispersive ability of the

I. Farago et al. (eds.), Advances in Air Pollution Modeling for Environmental Security, 285–295.

atmosphere. Moreover, during sea breeze conditions there is a possibility of both vertical and horizontal recirculations of pollutants (Steyn, 1995).

During the last two decades much has been written on the sea breeze and its profound influence on pollutant concentrations in Greater Athens Area (GAA) (Lalas et al., 1983, Moussiopoulos et al., 1993, Pilinis et al., 1993, Katsoulis, 1995, Ziomas et al., 1998, Melas et al., 1998a, to mention just a few examples). The strong link between the wind regime and photochemical pollution in the area is summarized in Mantis et al. (1992). Photochemical pollution episodes are largely restricted to the warm period of the year. During summer, a persistent circulation system that is called Etesians, consisting of a northern wind flow, gives maximum ventilation of Athens basin controlling pollution levels. However, the Etesians are occasionally interrupted giving space to the development of local circulation systems, such as sea/land breezes along the axis of the basin (NE to SW) and katabatic flows from the surrounding mountains. This can lead to severe photochemical pollution episodes in GAA.

Despite the multitude of the studies dealing with air pollution in GAA, there has not been any attempt to compare the boundary layer dynamics during an Etesian and a sea breeze day and provide a quantitative measure of air pollution potential during the two days. This paper aims to present a comparative study of pollution potential in GAA. For this purpose, numerical model results are analysed which are based on simulations performed by the MIUU model (Enger, 1990a). The model has been previously validated against an extensive data set from MEDCAPHOT-TRACE (MEDiterranean CAmpaign of PHOtochemical Tracers-TRAnsport and Chemical Evolution) (Melas et al., 1998a and b). The objectives of this paper are threefold:

a) To compare the flow field and the turbulence structure during two days with diverse air quality and meteorological conditions, namely an Etesian and a sea breeze day.

b) To provide quantitative measures of the ventilation and the recirculation in Athens basin during the aforementioned days.

c) To gain perspective for the relative importance of the two flow patterns on air pollution potential.

2. METHODOLOGY

2.1. Integral measures of air pollution potential

In this chapter integral quantities are defined in order to quantify the atmospheric conditions that are adverse to the dispersion of air pollutants. The definitions given below follow the methodology proposed by Allwine and Whiteman (1994) with two major modifications: (i) the usage of the layer-averaged wind in place of the single point wind used by the authors and (ii) the insertion of the boundary layer depth in the definition of the ventilation factor.

Given a time series of N data point pairs of south-north (u) and west-east (v) wind components (or equivalent wind speed, U, and wind direction, D) at each grid point, the ventilation factor is defined as:

$$V_i = T\sum_{j=i}^{i+p}(\langle U\rangle_j \cdot h_j) \tag{1}$$

where the angular brackets denote layer-averaged quantity, h is the boundary layer depth, T is the time interval of the wind data, i = 1,…, N-p and p = τ / T – 1, where τ is the desired transport time.

The wind run is then defined by:

$$S_i = T\sum_{j=i}^{i+p}\langle U\rangle_j . \tag{2}$$

The west-east (X_i), south-north (Y_i) and the resultant (L_i) transport distance are defined by:

$$\begin{gathered} X_i = T\sum_{j=i}^{i+p}\langle u\rangle_j , \\ Y_i = T\sum_{j=i}^{i+p}\langle v\rangle_j , \\ L_i = \sqrt{X_i^2 + Y_i^2} . \end{gathered} \tag{3}$$

The recirculation factor is defined as:

$$R_i = 1 - \frac{L_i}{S_i}. \tag{4}$$

From the above definition it emerges that $0 \leq R_i \leq 1$, where high values are indicating the presence of recirculations on time scales comparable to τ while $R_i = 0$ shows a straight line transport without any recirculation (Allwine and Whiteman, 1994). It should be mentioned that the recirculation factor calculated by (4) takes into account only the horizontal recirculation (Steyn, 1995).

2.2. Overview of the numerical model

The MIUU model is a 3-D, hydrostatic meso-γ-scale model with a consistent higher-order turbulence closure. The model is well documented in the international literature (Tjernstöm, 1987, Enger, 1990a, and Enger et al., 1993, to mention just a few publications). Only a brief outline will be given here. A terrain-influenced coordinate system is used to introduce the topography in the model. The turbulent exchange coefficients in this model are determined in terms of the turbulent kinetic energy, which is obtained from a prognostic equation. The turbulence closure is based on the approach developed by Yamada and Mellor's (1975) 'Level 2.5' model. Details about the parameterization of the higher order terms are found in Andrén (1990). The advection scheme used in the model has been corrected for atmospheric diffusion and is of third order both in time and space. The model has been employed and validated in a variety of applications. Melas et al. (1995, 1998a and b) and Svenson (1998) have previously performed a detailed evaluation of the model performance for Athens area (including the two days studied in the present paper).

2.3. Model setup and initialization

The simulation results discussed here were obtained for a domain 150 x 150 km^2. The central part of this domain is shown in Figure 1. The grid distance at the centre of the model was 2 km, gradually increasing towards the lateral boundaries. The model top was set at 7 km. The total amount of grid points is 39 x 39 x 23. Two 24-hour simulations were performed for the 12th and the 15th of September, 1994, respectively. For each simulation day the initial profiles of potential temperature and specific humidity were taken from the rawinsonde measurements at Athens airport, Hellinikon. The initial wind profile is the estimated geostrophic wind at 2.00 LST and below 150 m a logarithmic wind profile. The 3D initialization of the model lasted for 3 h. The real simulation started at 2.00 LST. The time step is 15 s.

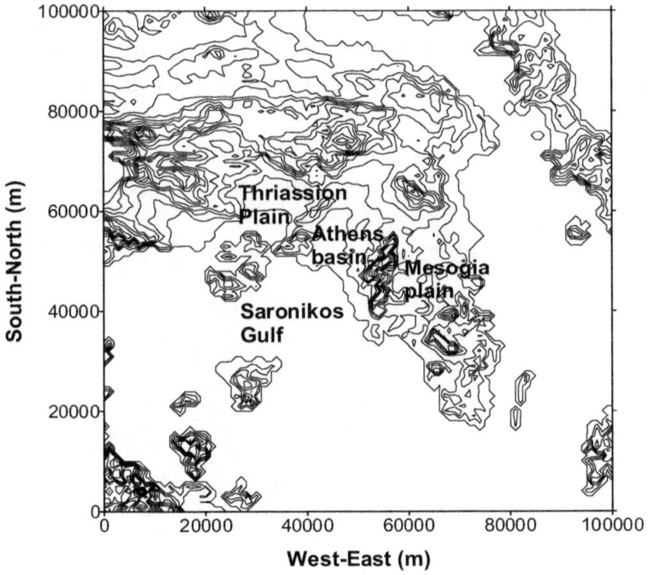

Figure 1: Topographic map of Attica peninsula. Contours are every 100 m.

3. RESULTS AND DISCUSSION

3.1. Boundary-layer dynamics

During the 12th of September, 1994, the synoptic flow pattern was largely determined from the combination of a low-pressure system situated to the east of Greece and an anticyclone, located over the central Mediterranean. The synoptic meteorological conditions were characterized by clear skies while a relatively strong northerly flow was prevailing over Athens area. On the other hand, during the 15th of September, 1994, Greece was covered by a high-pressure system. The pressure gradients were weak and clear sky conditions were prevailing. The diurnal variation of pollution levels at two monitoring stations is depicted in Figure 2.

The station of Patision is located in the center of Athens and it consistently shows the highest NO_2 levels. The Marousi station has a suburban character and is located at the northern part of Athens basin. As it is shown in Figure 2 the pollution levels were appreciably lower during the Etesian day compared to the respective levels during the sea breeze days. NO_2 levels at the central station of Patision peaked at ~450 μg m^{-3} during the sea breeze day while, during the Etesian day, it barely reached the alert limit (200 μg m^{-3}). The suburban station at Marousi shows a similar behavior but pollution levels are generally lower. During the sea breeze day ozone shows a sharp peak of ~300 μg m^{-3} at midday decreasing to negligible values during nighttime. On the contrary, the Etesian day is mostly characterized by a rather flat diurnal variation, especially during daytime hours.

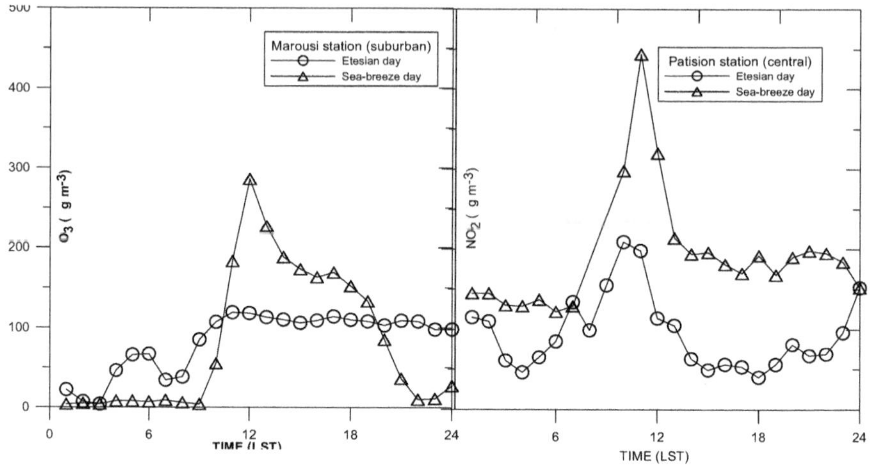

Figure 2: Pollution levels at a suburban (left) and a central (right) station during the two selected days.

During both days the nighttime wind pattern is strongly influenced by local topography, hence being very similar (not shown here). The stably stratified boundary layer is amenable to drainage flow and its most notable feature is the development of katabatic winds over the slopes of the major mountains of the area. These are converging over Athens basin to a light offshore wind. During morning hours, before the inversion break-up, the wind is weak from variable directions. The wind minimum appears at 9.00-10.00 LST and coincides with the maximum concentration of NO_2 (Ziomas et al., 1995 and 1998). The similarities between the two days end at this point. Subsequently, the stability change has a large impact on the wind field.

During the Etesian day (12/9) the wind field is more homogeneous and it is characterized by a rather strong flow from N-NE directions that prevails over Athens basin (Figure 3a). Melas et al. (1998b) attributed this flow to a sea breeze system embedded in the northeasterly synoptic flow. It is worth noticing that the winds over Saronikos gulf are from southerly directions. During the early morning hours of the 15th of September the air mass practically stagnates over Athens basin. The first indication of the sea breeze appears at ~9.00 LST while during the subsequent hours the sea breeze gradually penetrates over Athens basin. The simulation results indicate a sea breeze cell over Saronikos gulf which influences Athens basin and the industrialized Thriassion plain while another sea breeze system prevails over the eastern part of Attica peninsula (Figure 3b). The winds over Athens basin are weak from SW, generally not exceeding 4 m/s, decelerating downwind of the shoreline due to the urban roughness.

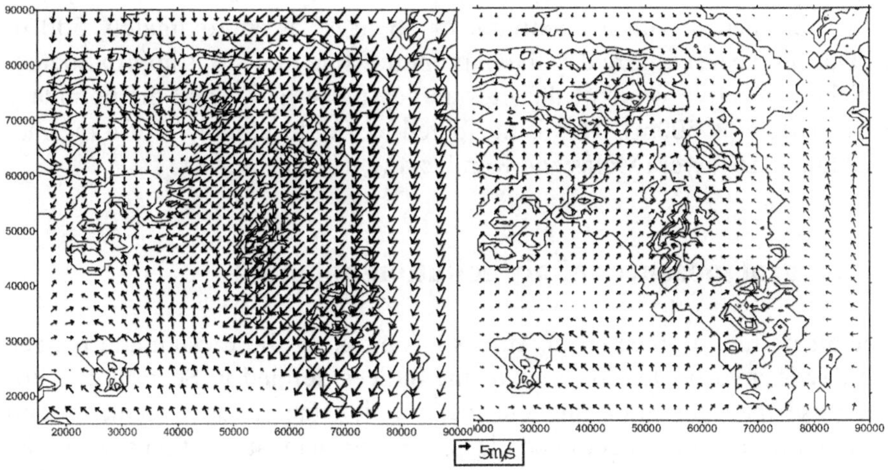

Figure 3: The wind field at ~25 m AGL at 14.00 LST; (a) 12/9/1994 (left), (b) 15/9/1994 (right).

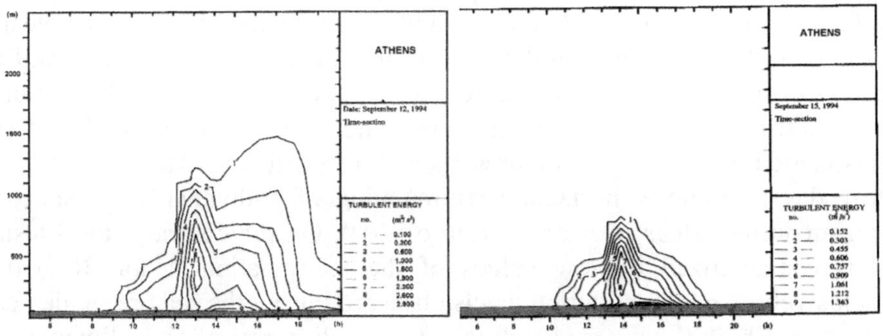

Figure 4: Time sections of turbulent kinetic energy at the National Observatory of Athens for the select days; (a) 12/9/1994 and (b) 15/9/1994.

Time-sections of the turbulent kinetic energy (TKE) for a grid point corresponding approximately to the location of the National Observatory of Athens (located in central Athens) are shown in Figures 4a-b for the Etesian and the sea breeze day, respectively. During both days, the TKE has very small values at nighttime. However, the development of the unstable boundary layer during morning hours has a dramatic effect on the TKE, which increases by 1-2 orders of magnitude. The maximum value appears at early afternoon but there are substantial differences between the two days. The Etesian day is characterized by a deep boundary layer (~1500 m) and intensive turbulence with a maximum of ~3 m^2/s^2 at approximately 500 m AGL. During the sea breeze day the maximum value of TKE is approximately half of the respective value of the Etesian day (~1.4 m^2/s^2) and the boundary layer is shallow (< 500m). This is attributed to the development of a thermal internal boundary layer (Melas and Kambezidis, 1992).

3.2. Integral measures of air pollution potential

The air pollution potential in Athens basin during the Etesian and the sea breeze day is investigated in terms of the indices defined in Section 2.1. For this purpose the ventilation and the recirculation factors are calculated and compared for the aforementioned days. The present study focuses on the influence of local flows on air quality and a complete cycle of these flows has a time scale of one day. For this purpose we have chosen $\tau = 24$ h and the values of the parameters calculated from equations (1) and (4) represent daily values. The model results are stored with time intervals of 1 h and consequently $T = 1$ h.

The results for the ventilation equation (1) and the recirculation equation (4) factors are shown in Figures 5 and 6, respectively. As it is clearly illustrated in Figure 5, the ventilation during the Etesian day is almost half an order of magnitude larger than the respective one during the sea breeze day. This is attributed to differences in both wind speeds and in depths of the atmospheric boundary layer (Figures 3 and 4, respectively). Moreover, the sea breeze day is prone to horizontal recirculation of pollutants, the respective factor attaining values in Athens basin of ~0.9. On the contrary, the Etesian day is characterized by low values of the recirculation factor, $R_i \sim 0.2$. Conclusively, the high pollution levels observed during the sea breeze day can be attributed both to adverse conditions for the dispersal of air pollutants and to the presence of local recirculations.

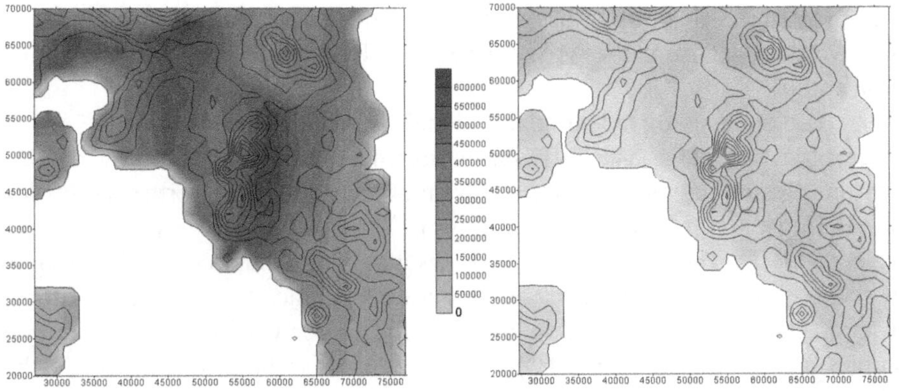

Figure 5: Geographic distribution of ventilation factor defined by equation (1); (a) 12/9/1994 (left) and (b) 15/9/1994 (right).

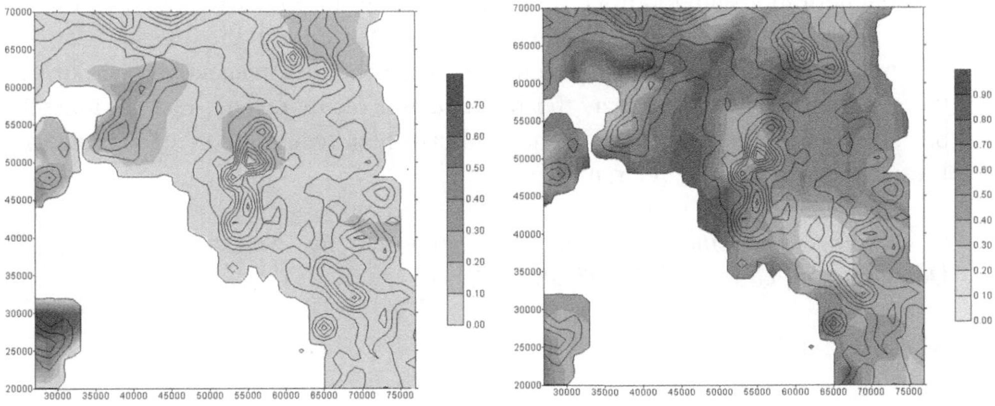

Figure 6: Geographic distribution of recirculation factor defined by equation (4); (a) 12/9/1994 (left) and (b) 15/9/1994 (right).

4. SUMMARY AND CONCLUSIONS

In the present study numerical model simulations are utilized to assess the pollution potential in Greater Athens Area during two days of diverse meteorological conditions. During the first day, the 12th September, 1994, the synoptic forcing was relatively strong and a moderate-to-strong northeasterly flow, known as Etesians, was prevailing in GAA. Anticyclonic conditions and the development of sea breeze circulations in the area characterized the second day, the 15th September, 1994. The numerical simulations were performed using a three-dimensional, higher order turbulence closure model

(Enger, 1990a). In a previous study the model was validated against the extensive data set from the MEDCAPHOT-TRACE campaign (Melas et al., 1998a and b).

The analysis of the simulation results reveals some distinct differences between the two days. The differences are more pronounced during daytime when the Etesian day is characterized by moderate-to-strong northeasterly flow while during the sea breeze day the flow in Athens basin is weak-to-moderate from SW. During both days the TKE shows a dramatic diurnal variation roughly being two orders of magnitude larger during daytime than during nighttime. However, during the Etesian day the TKE has larger values and by midday it occupies a layer of ~1500 m. The sea breeze day is characterized by a rather shallow boundary layer height (< 500 m).

In order to quantify the dispersive ability of the atmosphere, a ventilation and a recirculation factor are defined which provide integral measures of these terms. The mathematical definitions follow the methodology proposed by Allwine and Whiteman (1994). However, the original definitions are modified in order to account for the different type of data used in the present study. The ventilation factor during the Etesian day is approximately half an order of magnitude larger than the respective one during the sea breeze day. Moreover, the recirculation factor reveals that during sea breeze days Athens basin is prone to horizontal recirculation of pollutants. Etesian days have a low recirculation factor.

Acknowledgements: The work was supported by the Greek Ministry of Development, contract 4636/05-06-2002.

REFERENCES

Allwine, K. J., Whiteman, C. D., 1994, Single-station integral measures of atmospheric stagnation, recirculation and ventilation. *Atmos. Environ.* **28**, pp. 713-721.

Andrén, A., 1990, Evaluation of a turbulence closure scheme suitable for air-pollution applications. *J. Appl. Meteor.* **29**, pp. 224-239.

Enger, L., 1986, A higher order closure model applied to dispersion in a convective PBL. *Atmos. Environ.* **20**, pp. 879-894.

Enger, L., 1990a, Simulation of dispersion in moderately complex terrain-Part A. The fluid dynamic model. *Atmos. Environ.* **24**, pp. 2431-2446.

Enger, L., 1990, Simulation of dispersion in moderately complex terrain-Part B. The higher order closure dipersion model. *Atmos. Environ.* **24**, pp. 2431-2446.

Katsoulis, B. D., 1995, The relationship between synoptic, mesoscale and microscale meteorological parameters during poor air quality events in Athens, Greece. *The science of the total Environment* **181**, pp. 13-24.

Lalas, D. P., Asimakopoulos, D. N., Deligiorgi, D. G., Helmis, C. G., 1983, Sea breeze circulation and Photochemical Pollution in Athens, Greece. *Atmos. Environ.* **17**, pp. 1621-1632.

Mantis, H. T., Repapis, C. C., Zerefos, C. S., Ziomas, J. C., 1992, Assessment of the potential for photochemical air pollution in Athens: A comparison of emissions and air pollutant levels in Athens with those in Los Angeles. *J. Appl. Meteor.* **31**, pp. 1467-1476.

Melas, D., Kambezidis, H. D., 1992, The depth of the internal boundary layer at an urban area under sea breeze conditions. *Boundary-Layer Meteor.* **61**, pp. 247-264.

Melas, D., Ziomas, I. C., Zerefos, C. S., 1995, Boundary layer dynamics in an urban coastal environment under sea breeze conditions. *Atmos. Environ.* **29B**, pp. 3605-3617.

Melas, D., Ziomas, I. C., Klemm, O., Zerefos, C. S., 1998a, Anatomy of the sea breeze circulation in Athens area under weak large-scale ambient winds. *Atmos. Environ* **32**, pp. 2223-2237.

Melas, D., Ziomas, I. C., Klemm, O., Zerefos, C. S., 1998b, Flow dynamics in Athens area under moderate large-scale ambient winds. *Atmos. Environ.* **32**, pp. 2209-2222.

Moussiopoulos, N., Flassak, Th., Sahm, P. Berlowitz, D., 1993, Simulations of the wind field in Athens with the nonhydrostatic mesoscale model MEMO. *Env. Software* **8**, pp. 29-42.

Pilinis, C., Kassomenos P., Kallos, G., 1993, Modelling of photochemical pollution in Athens, Greece. Application of the RAMS-CALGRID modelling system. *Atmos. Environ.* **27B**, pp. 353-370.

Steyn, D., 1995, Air pollution in coastal cities. *Air pollution Modeling and Its Application XI*, edited by Sven-Erik Gryning and Francis A. Schiermeier, Plenum Press, New York and London, pp. 505-518.

Svensson, G., 1998, Model simulations of the air quality in Athens, Greece, during the MEDCAPHOT-TRACE Campaign. *Atmos. Environ.* (in press).

Tjernström, M., 1987, A study of flow over complex terrain using a three dimensional model. A preliminary model evaluation focusing on stratus and fog. *Annales Geophysicae* **5B**, pp. 469-486.

Yamada, T., Mellor, G. L., 1975, A simulation of the Wangara atmospheric boundary layer data. *J. Atmos. Sci.* **32**, pp. 2309-2329.

Ziomas, I., Melas, D., Zerefos, C., Paliatsos, A., Bais, A., 1995, Forecasting peak pollutant levels from meteorological variables. *Atmos. Environ.* **29**, pp. 3703-3711.

Ziomas, I., Tzoumaka, P., Balis, D., Melas, D., Klemm, O., Zerefos, C. S., 1998, Ozone Episodes In Athens, Greece. A modelling approach using data from the MEDCAPHOT-TRACE. *Atmos. Environ.* **32**, pp. 2313-2321.

DEVELOPMENTS AND APPLICATIONS IN URBAN AIR POLLUTION MODELLING

Clemens Mensink, Filip Lefebre, and Koen De Ridder
Vito - Centre for Integrated Environmental Studies, Boeretang 200, B-2400 Mol, Belgium

Abstract: An integrated assessment approach is presented to evaluate the impact of abatement strategies and mobility scenarios on transport emissions and urban air quality. The system consists of a set of flexible models that can be connected in a modular way within the framework of the impact pathway methodology. The MIMOSA model is used to evaluate road transport emissions. The AURORA model is used for urban air quality assessment and management. The evaluation of health impacts and environmental damage costs is carried out by applying the ExternE methodology.

Key words: Urban air pollution, emission modeling, air quality modelling.

1. INTRODUCTION

Europe is an urbanised continent. More than two thirds of the population is living in cities. The concentration of human activities on a relatively small area imposes large pressures on the urban environment and can lead to serious environmental problems. As a result cities experience increasing signs of environmental stress, notably in the form of poor air quality. A vast majority of the urban and suburban population is exposed to conditions that exceed air quality guidelines established by the World Health Organisation (WHO, 1999). It is obvious that people in urban areas are increasingly concerned about this, especially since the growing awareness of the possible health impacts of exposure to air pollution. The European Environmental Agency (EEA) considers urban air pollution as one of the major environmental concerns in Europe.

I. Farago et al. (eds.), Advances in Air Pollution Modeling for Environmental Security, 297–308.

A thorough knowledge of the present and future air quality in cities and of the parameters that determine the air quality is considered to be the necessary base for the development of urban air quality management policies and programmes. Reliable air quality models can help to provide this knowledge.

In the European framework directive on ambient air quality (96/62/EEC), the basic principles are given of a common strategy to define and establish objectives for ambient air quality in the European Community. This strategy is designed to avoid, prevent or reduce harmful effects on human health and the environment as a whole. Further objectives of this directive are to assess the ambient air quality in member states on the basis of common methods and criteria, to obtain adequate information on ambient air quality and ensure that it is made available to the public, inter alia by means of alert thresholds, and to maintain ambient air quality where it is good and improve it in other cases.

As a consequence of this directive, member states have to carry out air quality assessments and, depending on the outcome, have to develop action plans for zones where concentrations of pollutants in ambient air exceed the limit values. The directive also states that 'assessment of air quality means any method used to measure, calculate, predict or estimate the level of a pollutant in ambient air' and therewith recognises air quality models as assessment tools. Depending on the difference between limit value and representative actual levels, measurements can be complemented or replaced by models as assessment technique. Further requirements with respect to air quality modelling within the context of the EU directives is discussed in Section 2.

Besides their applications in air quality assessments, urban air quality models can be used for town planning, for determination of the environmental impact of future urban developments and activities (e.g. short term modifications in actual traffic situations or mobility plans on a long term), but also for finding the optimal locations for monitoring stations in cities in order to obtain a representative picture of the air quality in the city.

In Section 3 we discuss the integrated pathway methodology as a framework for the assessment of air quality impacts. Recently we developed a model for urban transport emissions (MIMOSA) and an integrated urban air quality model (AURORA) as tools within this framework. Section 4 describes the main characteristics of the MIMOSA model. Section 5 gives an overview of the main features related to the AURORA model. For the evaluation of health impacts and economic impacts (environmental damage costs) we follow the ExternE methodology, as will be discussed in Section 6.

2. EU AIR QUALITY DIRECTIVES

The European framework directive on ambient air quality (96/62/EEC) recommends that the assessment locations are to be sited at locations where the highest concentrations occur to which the population is likely to be exposed. At traffic-oriented sites the assessment point should be representative of the air quality in a surrounding area of at least 200 m². At urban background sites this should be several square kilometres. This implies that an air quality model should focus not only on the urban background, but also on calculating representative values at street level.

Information on the limit values and on the expected quality of the assessment results is provided in the so-called daughter directives. Accuracy for modelling is defined as the maximum deviation of the measured and calculated concentration levels, over the period considered by the limit value, without taking into account the timing of the events. Council directive 1999/30/EC of 22 April 1999 is related to limit values for sulphur dioxide, nitrogen dioxide and oxides of nitrogen, particulate matter and lead in ambient air. As a guideline the quality objective has an accuracy of 50% - 60% for hourly averaged values of nitrogen dioxide, nitrogen oxides and sulphur dioxide; an accuracy of 50% for their daily averaged values and an accuracy of 30% for their yearly averaged values. The guidelines for particulate matter and lead are restricted to the yearly averaged values, with an accuracy of 50%. A value for the accuracy of their daily averaged values has not been provided so far. Council directive 2000/69/EC of 16 November 2000 is related to limit values for benzene and carbon monoxide in ambient air. The guideline for data quality allows an uncertainty of 50%, both for the modelled yearly averaged benzene concentrations and for the 8-hourly averaged values for carbon monoxide.

3. INTEGRATED ASSESSMENT: THE IMPACT PATHWAY METHODOLOGY

Integrated assessment can be defined (Munn, 2002) as an interdisciplinary process of combining, interpreting and communicating knowledge from diverse scientific disciplines in such a way that the whole cause-effect chain of a problem can be evaluated from a synoptic perspective with two characteristics: (i) it should have added value compared to single disciplinary assessment; and (ii) it should provide useful information to decision makers. Evaluating the impact of potential emission reduction strategies on pollutant concentrations is one of the main concerns to policy makers in Belgium and elsewhere dealing with ambient air quality. Tools evaluating the impact of abatement strategies and mobility scenarios on ozone and other pollutants must provide fast assessments at low computational costs. In this way a large number of scenarios can be evaluated.

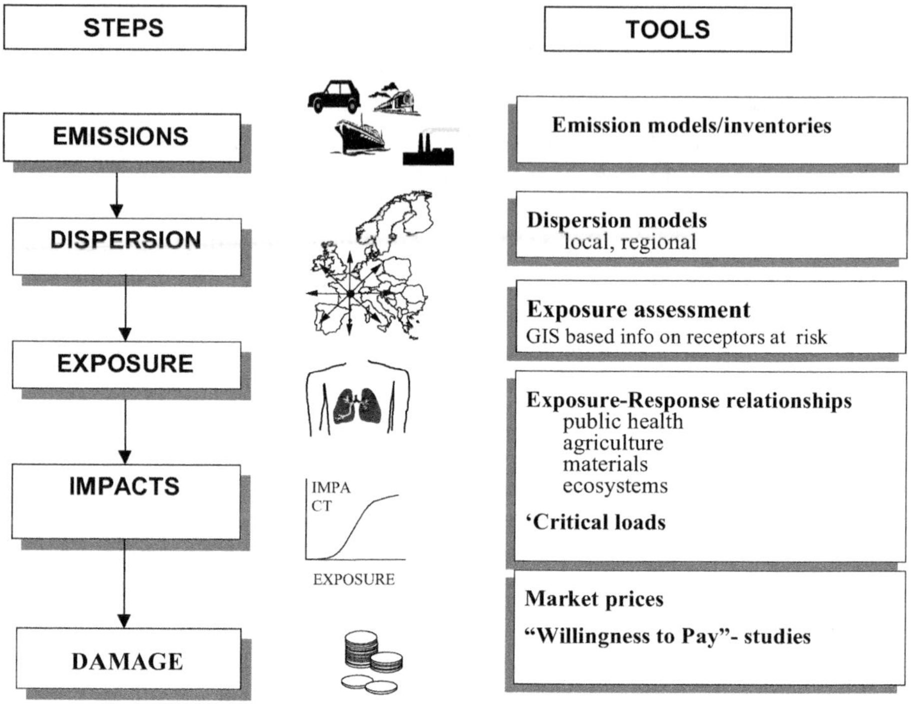

Figure 1: Integrated assessment framework: the impact pathway methodology.

Over the past years, we developed an integrated assessment approach to evaluate the impact of emissions from energy use and road transport emissions on air quality on an urban and regional scale. The framework for this assessment is the impact pathway methodology as shown in Figure 1. The impact pathway method follows the fate of pollutants along the steps in the DPSIR chain: Drive (human activities), Pressure (emissions), State (air quality and exposure), Impact (health, economic) and Response (policy).

As a tool for emission modelling we developed the MIMOSA model, as will be discussed in Section 4. MIMOSA provides road transport emissions for various pollutants on an urban or regional scale. For the evaluation of urban air quality, the output of MIMOSA is linked to the AURORA model. Air quality in street canyons is assessed by using a street box model. The AURORA model is further discussed in Section 5.

The applications of the integrated assessment methodology and its tools are very relevant for policy support. Some examples of applications are:

- Evaluation of effects of mobility actions and mobility plans
- Assessment of EU legislation (limit values)
- Optimal siting for monitoring stations
- Management of air quality
    - Short term measures (episodic)
    - Long term measures (structural)
- Cost/Benefit analysis (measures vs. effects)

4. THE URBAN TRANSPORT EMISSION MODEL MIMOSA

Urban emission inventories describe in a very detailed way the stationary and mobile emission sources that can contribute to any form of air pollution within the urban canopy. For urban air quality modelling an emission inventory should provide the input data with a sufficient time resolution and an adequate spatial resolution. For the urban situation this means that at least hourly variations in emissions can be described within a small grid domain (max. 1 x 1 km^2) or even at street level. Also the sensitivity of emissions with regard to temperature variations is important, not only for emissions stemming from spatial heating, but for traffic emissions as well.

The required dynamic character and spatial accuracy with which emission variations have to be described have led to the development of an urban transport emission model (Mensink et al., 2000). Initially, its design was based on the results of an urban traffic flow model (Nys, 1995). This model is actually in use by the Antwerp City authorities. The urban traffic flow model provides hourly traffic volumes for each element in a network of 1963 streets and road segments in Antwerp. It was implemented in a GIS environment. By combining the hourly traffic volumes computed per road segment with fleet statistics and the corresponding emission factor, the hourly urban transport emissions can be obtained.

The emission E(t) ($g \cdot h^{-1}$) calculated for pollutant i, vehicle class j, road type k and road segment n, can be expressed by:

$$E_{i,j,k,n}(t) = EF_{i,j,k} \cdot F_{m,d,h}(n, t) \cdot L(n) \qquad (1)$$

where EF is the emission factor ($g \cdot km^{-1}$) for pollutant i, vehicle class j and road type k. F is the time dependent traffic flow rate (h^{-1}) and L the road length (km) per road segment.

More recently, the model was updated and extended for the computation of road transport emissions on both urban and regional scales under the name

MIMOSA. It is used now by the Flemish administration and the Flemish Environmental Agency for the yearly evaluation of geographically distributed road transport emissions and for the assessment of mobility plans.

MIMOSA is based on the MEET/Copert-III methodology and has now been implemented both on a regional (Flanders) and on an urban scale (Antwerp). The model uses hourly averaged speeds of the different vehicles as defined by the output of a traffic flow model, like for example the TRIPS/32 road network. By means of the traffic flow models, a variety of mobility scenarios can be defined.

The emission factors used in MIMOSA are partially extracted from experimental data collected by Vito's VOEM system ('on-road' measurements) (De Vlieger, 1997) as well as from the Copert-III methodology. For missing data, emission factors from MEET were implemented. The model also distinguishes between hot emissions, cold start emissions and emissions due to evaporation. The number of pollutants that are evaluated has been extended to sixteen: CO, NO_x, VOC (61 species), PM, SO_2, CO_2, N_2O, CH_4, NH_3 and the heavy metals Pb, Cu, Cr, Ni, Se, Zn and Cd.

Five classes of vehicles are distinguished (Passenger Cars, Light Duty Vehicles, Heavy Duty Vehicles, Motorcycles and Mopeds) with a further subdivision into 105 classes depending on the age of the vehicle, cylinder capacity and weight. Different fuel types (gasoline, diesel, LPG and 2-stroke gasoline) are considered, with lead and sulphur contents depending on the year of the simulation. The road network is divided in six road classes (highways, national roads, main roads outside the city, main roads inside the city, secondary roads and harbour area). A specific year-dependent fleet composition is associated to each road class. The relative fleet composition has been updated for the years 1996 to 2002 using the most representative and actual data. Estimated vehicle park compositions for the years 2005 and 2010 were also implemented.

A scenario manager was implemented to allow the user to modify parameters and build his own scenario. A user-interface (Figure 2) developed under ARCVIEW 3.2 offers visualisation of the results in function of the selected pollutant, emission type (hot, cold, evaporation or total), road category and resolution (road segments or surface integrated emissions per km^2).

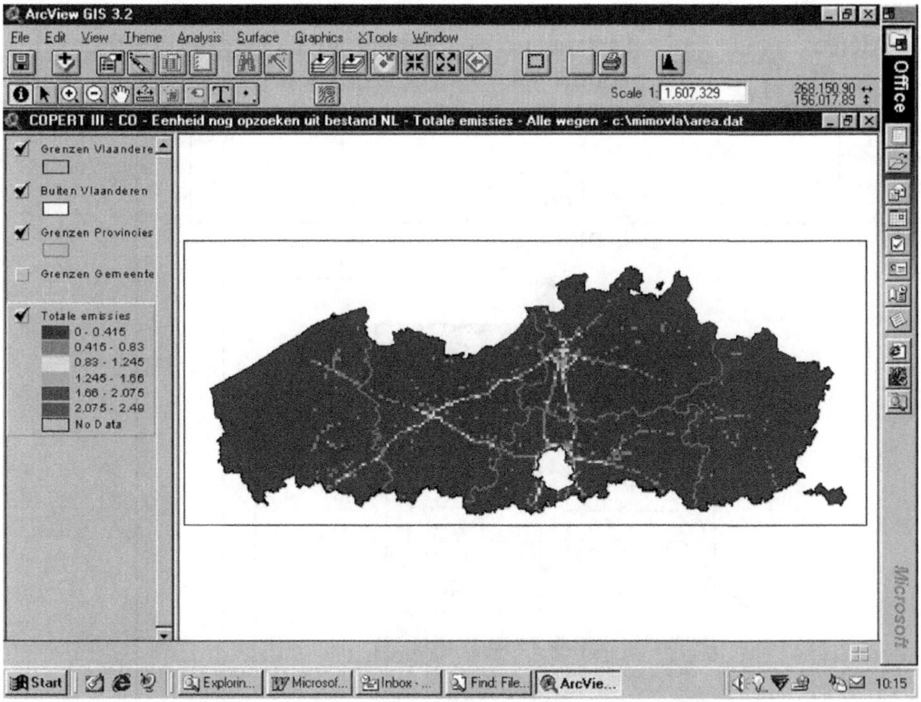

Figure 2: Total CO emissions per km² simulated for Flanders for a specific simulation period.

5. THE URBAN AIR QUALITY MODEL AURORA

For the assessment of air quality in cities, Vito introduced an *integrated model system*, known as AURORA (**A**ir quality modelling in **U**rban **R**egions using an **O**ptimal **R**esolution **A**pproach). This urban air quality management system has been designed for urban and regional policy support and reflects the state of the art in air quality modelling, using fast and advanced numerical techniques (Mensink *et al.,* 2001). The model input consists of terrain data (digital elevation model, land use, road networks) that are integrated in a GIS system. Meteorological input data are provided, with a resolution up to a few hundred meters, by a separate meteorological model (ARPS). The emission input data are resulting from a detailed inventory and acquisition of existing emission data in combination with emission modelling (Mensink *et al.*, 2000). In this way the emissions are described as a function of space, time and temperature.

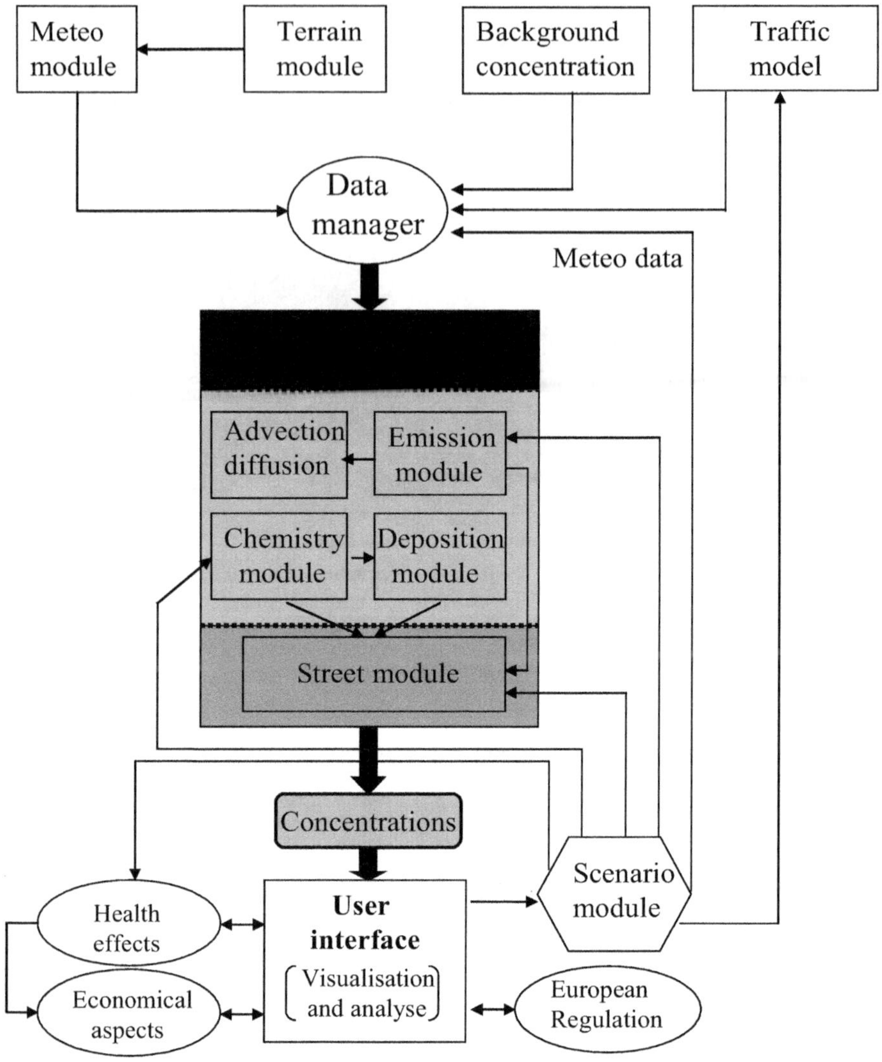

Figure 3: Modular structure of the AURORA model.

The physical and chemical processes are modelled in a modular way (see Figure 3), following the state of the art in urban air quality modelling and involving large-scale computations. Every module was tested and validated individually before integration in AURORA. The most important parts of the model system are the modules for physical transport phenomena (advection, diffusion, deposition, etc.) and the modelling of photochemical processes and chemical reactions. These modules include some advanced and improved numerical solvers.

Concentrations at the street level are estimated by means of the Street module (Mensink and Lewyckyj, 2001), using the 3D spatial configuration of the considered street and the related traffic information. Calculated pollutant concentrations are coupled to specific dose-response functions in order to assess both health effects and ecosystem (or material) degradations.

The AURORA model is successfully applied in the Cities of Antwerp and Ghent in Belgium, Budapest in Hungary and the Ruhr area in Germany. It also contributes to the EU 5th framework projects BUGS (De Ridder, 2003) and DECADE (Development and validation of highly accurate emission simulation tool capable of comparatively assessing vehicles operating under dynamic conditions) (Mensink et al., 2004).

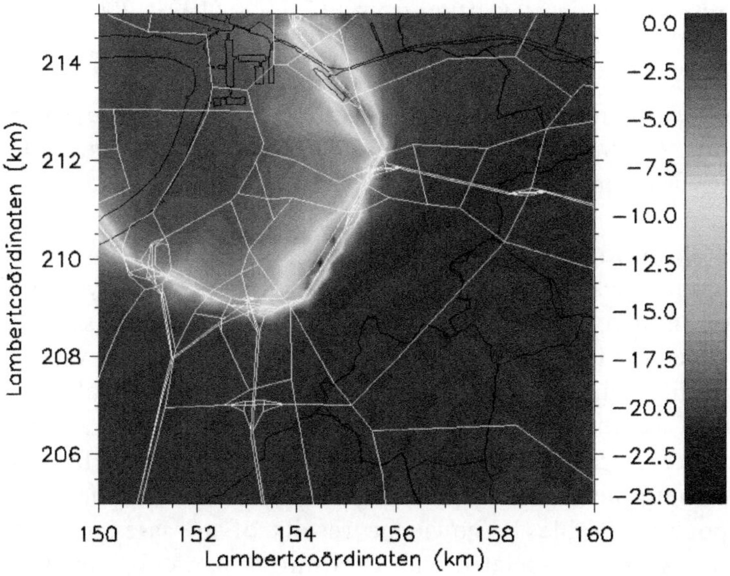

Figure 4: Modelling the impact of ring road traffic reductions on benzene concentrations in Antwerp.

Figure 4 shows an application of the AURORA model for the city of Antwerp. In this application the impact of tunnelling the ring road of Antwerp was investigated. In Figure 4 the reduction in benzene concentrations are shown (in percentage), associated with a covering of the ring road. From the figure it can be concluded that at least 75% from the benzene concentrations near the ring road are not caused by its emissions. This means that a rather high background concentration can be expected at this urban location.

6. THE EXTERNE METHODOLOGY

The evaluation of environmental impacts is based on the accounting framework of the European ExternE project. Using the ExternE methodology, estimations of the environmental damage costs related to the impacts can be provided. Environmental damage costs (e.g. health impacts) are not reflected in the prices of e.g. transport, fuels or electricity. They are not "internalized" in the product cost and therefore called "external costs".

The calculation framework for external costs results from a series of European research projects, starting from EC, Joule (1991- 2000); through the EU 5th and 6th framework programmes (EC, 1995; EC, 1998; Friedrich and Bickel, 2001). It is the results of a cooperation between +/- 50 multi-disciplinary teams from EU and the US. In the early years, the external costs of all forms of energy consumption were considered. Since 1996, the methodology is extended to transport.

In general, external costs of health effects of particulate matter (PM) are very important. These costs are based on a willingness-to-pay approach and are very useful in cost-benefit analyses of policy plans or emission reduction measures. The way to express or measure the health impact is through the so called Disability Adjusted Life Years or DALY's. These are already used in the context of health policy and enable a comparison of risks and impacts of PM with other disease factors. In this way, environmental factors can be incorporated into public health policies and prioritisation of health measures becomes possible. Table 1 shows the results of an assessment of health impacts (in DALY's) associated with exposure to particulate matter (PM) in Flanders and the related external costs.

	Attributable cases	DALY	External costs (1 000 EUR)
Acute cardiorespiratory mortality	750 (375 - 1 100)	131 (66 - 195)	31 000 (15 600 - 46 000)
Respiratory hospital admissons	1 400 (900 - 1 900)	34 (22 - 46)	6 000 (4 000 - 8 250)
Cardiovascular hospital admissions	1 700 (1 150 -2 300)	46 (31 - 61)	28 500 (19 000 - 38 000)
Bronchodilator use	1 900 (1 250 - 2 500)	2 (1,4 - 2,8)	75 (50 - 100)
Prevalence of cough	1 100 (550 - 1 300)	1 (0,5 - 1,3)	48 (24 - 60)
Lower respiratory symptoms	350 (250 - 600)	3 (2 - 5)	2 ,8 (1,9 - 4,7)
Respiratory hospital admissons for COPD[a]	160 (50 - 280)	37 (11 - 63)	700 (200 - 1 200)

Table 1: DALY's and external costs related to PM exposure, as calculated for Flanders in 2001.

7. SUMMARY AND CONCLUSIONS

For the assessment of air quality in cities, a modular integrated modelling system has been developed. The system consists of a set of flexible models and tools that can be connected in a modular way. The MIMOSA model is used to evaluate road transport emissions both on urban and regional scales. On an urban scale the AURORA model is used for air quality assessment and management. It has been designed for urban and regional policy support, using fast and advanced numerical techniques. Various modules in the system have been designed and refined, resulting in an air quality management tool that can provide reliable answers to policy makers and traffic planners. The AURORA model is successfully applied in various cities and is supported by the EU 5th framework projects BUGS and DECADE. The ExternE methodology is used to assess health effects and to calculate environmental damage costs in terms of external costs.

REFERENCES

De Ridder, K., 2003, BUGS: Benefits of Urban Green Spaces. http://www.vito.be/bugs/.

De Vlieger, I., 1997, On-board emission and fuel consumption measurement campaign on petrol-driven passenger cars. *Atmospheric Environment* **31**, pp. 3753-3761.

European Commission, 1995, DGXII, Science Research and Development, JOULE, Externalities of Fuel Cycles. Report numb0ers EUR 16520 EN to 16525 EN. (see website: http://externe.jrc.es/).

European Commission, 1998, DGXII, Science Research and Development, JOULE, Methodology Report, 2nd Edition, EUR 19083, EC.

Friedrich, R. and Bickel, P. (ed.), 2001, Environmental External Costs of Transport. Springer Verlag, Heidelberg.

Mensink, C., Cosemans, G. and Pelkmans, L., 2004, Dynamic modelling of transient emissions and concentrations from traffic in street canyons, in: P Suppan (Ed.): Proceedings of the 9th Int. Conference on Harmonisation within Atmospheric Dispersion Modelling for Regulatory Purposes, Vol 2, pp. 104-108, Garmisch-Partenkirchen.

Mensink, C., De Ridder, K., Lewyckyj, N., Delobbe, L., Janssen, L. and Van Haver, Ph., 2001, Computational aspects of Air quality modelling in Urban Regions using an Optimal Resolution Approach. In: S. Margenov, J. Wasniewski and P. Yamalov (Eds.) Large-Scale Scientific Computing, Lecture Notes in Computer Science **2179**, pp. 299-308.

Mensink, C. and Lewyckyj, N., 2001, A simple model for the assessment of air quality in streets, *Int. Journal of Vehicle Design* **27**, Nos. 1-4, pp. 242-250.

Mensink, C., De Vlieger, I. and Nys, J., 2000, An urban transport emission model for the Antwerp area. *Atmospheric Environment* **34**, pp. 4594-4602.

Munn, T., 2002, Encyclopedia of Global Environmental Change.

Nys, J., 1995, Computer program for Urban Traffic Planning, City of Antwerp, Antwerp (in Dutch).

World Health Organisation, 1999, Guidelines for air quality WHO, Geneva.

DEMANDS FOR MODELLING BY FORECASTING OZONE CONCENTRATION IN WESTERN SLOVENIA

Anton Planinsek
Environmental Agency of the Republic of Slovenia, Vojkova 1 b, SI-1000 Ljubljana, Slovenia

Abstract: The highest ozone concentrations in Slovenia occur in the southwest part of the country. This region has Mediterranean climate with a lot of sunshine during spring and summer. Ozone precursor emission in the western part of Slovenia is moderate. The main source of ozone precursor emission is traffic, emissions from other sources are small. Those emissions appear in the relatively narrow zone along the coast and the border with Italy. The area around the central part of the country is sparsely inhabited. The terrain there is rather hilly reaching 1500 m altitude. As opposed to this region, the emission of pollutants in the Po valley west from Gorica county is very high. This is one of the most industrialized regions in Europe. During the summer several episodes of photochemical smog are reported from big cities of North Italy where measures to abate pollution are introduced in case of such situations. Transport of pollutants across the border can be registered in the neighbouring countries. The main reason for the exceedances of the information threshold in the western part of Slovenia is the transport of pollutants. We have available good prediction of meteorological parameters in that area. To fulfil all demands of legislation regarding ozone concentration prediction, proper models should be chosen and adapted to these circumstances. More extensive monitoring will be performed for the validation of the models.

Key words: Air quality, weather, numerical forecast, atmospheric chemistry, atmospheric transport.

I. Farago et al. (eds.), Advances in Air Pollution Modeling for Environmental Security, 309–316.

1. INTRODUCTION

Slovenia lies between the Adriatic Sea, the Alps and the Pannonian plain. The area of the state is 20.000 km^2. The population is a little below two millions. Climatologically, Slovenia occupies three zones. The Mediterranean zone spreads along the coastal and southwest part of the country. The northwest and a part of the north side of the country belong to alpine climate. The east part of Slovenia belongs to the Pannonian plain, where the climate is continental. The central part of Slovenia has complex terrain with basins, valleys and hills reaching 1500 m altitude. The climate there has several transitions between three main climate types. The costal part is separated from Central Slovenia by a ridge that causes some local meteorological effects.

Figure 1: Map of Slovenia.

2. LEGISLATION DEMANDS

Directive 2002/3/EC of the European Parliament and of the Council of 12 February 2002 relating to ozone in the ambient air demands the following information when the information or alert threshold is exceeded or an exceedance is predicted:

Details to be supplied to the public on a sufficiently large scale as soon as possible should include:

1. information on observed exceedance(s):
   - location or area of the exceedance,
   - type of threshold exceeded (information or alert),

- start time and duration of the exceedance,
- highest 1-hour and 8-hour mean concentration;

2. forecast for the following afternoon/day(s):
   - geographical area of the expected exceedances of information and/or alert threshold,
   - expected change in pollution (improvement, stabilisation or deterioration);

3. information on type of population concerned, possible health effects and recommended conduct:
   - information on population groups at risk,
   - description of likely symptoms,
   - recommended precautions to be taken by the population concerned,
   - where to find further information;

4. information on preventive action to reduce pollution and/or exposure to it:
   - indication of main source sectors; recommendations for action to reduce emissions.

Demands of this directive, which are transposed into the Slovenian legislation, require several items.

- At first for each day the **prediction about information or alert threshold exceedance** must be made. At least it is necessary to make daily predictions for the monitoring sites for the present and next day. When one of those thresholds is exceeded, more information is needed.
- **Location** where exceedance occurs is clear, that is a monitoring site.
- A much more difficult task is to obtain the information about the **area** concerned. The nature of ozone as a secondary pollutant is different from that of primary pollutants, like sulphur dioxide or nitrogen oxides. If information about the emission source of an inert pollutant is known, regardless if it is a point, line or area source it is possible to calculate and predict the ambient concentration of the pollutant under consideration with certain accuracy. For ozone it is the chemistry and the weather that play the most important role. Ozone is also transported at long distances, so it is necessary to have information about ozone concentrations in the neighborhood. In the zones with dense traffic there are interactions between ozone and nitrogen oxides. All these facts must be taken into account before giving information.
- Another difficult task is to give the **expected change in the pollution level (improvement, stabilisation or deterioration)**. This information is very important for bodies which are responsible to introduce measures to abate pollution. The duration of the episode is especially important.

2.1. Monitoring of ozone in Slovenia

Monitoring of ozone is conducted at 11 monitoring sites in Slovenia. Locations are shown in Figure 2.

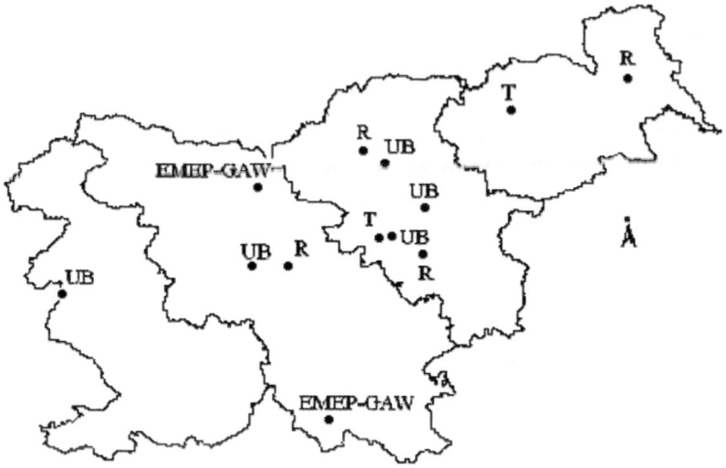

Figure 2: Monitoring sites for ozone in Slovenia from different networks. Station type is added: UB..urban background, R..rural, T..traffic.

At some locations there are series of measurements from 1992. In 2001 the renewed state network financed through EU programme PHARE started to operate. A new station was set in Nova Gorica, a town at the Slovenian border. This station is of urban background type and is partly influenced by emissions from road traffic.

Before 2001 there were no fixed measurements in the west part of Slovenia. The first year of monitoring showed that the level of ozone concentrations is already the highest there in Slovenia. In 2002 parallel measurements of the fixed station in Nova Gorica and the mobile station located at the Slovenian coastal zone took place. The distance between the two locations is about 50 km. A comparison showed similar ozone concentration levels. The weather in summer 2002 was not advantageous for ozone formation, so the ozone levels were not extremely high.

Date	Monitorig sites with exceedance of ozone information threshold
4.6.2003	Nova Gorica
9.6.2003	Nova Gorica
10.6.2003	Nova Gorica
11.6.2003	Nova Gorica, Ljubljana
12.6.2003	Nova Gorica
13.6.2003	Nova Gorica
25.6.2003	Nova Gorica
8.7.2003	Nova Gorica
16.7.2003	Trbovlje
21.7.2003	Nova Gorica, Ljubljana, Celje, Trbovlje, Hrastnik,
22.7.2003	Nova Gorica
26.7.2003	Nova Gorica
27.7.2003	Nova Gorica
28.7.2003	Nova Gorica, Iskrba
5.8.2003	Nova Gorica
6.8.2003	Nova Gorica
8.8.2003	Nova Gorica
9.8.2003	Nova Gorica
10.8.2003	Nova Gorica, Trbovlje
11.8.2003	Nova Gorica
12.8.2003	Nova Gorica, Ljubljana
13.8.2003	Nova Gorica, Ljubljana, Celje, Murska Sobota, Iskrba
14.8.2003	Ljubljana, Iskrba
23.8.2003	Nova Gorica

Table 1: Exceedances of ozone information threshold (180 $\mu g/m^3$) in Slovenia for the period April – August 2003.

A much different situation took place in summer 2003. Long periods of sunny weather accompanied with high temperatures and weak winds were the reason of high ozone concentrations. The number of exceedances of information threshold is shown in Table 1.

The highest number of exceedances occurred at the monitoring station Nova Gorica. In 2003 the Slovenian Order on ozone in the ambient air entered into force. It demands informing the population and relevant public institutions when the information or alert threshold for ozone is exceeded. Informing started in May 2003. There were no difficulties with information about the ozone concentration because data are refreshed every hour. Up-to-date data are also available on the web-site of the Environmental Agency of the Republic of Slovenia and on the teletext of the national television. The ozone forecasting was more problematical because Slovenia does not have models for its prediction. Through programmes financed by accession help from the EU we got enough information about ozone forecasting. The following information for the forecast of ozone concentrations based on expert knowledge was available:

- Concentration of ozone in Nova Gorica
- Concentration of ozone in other towns in Slovenia
- Meteorological parameters from regional forecast model ALADIN for 48 hours in advance for location Nova Gorica
- Forecasted wind field 48 hours in advance with a resolution of 3.5 km

The required forecast was done on the basis of the knowledge we got through the analysis of measured data. Experience from forecasting showed the following findings:

- Exceedance of information and/or alert threshold is strongly related with clear weather.
- On the first day after the passage of a weather front the exceedance is not likely to happen in spite of strong solar radiation.
- The concentration depends mostly on the wind direction.

The area with the exceedance of information threshold was estimated on the base of personal experience.

2.2. Description of local meteorology

The monitoring site in Nova Gorica lies at a low altitude in the suburb of the city. Nova Gorica has 36.000 inhabitants. City Gorica in Italy has 40.000 inhabitants. The two towns are located next to each other and are separated only with the state border. The altitude of the city is 90 m. In north and east directions of Nova Gorica there are hills and mountains. On the south side there is a ridge of hills which separate the Gorica province from the Gulf of Trieste. The distance from Nova Gorica to the nearest coast is about 20 km. During fair weather there are two local phenomena to be mentioned. The first is daily wind change caused by heating and cooling of the slopes of the mountains respective to the time of the day. This effect is strengthened with the sea breeze because it acts in the same direction. The other phenomenon is the local wind bora. It blows from the Pannonian plain to the North Adriatic Sea. During spring and summer it is connected with the retreat of a cold weather front or with a low pressure area on the North Adria.

As mentioned before, high ozone concentrations occur primarily in a few-day long dry weather situation with strong solar radiation. During the night, east to north wind directions prevail over Nova Gorica and its surroundings. Ozone concentrations normally decrease during the night. There are no significant ozone precursor emissions in these directions. During the day the prevailing wind turns to south and west directions. Soon after this, ozone concentrations increase significantly. Such a case is shown in Figure 3.

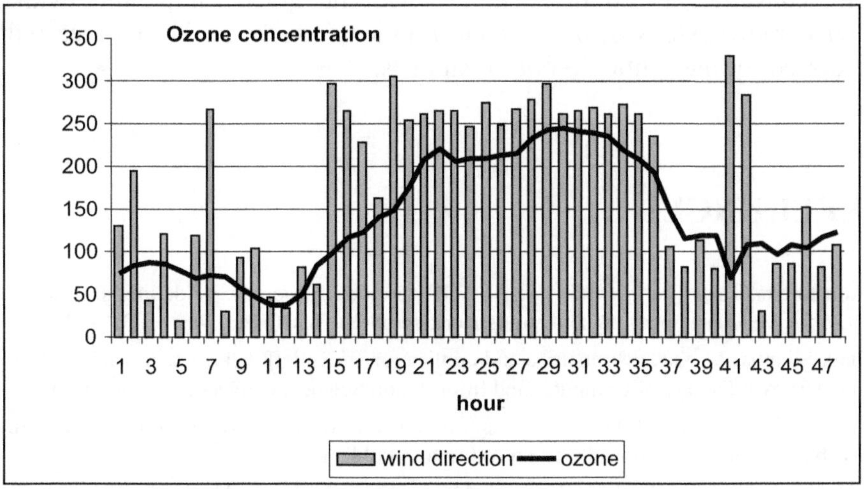

Figure 3: Daily course of ozone concentrations on June 6, 2003.

3. CONCLUDING REMARKS

How to predict ozone concentrations taking into account the above mentioned facts? Reliable forecasts of the meteorological parameters are available for 48 hours. The first step is to find a relation between the meteorological parameters and the high ozone concentration. This means a statistical model and also a "machine learning model". A part of that work has already been done. The main problem is caused by short series of data because only two years of reliable data are available for this area. Predicted concentrations are valid for the surroundings of the monitoring sites taking into account their radium of representativity. This kind of model cannot solve the whole problem. The legislation demands to define the area where information or alert threshold is exceeded. The location and strength of the emission sources cannot be defined exactly. Definition of the area with exceedance of information of alert threshold can be done with the help of calculating trajectories for a larger area. They are also available.

A project was already launched to solve some of these problems. It is a collaboration among the Faculty of Chemistry and Chemical Technology, the Faculty of Mathematics and Physics and the Environmental Agency of the Republic of Slovenia. The first phase will in short time give a machine learning model for the prediction of concentrations based on forecasts of the regional meteorological model ALADIN.

In the second phase more emphasis will be laid on defining emission sources using 3D trajectories. A contribution of several regions to the ozone concentration in Western Slovenia will be estimated. The program is still under consultation and the details cannot be given. In parallel with this activity, more extensive ozone concentration monitoring in this area will be performed for the calibration of the models.

REFERENCES

Aladin International Team, 1997, The ALADIN project: Mesoscale modelling seen as a basic tool for weather forecasting and atmospheric research. WMO Bull. 46, pp. 317-324.

Robnik Sikonja, M., 2001, Properties and application of ReliefF heuristic in machine learning, PhD Thesis, Faculty of Computer and Information Science, University of Ljubljana.

Witten, I. H., Frank, E., 2000, Data Mining: Practical Machine Learning Tools and Techniques with Java Implementations, Morgan Kaufmann Publishers.

Zabkar, J., Indihar, A., Vrabec, B., Petek M., Suc, D., Bratko, I., Merse, J., 2002, Improving ALADIN's temperature forecast with machine learning. Proc. of the 11th International Electrotechnical and Computer Science Conference, Volume B, ed. B. Zajc, pp. 459-460.

A PILOT SYSTEM FOR ENVIRONMENTAL IMPACT ASSESSMENT OF POLLUTION CAUSED BY URBAN DEVELOPMENT AND URBAN AIR POLLUTION FORECAST

Ion Sandu, Constantin Ionescu, Marian Ursache
National Intitute of Meteorology and Hydrology,Sos bucuresti Ploiesti, No. 97, Sector1, Bucharest, Romania

Abstract: In the framework of the LIFE –Environment Program, The National Institute of Meteorology and Hydrology (NIMH) carries out a demonstrative project concerning with the urban impact of air pollution. The project Life ASSURE has focused on impacts of long term estimation procedures of pollution levels in the atmosphere, and in the surface and underground hydrologic environment in the context of urban panning. Thus, the ASSURE system is able to provide data for evaluating the pollution potential of each objective of urban planning, and to encourage or discourage those urban or industrial scenarios which could constitute a pollution risk for environment, and consequently to the population health. In the first part of this paper the way in which dispersion model results are integrated in the operational final system on the Open–GIS platform is described. In the second part of the paper the statistical performances of two dispersion models are presented.

Key words: Air quality, numerical forecast, dispersion model, environmental impact.

1. INTRODUCTION

In the first section of this paper a new technical method of evaluation of the environmental impact caused by urban land-use planning is presented. The methodology regards impacts on the air quality as well as impacts on surface and underground water.

The ASURE Project, financed by the European Commission, have had the following main objectives:

I. Farago et al. (eds.), Advances in Air Pollution Modeling for Environmental Security, 317–326.

- to implement a demonstrative numerical system for integrated assessment of the probable pollution caused by implementation of urbanistic development;
- to demonstrate the capabilities for modeling of the environmental parameters in air, surface and underground water for the local authorities;
- the first attempt to develop an integrated Open-GIS-on-line allowing remote operation of the system, in the benefit of the local authorities;
- daily forecast of air pollution, pollution caused by the main metallurgical industries in the area of Baia Mare.

The Life ASSURE Project relies on the Open-GIS system and on dispersion models for estimation of pollution levels in all three domains: air, surface water and underground water. In successfully implementing the Project, top IT technologies have been used: Geographical Information Systems (GIS), Open Geographical Information Systems (Open-GIS), Computer Aided Design (CAD), Global Positioning Systems (GPS), Object Oriented Programming (O-O Programming), CAD Programming, WEB-GIS-programming in vector and raster GIS environments.

In the first section of the paper, the Open-GIS-on-line system which is implementing an Internet computing environment is described. The system contains and is able to use deterministic models: digital terrain model, surface water pollution model, underground vulnerability model and air pollution dispersion and transport models as well as pollutant direct/backward trajectory models. The whole system incorporates three main modules (or sub-systems) for pollution assessment: atmosphere, surface water and underground water, all interacting through the Open-GIS-on-line (operable in the Internet) integrator system. Starting from it the user is able to use special environmental functions as to: - visualize, optionally separate model results; - optionally dispose any post-processed diagnostic index, ean /daily pollutant trajectories; - dispose the integrated system of cumulated impacts as tool for an optimal decision choice for a given event of land-use design; - is able to include designed data into an Expert system for an objective decision indicator.

In the second section of this paper the daily mean air concentrations of SO_2 calculated with OML and ADMS dispersion models are compared with measurements made in a point of Baia Mare city for a three-month interval. The statistical performance of the mentioned dispersion models and the conclusions concerning the way in which the integrated system developed in this project can be used by the local authorities are presented.

2. DESCRIPTION OF TECHNICAL / METHODOLOGICAL SOLUTION; ROLE OF OPEN-GIS TECHNOLOGIES IN THE IMPLEMENTATION

One of the main achievements of the ASSURE Project consists in developing and implementing an **integrated computerized environmental protective system**, a tool able to assess for urbanistic planned land-use triggered environmental impacts concerning: atmosphere, surface water and underground water pollution.

At the final phase of the Project, the system has been developed in all of its components, such as: Knowledge Database; Digital georeferenced GIS-based databases; User input interface in GIS environment; Environmental protection models development; Models interface with GIS (air, surface water and underground water); System functioning automation; Output georeferenced for user analysis; Open-GIS on-line available to users with impact assessment on air pollution, surface water pollution, underground water pollution.

The proposed system aims at concurrent description of the environmental parameters on air and water at the town scale and beyond, in its peri-urban areas. The project is designed as a pilot system on behalf of Baia Mare City Council, a town which has one of the largest and broadest industrial display of activities, with emphasize on overlapping of environmental impacts. For that reason, the most suitable tool for the assessment is the Open-GIS technology, as the cutting edge of the GIS methodologies for environmental impact assessment.

2.1 The methodology for system integration: Open-GIS – as an integration tool for the system

The Open-GIS environment ensures the easy access to various types of formats used by the land-use designers operating at the urbanistic bureaus. Thus, in the case of the Baia Mare City Hall urbanistic bureau, Autodesk is the deployment tool for the land-use planning. Using an Open-GIS tool (such as Intergraph's GeoMedia Open-GIS) offers the open architecture environment in order to allow georeferenced input from the urbanistic bureaus, to the environmental modeling.

The Open-GIS on line system, which has been implemented for the application, relies on Intergraph's WebMap application. On top of the product, which functions as a Web Mapping Service (WMS), several functions have been programmed such that they insure additional functionality to the product, such as integrating raster data. Intergraph's WebMap version 5 has been used as base for developing the additional functionality. In this respect, since GeoMedia and WebMap versions 4 and 5 are built in accordance to Open-GIS Consortium's regulations for vector processing, the developed system insures the implementations of the Open-GIS Consortium to the same degree as Intergraph's base product – WebMap version 5. The additional features which have been programmed by the authors of this article, involve mainly raster visualisation and querying of data. Since these functionalities have not been submitted to the Open-GIS Consortium for consideration, one may conclude that these functions do not obey Open-GIS Consortium's regulations.

The system functions due to the integration of each model into the Open-GIS. The input of pollution emmissions is constructed by the users in Baia Mare in the urbanistic GIS that they have at their disposal in Baia Mare's case, they are using the AutoCAD. The input is developed automatically by using a special interface operating into AutoCAD environment, the interface being developed by NIMH for the project. The input is sent by the user via FTP protocol in the internet, towards the Open-GIS data server at NIMH. An automation software triggers the models respectively to the type of input which has been sent by the user. The models run automatically and their input is integrated in the same automatic way into the Open-GIS-on-line server. The user in Baia Mare then opens an Internet Explorer (or a NetScape) window and views the results in a GIS environment. The same window has a set of buttons constructed by NIMH, which allows the loading of special analysis functions that allow the display of each type of pollution, the interogation of the punctual pollution over a GIS element or the pollution simulated over entire areas over the city of Baia Mare. The buttons are specific for: 1) distribution of air pollution; 2) distribution of frequency of pollution surpassing the sanitary levels 3) distribution of air pollution climatological pathways; 4) distribution of retrotrajectory of air pollution pathways; 5) areas where pollution exceeds the sanitary level; 6) surface water pollution, where the surface water exceeds defined levels; 7) level of aquifer; 8) areas where the pollution of the aquifer exceeds the sanitary levels; 9) distribution of levels of aquifer pollution (Ionescu et al., 2002).

The user of the system may visualise, query and analyse the modelled data against the background geography. The way the user may get all this information is by using its usual Internet Explorer or Netscape web browsers. In order to enable their browsers to access the WMS, the user has to install an Active CGM browser ActiveX Control available free of charge on-line from MicroGraphix Company. In this respect, any user that is authenticated to access the server, is able to use the system for free. The only restriction is installed in order to protect the safety of the system against hacking attack. As such, only specific users are allowed to send input files (and thus to remotely operate and run the system), the access being restricted to specific IP addresses.

The models which are linked into the system are activated through a TSR (Task_and_Stay_Resident) program which has been programmed to monitor the occurence of any change in specific input data files. As such, the monitoring program (called „NIMH File Monitor") searches for arrival of new input data for the models each 10 seconds. Whenever a new file is sent by the user via FTP to the system, the monitoring program triggers a sequence of operations within the system, resulting in the production of GIS data and the submission of these data to the WMS.

The two figures presented are illustrating the kind of graphical output that may be available in GIS context and showing the spatial extension of pollutant simulations using two separate models. The following figures present the output in WMS context. the output in WMS context.

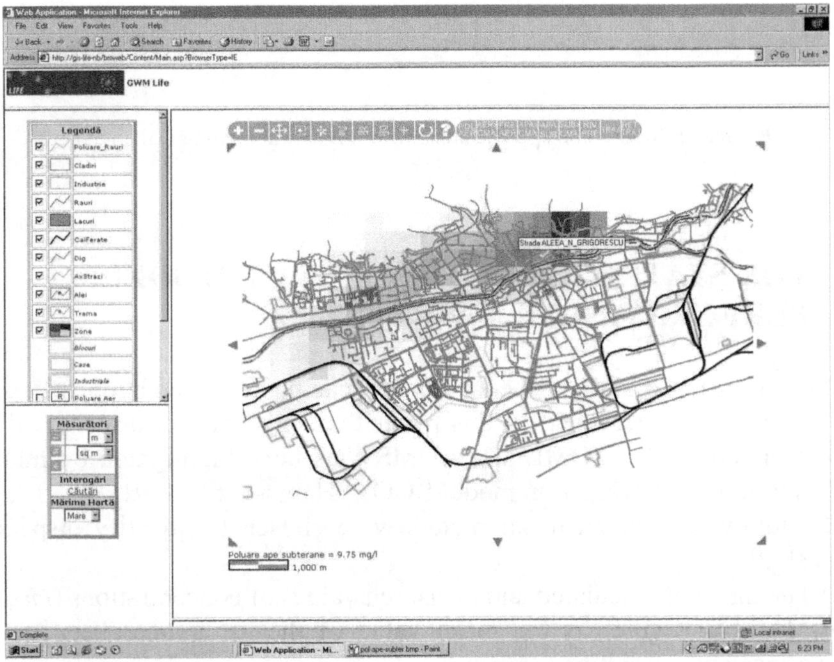

Figure 1: Snap-shot of screen display on the Open-GIS-on-line user output within Internet Explorer of a dispersion plume.

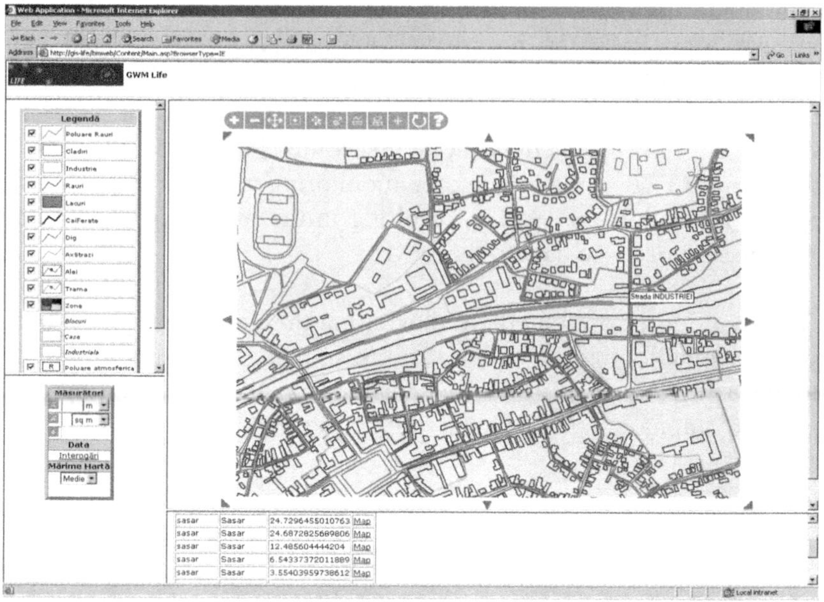

Figure 2: Zoom-in on specific areas within Baia Mare municipal coverages.

3. THE STATISTICAL MEASURES OF MODEL PERFORMANCES

The performance of the dispersion model can be described by means of the relevant statistical measures. In this paper the statistical performance of the two dispersion models (OML and ADMS) has been highlighted by making use of the statistical evaluation model BOOT (Hanna et al., 1991).

The following statistical parameters were chosen to quantify dispersion model skill:

(i) The mean of calculated and measured values of concentrations (*mean*);

(ii) The bias, which is the mean value of the differences between the observations and predictions *(bias)*:

$$bias = C_O - C_P .$$

In a perfect model the bias is equal to zero. Positive (negative) value of bias indicate that, on the average, the model is underpredicting (overpredicting) the observations.

(iii) Standard deviation of the observations and of the predictions

$$\sigma_o = \sqrt{\overline{(C_o - \overline{C_o})^2}}\ ;\quad \sigma_p = \sqrt{\overline{(C_p - \overline{C_p})^2}}$$

(sigma):

Ideally σ_p would equal the corresponding statistic for the observations.

(iv) Normalized mean square error (*nmse*):

$$nmse = \frac{\overline{(C_o - C_p)^2}}{\overline{C_o} \bullet \overline{C_p}}$$

The *nmse* is a fundamental statistical performance parameter, since it gives information on the actual value of the error produced by the model. The normalization assures that in most applications the *nmse* will not bias toward models that overpredict or underpredict.

(v) Correlation coefficient (*cor*):

$$cor = \frac{\overline{(C_0 - \overline{C_o}) \bullet (C_p - \overline{C_p})}}{\sigma_{C0} \bullet \sigma_{Cp}}$$

The correlation coefficient can describe proportional change with regard to the means of the two quantities in question, but cannot distinguish the type or magnitude of possible covariance.

(vi) normalized bias used by EPA (*fb*):

$$fb = \frac{\overline{C_o} - \overline{C_p}}{0.5(\overline{C_o} + \overline{C_P})}$$

(vii) Fraction of prediction within a factor two of observation (fa2): $fa2 : 0.5 \le C_p / C_o \le 2$.

4. RESULTS AND DISCUSSION

Using the OML and ADMS dispersion models, the daily mean concentrations at the ground level have been estimated in the same point in which the measurements were carried out for a three-month interval. The meteorological input data for dispersion models were obtained from the meteorological station of Baia Mare city.

The statistical performances of the two dispersion models are presented in Table 1-3, where C_{obs} stands for the observed concentration and C_{mod1}, C_{mod2} stand for the simulated concentration using the OML and the ADMS dispersion models.

Model	Mean	Sigma	Bias	nmse	cor	fa2	Fb
Cobs	209.03	132.28	0.00	0.00	1.000	1.000	0.000
Cmod1	240.71	92.73	-31.68	0.53	0.318	0.613	-0.141
Cmod2	80.97	56.53	128.06	2.42	0.256	0.419	0.883

Table 1: Statistical analysis for daily mean concentration (μg/m^3) – January.

Model	Mean	Sigma	Bias	nmse	cor	fa2	Fb
Cobs	145.71	134.81	0.00	0.00	1.000	1.000	0.000
Cmod1	236.14	95.88	-90.43	0.75	0.378	0.564	-0.474
Cmod2	94.75	53.90	50.96	2.01	0.285	0.536	0.424

Table 2: Statistical analysis for daily mean concentration (μg/m^3) – February.

Model	Mean	Sigma	Bias	nmse	cor	fa2	Fb
Cobs	166.13	86.42	0.00	0.00	1.000	1.000	0.000
Cmod1	224.35	69.71	-58.23	0.26	0.494	0.710	-0.298
Cmod2	122.90	69.04	43.23	0.59	0.180	0.613	0.299

Table 3: Statistical analysis for daily mean concentration (μg/m^3) – March.

With regard to the global statistics, the results indicate a positive bias between the measured and modelled concentrations of approximately 61%, 35%, 26% (against the mean value) for the ADMS model in January, February and March, respectively. The OML model performs with a negative bias of 15%, 62%, 35% for January, February and March. The normalized mean square error *(nmse)* is best for the OML dispersion model (0.53, 0.75, 0.26). The best value of *fa2* (0.613, 0.564, 0.710) is also provided by the OML dispersion model for all statistics. In Figures 1-2 an example of the predicted concentration values for the two dispersion models are presented.

The models, linked into the described system were applied successfully at Baia Mare – an industrial town in the northern part of Romania which is heavily affected by air pollution, surface and underground water pollution. The town, hosting some 140 000 inhabitants is situated at the linkage between high-land and mountainous region. Several mining industries are impacting both air and water quality. As such the municipality together with the Environmental Protection Agency in Baia Mare are currently operating the system in assistance for Environmental Impact Assessments related to urban developments.

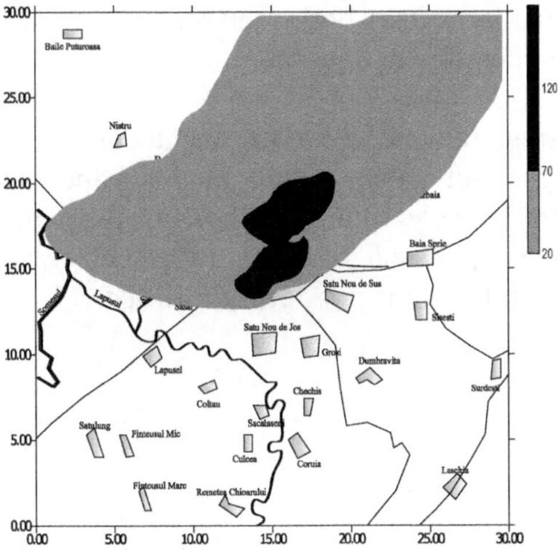

Figure 3: Dealy mean concentrations SO_2 ($\mu g/m^3$)– OML.

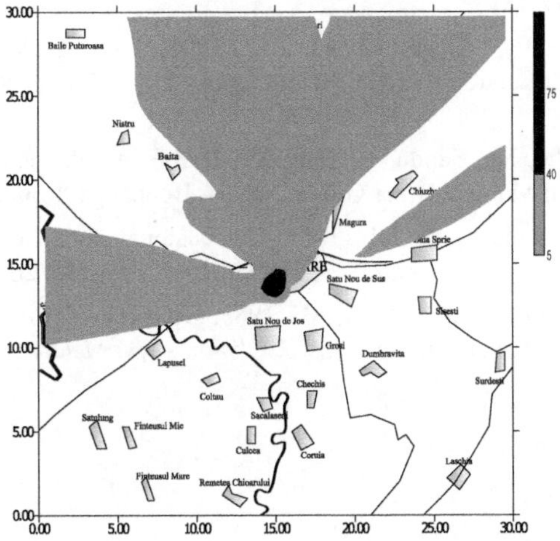

Figure 4: Dealy mean concentrations SO_2 ($\mu g/m^3$) – ADMS.

In conclusion, the OML dispersion model overpredicts the measured values for all statistics, and the ADMS dispersion model underpredicts the analysed measured values. All schemes produce poor correlation coefficients.

In our opinion, the differences between the two models that have been presented and are regarding air quality, is as follows: the ADMS model used in this paper is a Gaussian type model which uses the surface meteorological data, while the OML is a "new generation model" in which the dispersion coefficients are estimated as continuous functions of parameter of boundary layer and it particularly uses both meteorological surface data and vertical profiles of dry temperature and pressure.

As a concluding remark, we note that the level of agreement between predicted and observed values found in this paper show the fact that the performance statistics reflect the limitations of the Gaussian model related to the real atmospheric physical processes, especially when a Gaussian model assumes that the flat terrain is not deeply fulfilled.

REFERENCES

Berkowicz, R. and Olesen, H. R., 1993, Danish point source dispersion model - OML, MST Luft-A94, NERI, Roskilde, Denmark.

Businger, J. A., Wyngaard, J. C., 1971, Flux-profile relationships on the atmospheric surface layer. *J. Atm. Sci.*, 28, pp. 181-189.

Golder, D., 1972, Relations among stability parameters in the surface layer, *Boundary Layer Meteor.*, 3, pp. 47-58.

Hanna, S. R., Strimaitis, D.J. and Chang, J. C., 1991, User's guide for software for evaluating hazardous gas dispersion models. Final Report Vol I, Sigma Research. Corporation, MA 01886, USA.

Ionescu, C., Ursache, M., Sandu, I., Trifu, C., Barbu, A., Pandele, A., 2002, Life-ASSURE Integrated Sistem Manual - User Guide. NIMH, Bucharest, Romania.

Monin, A. S and Obukhov, A. M., 1954, Dimensional characteristics of turbulence in the atmospheric surface layer. *Doklady A. N. USSR*, 93, pp. 223-226.

Van Ulden, A. P. and Holtslag, A. A. M., 1985, Estimation of Atmospheric Boundary Layer Parameters for Diffusion Applications. *J. Climat. Appl. Meteor.*, 24, pp. 1196-1207.

THE USE OF MM5-CMAQ AIR POLLUTION MODELLING SYSTEM FOR REAL-TIME AND FORECASTED AIR QUALITY IMPACT OF INDUSTRIAL EMISSIONS

R. San José[(1)], Juan L. Pérez[(1)], and Rosa M. González[(2)]

[(1)]Environmental Software and Modelling Group, Computer Science School, Technical University of Madrid (UPM), Boadilla del Monte, 28660 Madrid (Spain); [(2)]Department of Geophysics and Meteorology, Faculty of Physics, Complutense University of Madrid (UCM), Ciudad Universitaria 28004 Madrid (Spain)

Abstract: The sophistication of air pollution modeling systems has continued to increase in the last decade. Nowadays, the complex numerical mesoscale meteorological and chemical dispersion models have become very robust tools which are quite sensitive to small changes in the atmospheric process. The use of these tools as air quality management systems is becoming more and more common at urban and regional levels in many parts of the world. In this contribution we have applied the MM5-CMAQ modeling system in the framework of OPANA V3 with emission data sets produced by EMIMO to evaluate the capability of the system to be used as a tool to determine in real-time and forecasting mode the individual impact of five different virtual industrial sources located at the south-east direction of Madrid (Spain) city. The system has been developed for two different scenarios for each single industrial source: OFF (complete switching off the emissions, 100%) and OFF50 (reducing the expected emissions from each single industrial source by 50%). The simulations are carried out over 120 hours, and the emission reduction strategies are applied for the last 48 hours in a way that when the system is used in real-time and forecasting mode in daily operational use, the information is ready to be applied 12 – 20 hours before the hour 96 of the simulation. So, implementation of the emission reduction strategy – in case of need – can be done for such a single industrial plant in such a period of time. The results of 11 simulations (ON + 5xOFF + 5xOFF50) are analyzed by developing a robust post-processing tool which will report us about the time, location (grid cell) and eventually the industrial plant (or plants) which should

I. Farago et al. (eds.), Advances in Air Pollution Modeling for Environmental Security, 327–336.

apply the reported emission reduction scenarios. The modeling system has been mounted with two domains for MM5 with 81 and 27 km spatial resolutions and also with other two domains with 9 km and 3 km spatial resolutions. The CMAQ model has been mounted according to MM5 architecture just for the domains with 9 and 3 km spatial resolutions. Analysis of the air quality impact of industrial sources is prepared for the domains with 9 km and 3 km spatial resolutions. The system has proved to be robust and sensitive enough to be used in a reliable way. The computer time for all exercises is about 140 hours on a PIV-3,06 Ghz. In real operational mode, it can be reduced to 14 hours or less (daily operation) with a 20 PC cluster.

Key words: Air quality, modeling, Eulerian models, cluster computing, industrial plants, meteorology, emissions, atmosphere.

1. INTRODUCTION

The new generation of air quality modelling systems are capable of simulating the atmospheric processes in detail, which a few years ago was quite a difficult task. Nowadays, the advances in computer capabilities and the substantial increase in the knowledge of the atmospheric processes have conducted to new possibilities. In this contribution we have used the OPANA V3.0 air quality modelling system which is a framework of different air quality models and utilities. The MM5 model (PSU/NCAR) (Cheng and Dudhia, 2001) for producing meteorological fields in 4D mode and the CMAQ model (Community Multiscale Air Quality Modelling System / Models-3 / EPA, U.S.) for the chemical mechanism and dispersion of pollutants are used as part of OPANA V3 tool (San José et al., 2000). CMAQ is a representative of the third generation of air quality models which include clouds, aqueous and aerosol chemistry.

Industrial plants (including power generators) produce a considerable amount of pollution, although in the last decade the technology has improved significantly and the total emissions have been reduced substantially. In spite of that, industrial sources are an important cause of air pollution in the surrounding areas. In this contribution we will use state-of-the-art meteorological and chemical dispersion models (MM5/CMAQ) (Warner and Hsu, 2000; Stensrud et al., 2000; Ge et al., 1997) for obtaining the impact in time (120 hours) and space (4D and two nesting levels) of the industrial emissions in the air pollution concentrations. The complete software tool, OPANA V3, shows the portion of air pollution concentration (for all pollutants), in absolute and relative value, due to the industrial plant emissions. Furthermore, we have simulated five different virtual industrial plants in a domain located in the south-east area of Madrid city. Simulations involving the impact of the five sources and the impact of 50% emission reductions are reported. Results show that the tool is capable of controling the impact of several industrial plants simultaneously even for different emission reduction strategies.

2. METHODOLOGY

Traditionally Eulerian models have not been used for evaluating the air quality of industrial plants because of the complexity of the models and the low sensitivity to determine the portion in the immission due to industrial plant emissions. Local meteorology has a significant influence on the impact of industrial sources. MM5/CMAQ has the capability to quantify and qualify the impact on air pollution concentrations of different emission sources because of the high sensitivity of the architecture and the atmospheric science included. Chemical mechanisms (CBM-IV, RADM, SAPRAC, etc.) can trace the pollution concentration changes with a high degree of accuracy.

The methodology involved in this contribution is to simulate the atmospheric process over a domain (two nesting levels) with MM5-CMAQ and an emission model (in our case, EMIMO) to obtain a 4D dataset with the pollution concentrations in time and space for the simulation period. Five different virtual industrial sources have been implemented in the model area domain. This is called ON mode. The model has been implemented over one 9 km spatial resolution domain with an area of 405 km and over one 3 km spatial resolution domain with the dimensions of 81 km x 99 km. These are CMAQ model dimensions. In all cases (MM5 and CMAQ) (Byun et al., 1998; Colella and Woodward, 1984) 23 vertical layers are used with 100 mb as the top level height.

Five different vertical industrial emission sources have been implemented in the 3 km model domain (81 km x 99 km). Table 1 shows the UTM coordinates for the five virtual industrial sources. Figure 2 shows the architecture of the software and hardware design for five different virtual industrial sources (although with the same geometrical and emission characteristics).

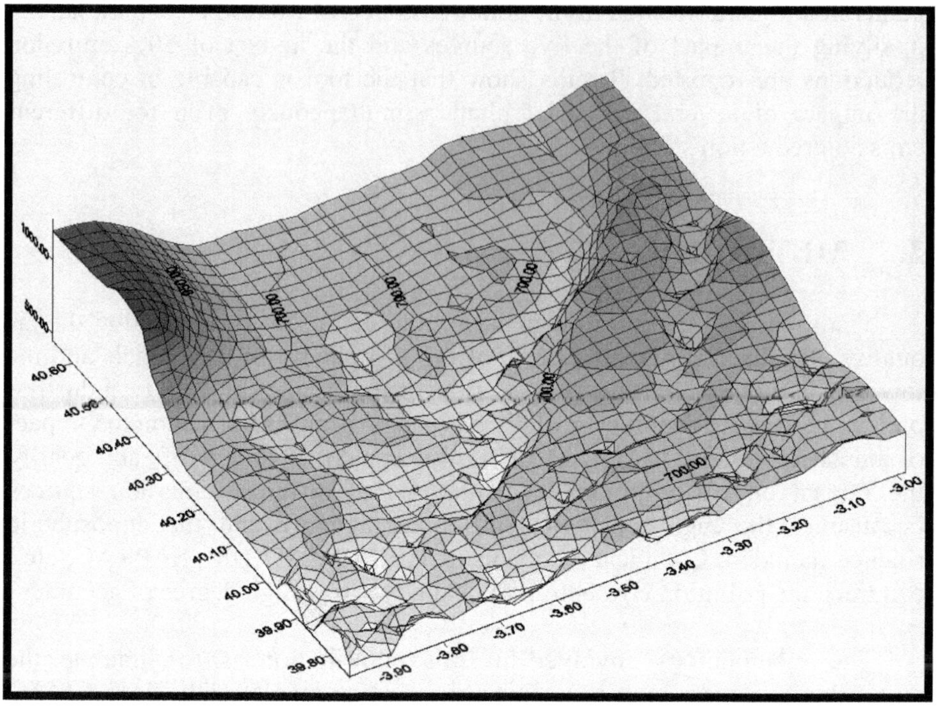

Figure 1: Topography of the area for the 3 km domain.

The full system was designed to run the ON scenario as reference case and the two different emission reduction scenarios (50, 100%) for each virtual industrial source for 120 hours, July, 8-12, 2002. The calibration phase consisted in running the full system without emissions from the five different virtual industrial plants and comparing the results with the air quality monitoring network in Madrid domain.

Figure 3 shows a comparison between observed and modelled ozone concentrations at Coslada air quality monitoring station (located at the eastern area of Madrid). Figure 4 shows the linear regression analysis for the same data sets as those used to produce Figure 3.

VIRTUAL EMISSION SOURCES Central Point: -3.42385, 40.24396 (Lat, Long)						
Coordinates	**Xlambert Conformal)**	**Y(Lambert Conformal)**	**Chimney Characteristics**	**All chimneys have the same characteristics**	**Emissions**	**All virtual industrial plants have the same emissions**
A	**500**	**500**	**Diameter**	**6.5 m**	**PM**	**1.9 g/s**
B	**20000**	**30000**	**Height**	**60 m.**	**VOC**	**2.4 g/s**
C	**20000**	**-40000**	**Temperature**	**377 K**	**CO**	**8.9 g/s**
D	**0**	**-40000**	**Gas velocity**	**21.59 ms^{-1}**	**NOx**	**34.56 g/s**
E	**-20000**	**-20000**	**Flux**	**716.38 m^3s^{-1}**	**SO2**	**2.07 g/s**

Table 1: Characteristics of the pollution emissions from the five different virtual industrial plants and geographical locations in the 3 km resolution domain (81 km x 99 km).

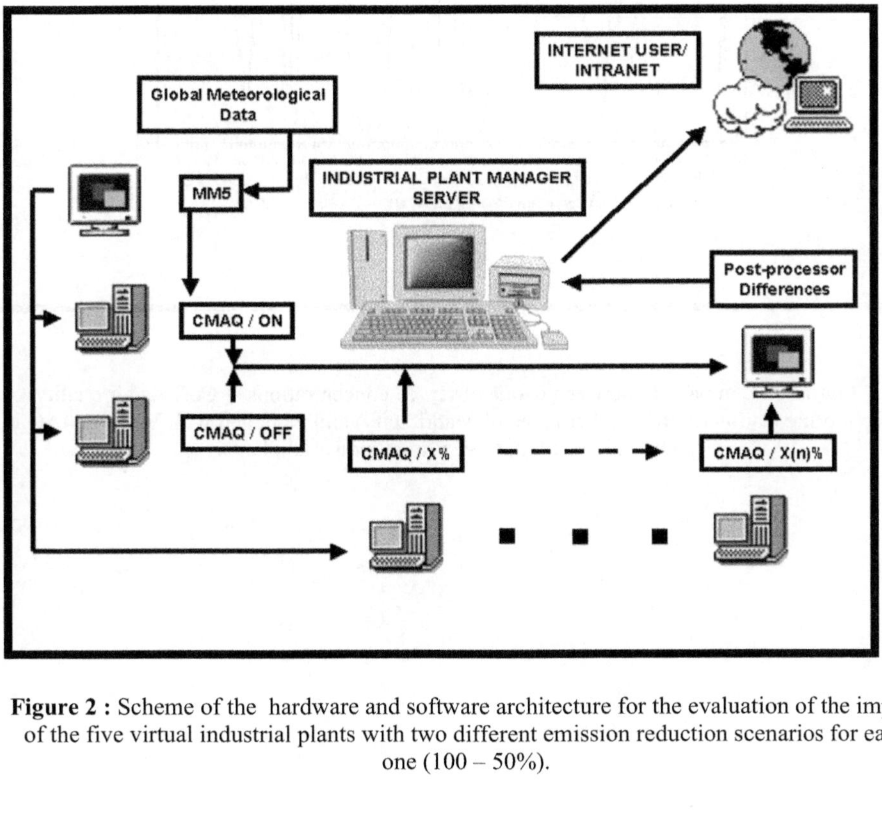

Figure 2 : Scheme of the hardware and software architecture for the evaluation of the impact of the five virtual industrial plants with two different emission reduction scenarios for each one (100 – 50%).

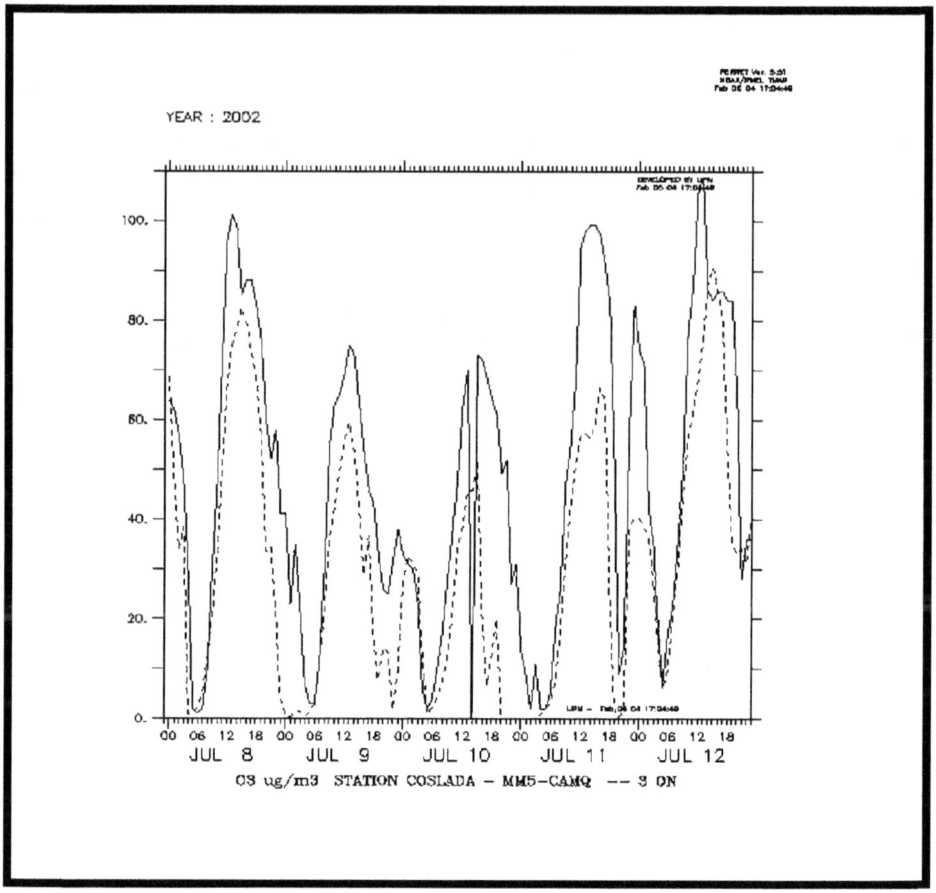

Figure 3: Comparison between ozone observed concentrations at Coslada air quality monitoring station (in the southern area of Madrid city) and modelled with MM5-CMAQ in ON mode with the five virtual industrial plants.

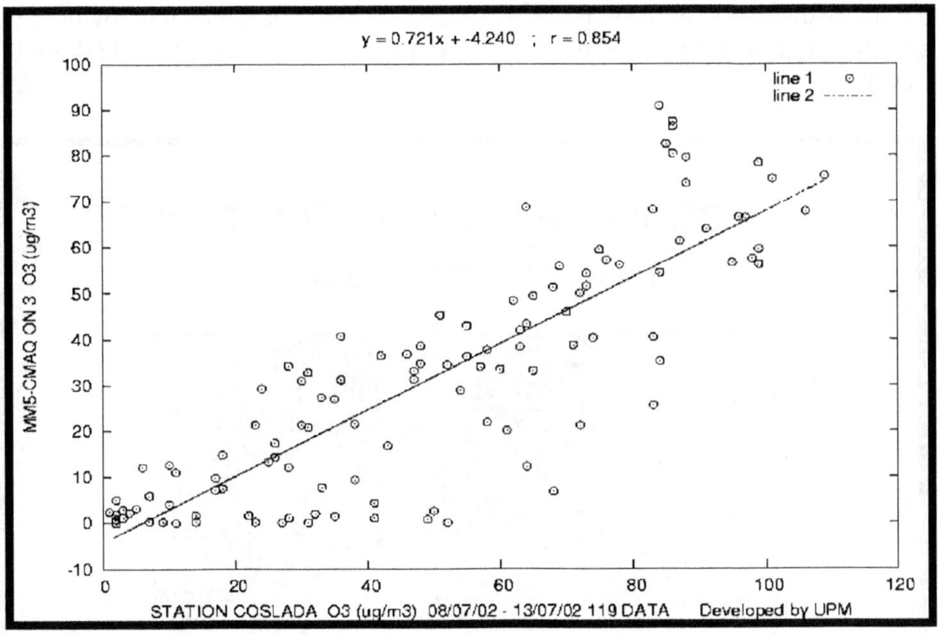

Figure 4: Regression analysis between observed data and simulated data during this experiment when all virtual industrial plants are emitting (ON mode).

3. RESULTS

The system took 140 CPU hours in a Pentium IV, 3.06 GHz, 1 Gbyte RAM. The 12 CMAQ simulations – two emission reduction scenarios (50 and 100%) for each virtual industrial plant and one for ON mode (all virtual industrial plants emitting) and CAL mode (the five virtual industrial plants are not emitting at all) for 9 km and 3 km spatial resolutions took 70% of the total CPU time. The post-processing analysis took 25 % of the total CPU time. The system could be mounted in a PC cluster with 20 PC's to obtain about 10 times speed-up in the CPU time which will enable the system to be "operational" under daily basis operations since the total CPU time will be about 14 hours.

The system generated a total of 134139 files which are mostly hourly images of the different combinations between OFF, OFF50 and ON (excluding switching off more than one virtual industrial plant). A total of 2.91 Gbytes were generated. Figure 5 shows – as an example – the boundary layer height (in geographical projection) at 05h00 GMT on July 10, 2002.

Figure 6 shows the percentage differences between ON mode and switching off all virtual industrial plants (mode CAL) for NO_2 on July 12, 2002, 03h00 GMT.

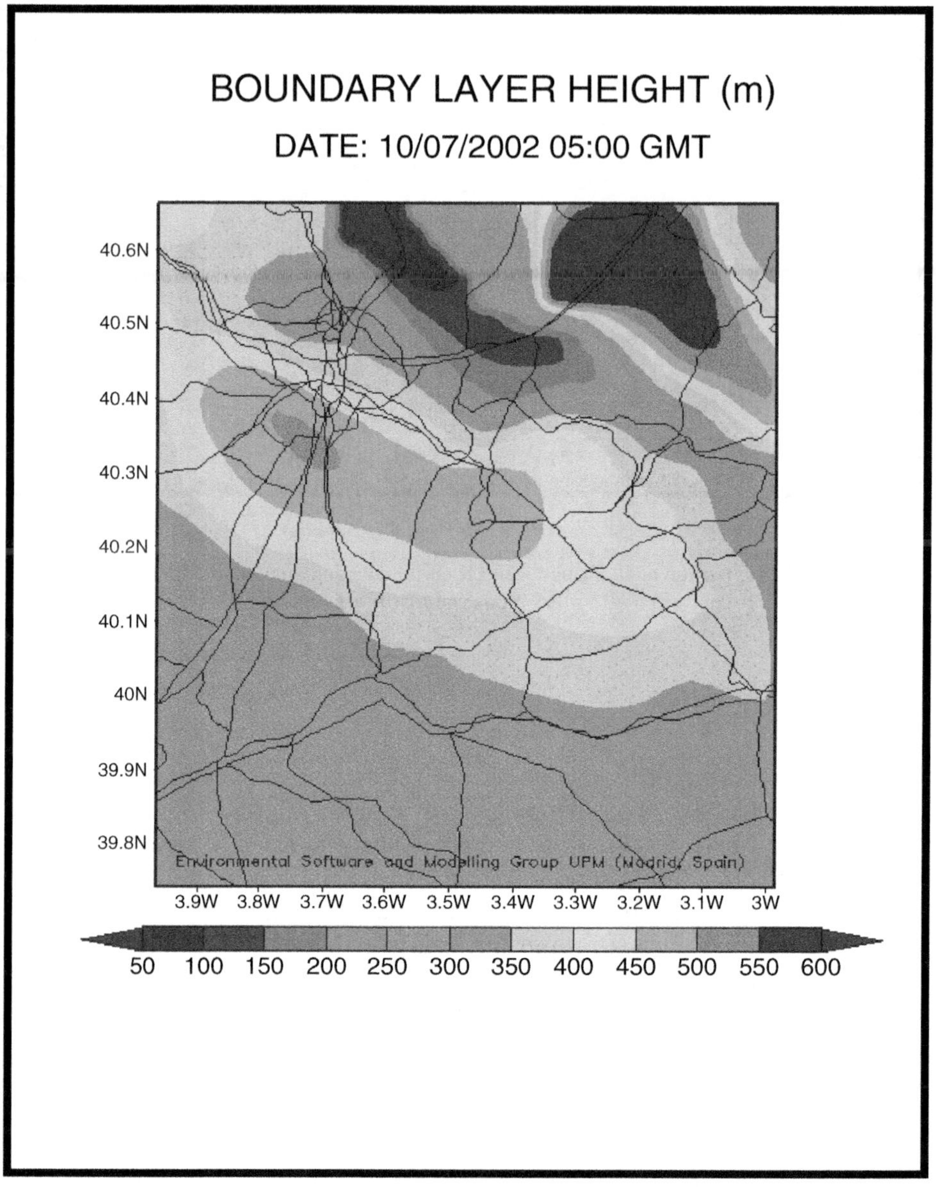

Figure 5.: Boundary Layer height surface pattern at 05h00 GMT on July 10, 2002 by using the MM5-CMAQ air quality modelling system.

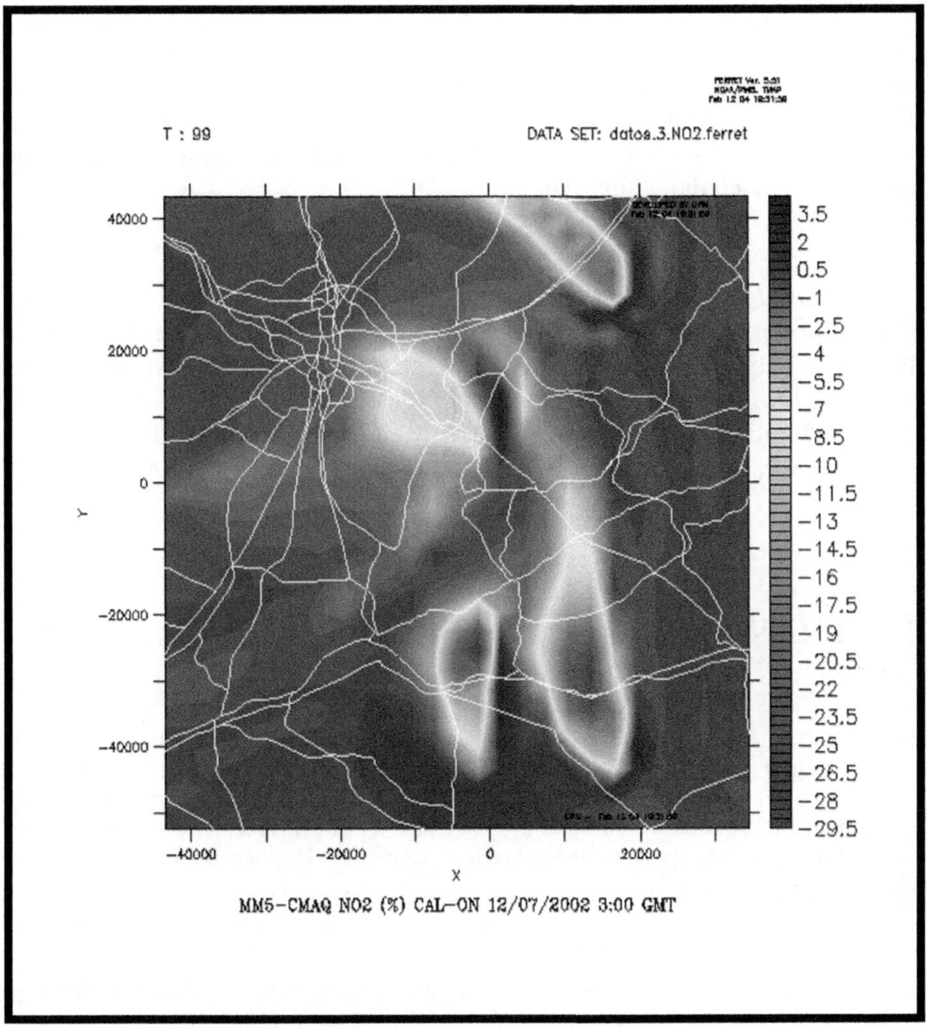

Figure 6: Percentage of change for NO_2 at 03h00 GMT on July 12, 2002 between ON mode and CAL mode (switching off the five virtual industrial plants) ((ON-OFF)/ON*100).

4. CONCLUSIONS

We have built a tool which implements an adaptation of the MM5-CMAQ modelling system named OPANA V3. The system shows an extraordinary capability of simulating all changes produced when five different virtual industrial plants are implemented in the 3 km resolution and 9 km resolution model domains. We have analysed in more detail 11 scenarios corresponding to OFF and 50% emission reduction for each virtual industrial plant and CAL scenario (all virtual industrial plants are switched off). The results show that the system can be used as a real-time and forecasting air quality management system for industrial plants. The system requires to be mounted in a cluster platform in order to handle the CPU times required for such daily simulations.

Acknowledgement: T We would like to thank Professor Dr. Daewon Byun formerly at Atmospheric Modeling Division, National Exposure Research Laboratory, U.S. E.P.A., Research Triangle Park, NC 27711 and currently Professor at the University of Houston, Geoscience Department for providing full documentation of CMAQ and help. We also would like to thank to U.S.E.P.A. for the CMAQ code and PSU/NCAR for MM5 V3.0 code.

REFERENCES

Byun, D., Young, J., Gipson, J., Godowitch, J., Binkowski, F., Roselle, S., Benjey, B., Pleim, J., Ching, J., Novak, J., Coats, C., Odman, T., Hanna, A., Alapaty, A., Mathur, R., McHenry, J., Sankar, U., Fine, S., Xiu, A. and Jang, C., 1998, Description of the Models-3 Community Multiscale Air Quality (CMAQ) model. Proceedings of the American Meteorological Society 78th Annual Meeting, January 11-16, Phoenix, AZ.

Chen, F., and Dudhia, J., 2001, Coupling an advanced land-surface/hydrology model with the Penn State/NCAR MM5 modeling system. Part I: Model implementation and sensitivity. *Mon. Wea. Rev.* 129, pp. 569-585.

Colella, P. and Woodward, P. R., 1984, The piecewise parabolic method (PPM) for gas-dynamical simulations. *J. Comp. Phys.* 54, pp. 174-201.

Ge, Xiaozhen, Li, Feng, and Ge, Ming, 1997. Numerical analysis and case experiment for forecasting capability by using high accuracy moisture advectional algorithm. *Meteorology and Atmospheric Physics*, Vienna, Austria 63(3-4), pp. 131-148.

San José, R., Rodríguez, M. A., Salas, I. and González, R. M., 2000, On the use of MRF/AVN global information to improve the operational air quality model OPANA. *Environmental Monitoring and Assessment* 65, pp. 477-484., Kluwer Eds.

Stensrud, D. J., Bao, J. –W., and Warner, T. T., 2000, Using initial condition and model physics perturbations in short-range ensemble simulations of mesoscale convective systems. *Mon. Wea. Rev.* 128, pp. 2077-2107.

Warner, T. T. and Hsu, H. M., 2000, Nested-model simulation of moist convection: The impact of coarse-grid parameterized convection on fine-grid resolved convection through lateral-boundary-condition effects. *Mon. Wea. Rev.* 128, pp. 2211-2231.

REGULATORY MODELLING ACTIVITY IN HUNGARY

Roland Steib
Hungarian Meteorological Service, 1675 Budapest, P.O. B. 39, Hungary

Abstract: Last year we adopted a second-generation local-scale dispersion model, called AERMOD, at the Hungarian Meteorological Service in order to make better regulatory modelling for concentration calculations than with our old model TRANSMISSION 1.0. In this article the main features of the AERMOD modelling system are mentioned. We summarize our experiences in connection with the model. We also examined the sensitivity of the model to the meteorological and surface parameters, used by this model.

Key words: Regulatory modelling, TRANSMISSION 1.0, AERMIC, ISC3, AERMOD, AERMET, AERMAP, INTERFACE, planetary boundary layer (PBL), SBL, CBL, Gaussian, bi-Gaussian, DEM, flat terrain, elevated terrain, terrain height scale, dividing streamline height.

1. INTRODUCTION

In 1991, the American Meteorological Society (AMS) and The U.S. Enviromental Protection Agency (EPA) initiated a formal collaboration with the designed goal of introducing current planetary boundary layer (PBL) concepts into regulatory dispersion models. A working group (AMS/EPA Regulatory Model Improvement Committee, AERMIC) comprised of AMS and EPA scientists was formed for this collaborative effort.

EPA's regulatory platform for near-field modelling, during 25 years has, with few exceptions, remained fundamentally unchanged. During this period, the Industrial Source Complex Model (ISC3) was the workhorse regulatory model. Therefore, AERMIC selected the EPA's ISC3 Model for a major overhaul. AERMIC's objective was to develop a complete replacement for

I. Farago et al. (eds.), Advances in Air Pollution Modeling for Environmental Security, 337–347.

ISC3 by: 1) adopting ISC3's input/output computer architecture; 2) updating, where practical, antiquated ISC3 model algorithms with newly developed or current state-of-the-art modelling techniques; and 3) insuring that the source and atmospheric processes presently modeled by ISC3 will continue to be handled by the AERMIC Model (AERMOD), albeit in an improved manner.

AERMOD underwent a comprehensive performance evaluation (Brode, 2002) designed to assess how well AERMOD's concentration estimates compare against a variety of independent data bases and to assess the adequacy of the model for use in regulatory decision making. AERMOD was evaluated against five independent data bases (two in simple terrain and three in complex terrain), each containing one full year of continuous SO_2 measurements. Additionally, AERMOD's performance was compared against the performance of four other applied, regulatory models: ISC3 (U. S. Environmental Protection Agency 1995), CTDMPLUS (Perry, 1992), RTDM (Paine and Egan, 1987) and HPDM (Hanna and Paine, 1989; Hanna and Chang, 1993). The performance of these models against AERMOD has been compared using the procedures in EPA's "Protocol for Determining the Best Performing Model" (U.S. Environmental Protection Agency, 1992).

On 21 April 2000 EPA proposed that AERMOD be adopted as a replacement to ISC3 in Appendix A of the Guideline on Air Quality Models (Code of Federal Regulations, 1997). As such, upon final action, AERMOD would become EPA's preferred regulatory model for both simple and complex terrains. Performance of the final version of AERMOD is documented in Perry et al. (2002) and Brode (2002).

2. AERMOD MODELLING SYSTEM

The AERMOD modelling system consists of two pre-processors and the dispersion model. The AERMIC meteorological pre-processor (AERMET) provides AERMOD with the meteorological information it needs to characterize the PBL. The AERMIC terrain pre-processor (AERMAP) both characterizes the terrain, and generates receptor grids for the dispersion model (AERMOD). Figure 1 shows the flow and processing of information in AERMOD.

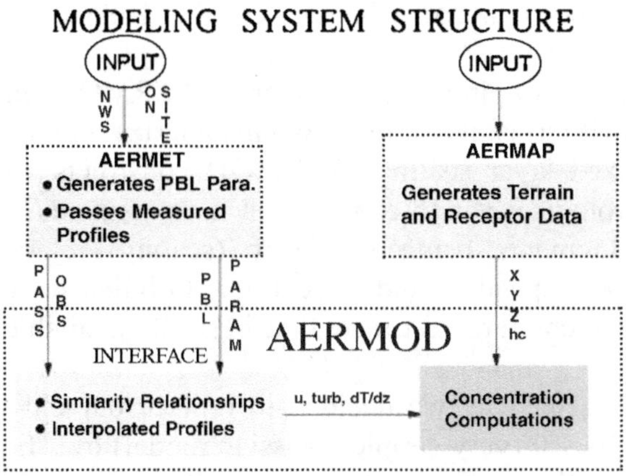

Figure 1: Data flow in the AERMOD modelling system.

2.1 AERMOD

AERMOD is a steady-state plume model. In the stable boundary layer (SBL), it assumes the concentration distribution to be Gaussian in both the vertical and horizontal. In the convective boundary layer (CBL), the horizontal distribution is also assumed to be Gaussian, but the vertical distribution is described with a bi-Gaussian probability density function (pdf). This behavior of the concentration distributions in the CBL was demonstrated by Willis and Deardorff (1981) and Briggs (1993). Additionally, in the CBL, AERMOD treats "plume lofting," whereby a portion of plume mass, released from a buoyant source, rises to and remains near the top of the boundary layer before becoming mixed into the CBL. AERMOD also tracks any plume mass that penetrates into the elevated stable layer, and then allows it to re-enter the boundary layer when and if appropriate. For sources in both the CBL and the SBL AERMOD treats the enhancement of lateral dispersion resulting from plume meander.

Using a relatively simple approach, AERMOD incorporates current concepts about flow and dispersion in complex terrain. Where appropriate, the plume is modelled as either impacting and/or following the terrain. This approach has been designed to be physically realistic and simple to implement while avoiding the need to distinguish among simple, intermediate and complex terrain, as required by other regulatory models. As a result, AERMOD removes the need for defining complex terrain regimes. All terrain is handled in a consistent and continuous manner while considering the dividing streamline concept (Snyder et al., 1985) in stably stratified conditions.

One of the major improvements that AERMOD brings to applied dispersion modelling is its ability to characterize the PBL through both surface and mixed layer scaling. AERMOD constructs vertical profiles of required meteorological variables based on measurements and extrapolations of those measurements using similarity (scaling) relationships. Vertical profiles of wind speed, wind direction, turbulence, temperature, and temperature gradient are estimated using all available meteorological observations.

That is the reason why we decided to replace our old dispersion model TRANSMISSION 1.0 (very simple Gaussian model) by AERMOD. We hope that AERMOD will give us much more results in our regulatory modelling calculations.

2.2 AERMET

Surface characteristics in the form of albedo, surface roughness and Bowen ratio, plus standard meteorological observations (wind speed, wind direction, temperature, and cloud cover), are input to AERMET. AERMET then calculates the PBL parameters: friction velocity (u*), Monin-Obukhov length (L), convective velocity scale (w*), temperature scale (θ*), mixing height (z_i), and surface heat flux (H). These parameters are then passed to the INTERFACE (which is within AERMOD) where similarity expressions (in conjunction with measurements) are used to calculate vertical profiles of wind speed (u), lateral and vertical turbulent fluctuations (σ_v, σ_w), potential temperature gradient ($d\theta/dz$), and potential temperature (θ).

2.3 AERMAP

The AERMIC terrain pre-processor AERMAP uses gridded terrain data to calculate a representative terrain-influence height (h_c), also referred to as the terrain height scale. The terrain height scale h_c, which is uniquely defined for each receptor location, is used to calculate the dividing streamline height. The gridded data needed by AERMAP is selected from Digital Elevation Model (DEM) data. AERMAP is also used to create receptor grids. The elevation for each specified receptor is automatically assigned through AERMAP. For each receptor, AERMAP passes the following information to AERMOD: the receptor's location (x_r, y_r), its height above mean sea level (z_r), and the receptor specific terrain height scale (h_c).

3. TEST RUNS OF AERMOD

In all our tests we used the following parameters:

- One point source (stack)
- Location of the source: 46.06N, 18.26E (south-western part of Hungary)
- Type of pollutant: NO_X
- Receptor resolution: ~ 500 m
- Numbers of receptors: 59x59
- Domain size: ~ 15x15 km
- Emission rate: 28 g/s
- Release height above ground: 82 m
- Stack gas exit temperature: 373 K
- Stack gas exit velocity: 13 m/s
- Stack inside diameter: 6 m

3.1 Test 1

In our first test we wanted to examine AERMOD's behaviour in case of flat terrain, because our former dispersion model TRANSMISSION 1.0 could only treat flat terrain. We wanted to see the differences or similarities in the concentration distributions between the two models. The two models gave almost the same results (concentration fields) when the yearly averaging period was used.

In Figure 2 the yearly average NO_X concentration field of AERMOD can be seen on the above mentioned domain. When we used a short averaging period (1 hour), the results between the two models were much more significant. In our opinion, the differences between the two models in the case of short averaging periods came from the different structures of the two models. TRANSMISSION 1.0 uses empirical values to calculate the planetary boundary layer parameters, while AERMOD calculates unique parameters of the planetary boundary layer for every hour.

In Figure 3 the 1-hour maximum NO_X concentration field of AERMOD is shown in the domain. The results are not surprising. When a long averaging period was used, we learnt that the concentration field was mostly governed by the average wind direction of the domain. In this area the wind usually blows from north-west so we could find the highest concentration values to the south-east direction of the source. In case of the 1-hour averaging period the situation was different. We could recognize high concentration values almost in all directions. It was not surprising either, because any direction of the wind can result in high concentration values. The model usually gave high concentration values in calm wind situations.

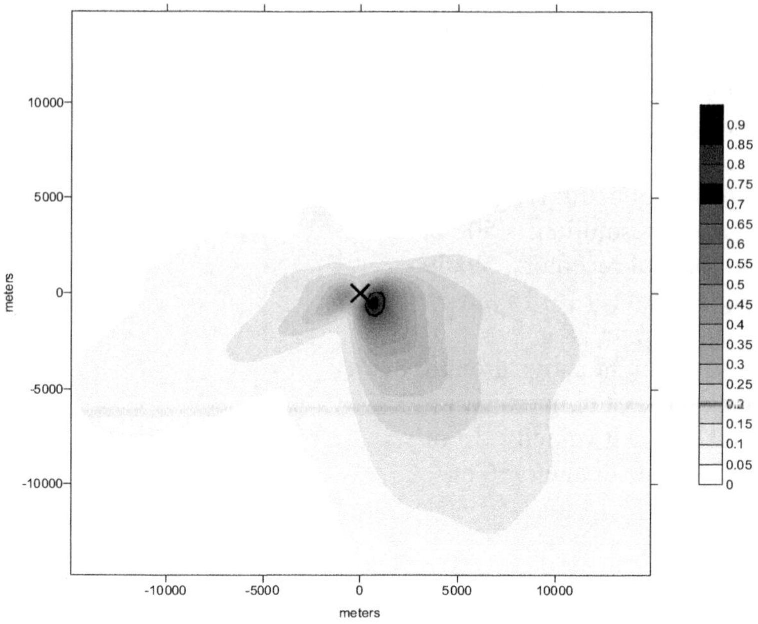

Figure 2: Yearly average NO_X concentration with flat terrain. Units: ug/m3.

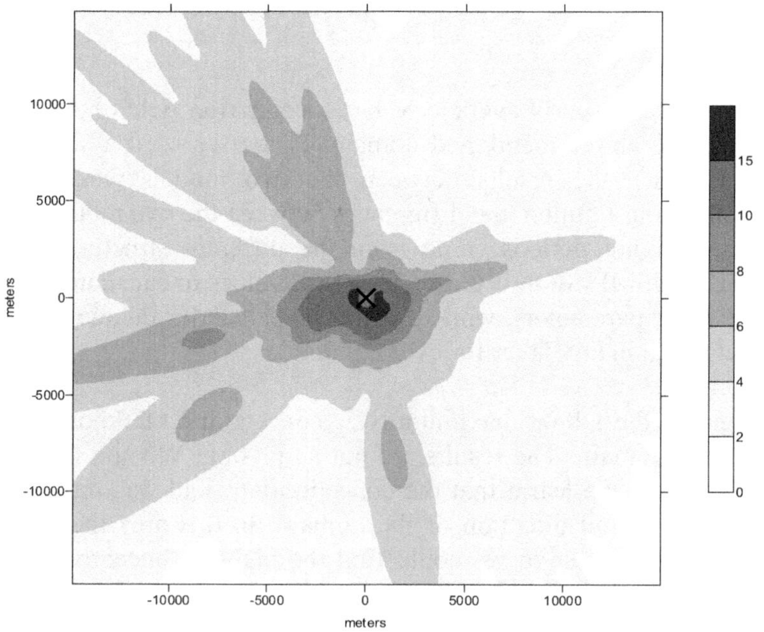

Figure 3: 1-hour max NO_X concentration with flat terrain. Units: ug/m3.

3.2 Test 2

In this test we wanted to see the changes in the concentration field when the model used elevated terrain. By short and also by long averaging periods we experienced big differences in the concentration fields between calculations with elevated terrain and with flat terrain. In Figure 4 the concentration distribution in the case of the yearly averaging period can be seen and in Figure 5 the same but with 1-hour averaging period. By long time averaging period we can notice two maximum places of the concentration distribution. The first maximum is at the same place as it was at the flat terrain. But also a second maximum can be seen at the place where the terrain begins to elevate at the feet of the mountains. This second maximum can be found near the dividing streamline height of the plume. We can also see that the plume flows around the mountains, and at the top of the mountains (500-600 m) the concentration values are lower again.

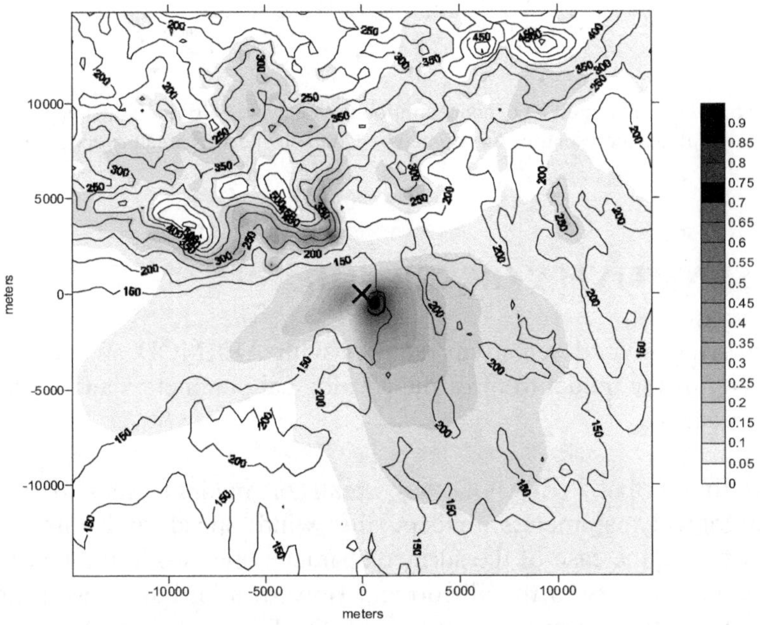

Figure 4: Yearly average NO_X concentration with elevated terrain. Units: ug/m3. The black isolines represent the terrain height. Contour intervals are at every 50 m.

The biggest concentration values can be found at the feet of the nearest mountain to the source by the 1-hour averaging period. So when short averaging periods were used, the elevated terrain had even more influence on the concentration distribution than by the yearly averaging period.

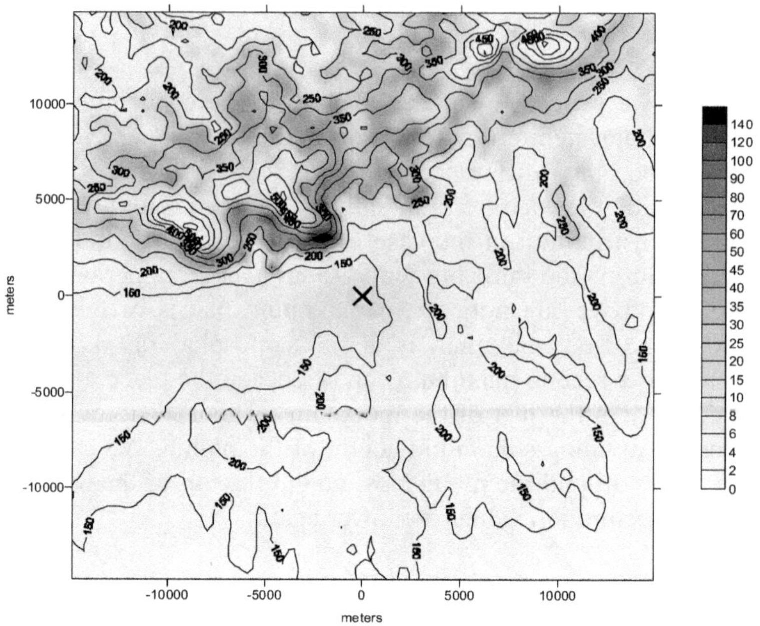

Figure 5: 1-hour maximun NO_X concentration with elevated terrain. Units: ug/m3. The black isolines represent the terrain height. Contour intervals are at every 50 m.

4. SENSITIVITY OF AERMOD

We made several sensitivity studies with AERMOD. We examined the sensitivity of the model to most meteorological parameters and to the surface characteristics.

We found that AERMOD has small or medium sensitivity to most meteorological parameters, except the wind speed and direction (high sensitivity). In the case of the surface characteristics we found that the model has small sensitivity to the albedo and Bowen ratio, but it has a rather high sensitivity to the surface roughness length. The next three figures show us how the change of the surface roughness length influences the concentration distribution. We made three model runs, all with flat terrain. In the first run the roughness length was 0.001 m (typical roughness length over water). In the next run it was 0.1 m (typical roughness length over grassland) and in the third model run it was 1 m (typical roughness length over forest/city).

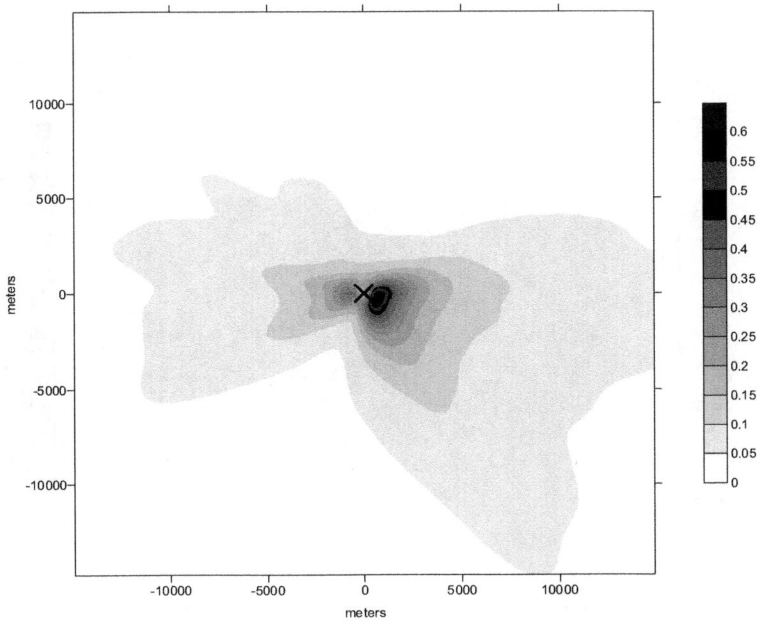

Figure 6: Yearly average NO_X concentration with flat terrain when $z_0 = 0.001$ m. Units: ug/m3.

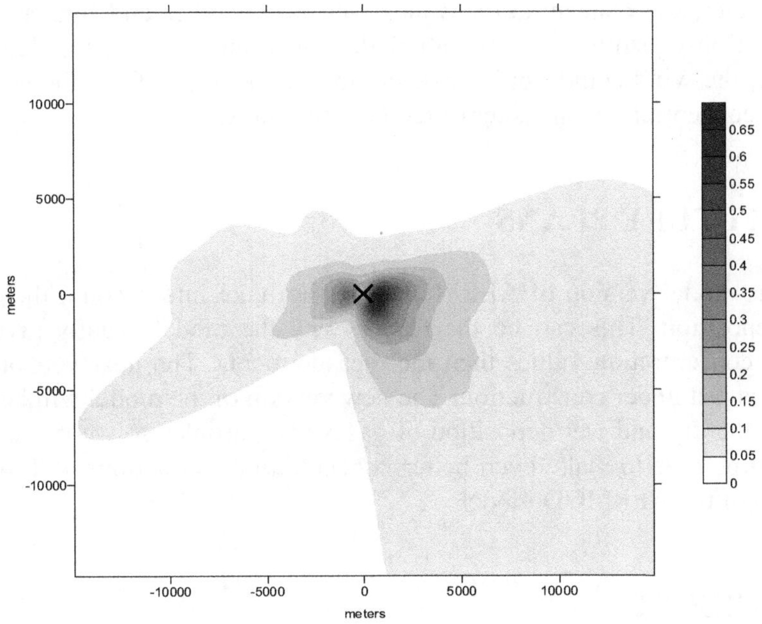

Figure 7: Yearly average NO_X concentration with flat terrain when $z_0 = 0.1$ m. Units: ug/m3.

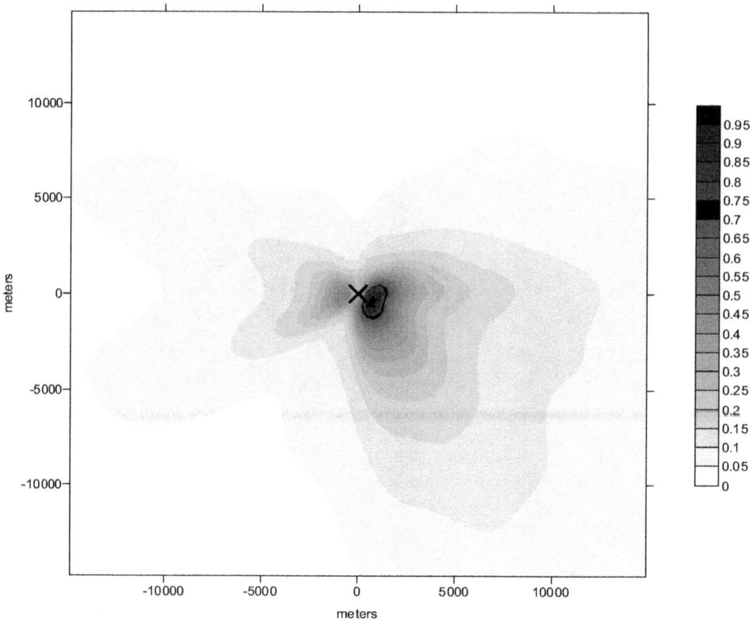

Figure 8: Yearly average NO_X concentration with flat terrain when $z_0 = 1$ m. Units: ug/m3.

We could see that, when the surface roughness length was getting greater and greater, the concentration values were also higher and higher near the source. Consequently we could conclude that when the roughness length was greater, the wind could not blow away the plume easily from the source, so higher concentration values appeared near the stack.

5. FUTURE PLANS

This model version of AERMOD does not take into account the dry and wet deposition. This can be the reason why the model usually gives a bit higher concentration values than the measurements. The next version of the model is just under construction. The new version of the model will be able to handle the dry and wet deposition of gases and particles so we can hope that we will be able to make even better concentration estimations with this new version of the AERMOD model.

REFERENCES

Briggs, G. A., 1993, Plume dispersion in the convective boundary layer. Part II: Analysis of CONDORS field experiment data. *J. Appl. Meteor.* **32**, pp. 1388-1425.

Brode, R. W., 2002, Evaluation of the AERMOD Dispersion Model.

Hanna, S. R. and Chang, J. S., 1993, Hybrid Plume Dispersion Model (HPDM), improvements and testing at three field sites. *Atmos. Environ.* **27A**, pp. 1491-1508.

Hanna, S. R. and Paine, R. J., 1989, Hybrid Plume Dispersion Model (HPDM) development and evaluation. *J. Appl. Meteor.* **28**, pp. 206-224.

Paine, R. J. and Egan, B. A., 1987, User's guide to the Rough Terrain Diffusion Model (RTDM) - Rev. 3.20. ERT Document PD-535-585, ENSR, Acton, MA.

Perry, S. G., 1992, CTDMPLUS: A dispersion model for sources in complex topography. Part I: Technical formulations. *J. Appl. Meteor.* **31**, pp. 633-645.

Perry, S. G., Cimorelli, A. J., Weil, J. C., Venkatram, A., Paine, R. J., Wilson, R. B., Lee, R. F. and Peters, W. D., 2002, AERMOD: A dispersion model for industrial source applications Part II: Dispersion formulations and model performance evaluation. *J. Appl. Meteor.*. (submitted).

Snyder, W. H., Thompson, R. S., Eskridge, R. E., Lawson, R. E., Castro, I. P., Lee, J. T., Hunt, J. C. R. and Ogawa, Y., 1985, The structure of the strongly stratified flow over hills: Dividing streamline concept. *J. Fluid Mech.* **152**, pp. 249-288.

U. S. Environmental Protection Agency, 1992, Protocol for Determining the Best Performing Model. EPA-454/R-92-025, U.S. Environmental Protection Agency, RTP, NC.

U. S. Environmental Protection Agency, 1995, User's Guide for the Industrial Source Complex (ISC3) Dispersion Models (revised) Volume I - User Instructions. EPA-454/b-95-003a, U.S. Environmental Protection Agency, Research Triangle Park, NC.

Willis, G. E. and Deardorff, J. W., 1981, A laboratory study of dispersion in the middle of the convectively mixed layer. *Atmos. Environ.* **15**, pp. 109-117.

CREATION AND TESTING OF FLUX-TYPE ADVECTION SCHEMES FOR AIR POLLUTION MODELING APPLICATION

Dimiter Syrakov, Hristina Kirova, Maria Prodanova

National Institute of Meteorology and Hydrology 66, Tzarigradsko chaussee, Sofia 1784, Bulgaria

Abstract: The advection scheme TRAP (from TRAPezium) was elaborated for building in the Bulgarian air pollution dispersion model EMAP, which is a 3D PC-orientated Eulerian multi-layer model. The TRAP scheme is a Bott-type one and possesses all good properties of this scheme: it is explicit, positive definite, conservative with limited numerical dispersion and good transport ability being at the same time much faster than the Bott scheme. Instead of integrating the polynomial fit over the neighboring grid values as in the Bott scheme, the flux area in the TRAP scheme is supposed trapezoidal and its value is determined as a product of the Courant number and a single value of the polynomial referring to the middle of the passed distance. In this paper some new and faster schemes built on the base of the TRAP scheme are described and tested. They are obtained by optimizing Bott's "normalization" procedure. The performance quality is determined by exploiting the rotational test: an instantaneous release is rotated in a 101×101 grid point field. A set of criteria is established for quantitative estimates of scheme properties.

Key words: Air pollution modeling, advection schemes, rotational test.

1. INTRODUCTION

The description and prediction of local and transboundary air pollution is one of the key problems of environmental agencies and meteorological services. A state-of-the-art method for this is the numerical modeling which requires to account for the dispersion processes at transport (advection and convection); diffusion (horizontal and vertical); physical and chemical transformations, wet and dry deposition. Very often such kinds of numerical models include one or more advection schemes.

I. Farago et al. (eds.), Advances in Air Pollution Modeling for Environmental Security, 349–359.

The major problems in the numerical modeling of advection are the numerical diffusion and the non-monotonicity. In case of constant in space velocity, a scheme is monotone if it does not generate new extremes in the pollutant field. Different techniques are applied to assure monotonicity and very often they are not related to the physics of the problem. In case of ideal advection the pollution field is transported without any dispersion of existing peaks. None of the discrete schemes can do this; the peaks are always dispersed, a phenomenon known as "numerical diffusion". The main efforts of the modelers are aimed at decreasing this numerical diffusion.

Many schemes are described in the literature but none of them posses all properties of the exact solution. One of the most widely used schemes is developed by Bott (*Bott, 1989)* and further improved by him. It is an explicit flux scheme. The in/out fluxes are calculated for each grid cell and the change of concentration is determined as their difference. In the Bott scheme, the advective fluxes are computed utilizing the integrated flux concept of Tremback. Upper and lower limitation and normalization are applied as well. The produced scheme is conservative and positive definite with small numerical diffusion. These properties make the Bott scheme very attractive for further improvements and optimizations. The TRAP scheme (Syrakov, 1995; Syrakov and Galperin, 1997) is a kind of a daughter scheme. It is supposed there that the shape of the so called "flux area" is a trapezium and its value is defined as a product of the Courant number and the concentration in the middle of the trapezium. The last one is determined from the polynomial fitting concentrations of the nearest neighboring grid points. Some variants of the TRAP scheme are elaborated and tested, decreasing the order of the fitting polynomial (Syrakov, 2003).

In the paper, two new variants of the TRAP scheme (*Tn1* and *Tn2*) are presented. They are obtained by optimization of the so called "normalization procedure" introduced by Bott. After that the best third order TRAP variant (*Tb*) and the newly created self-normalizing modifications *Tn1* and *Tn2* are tested using the two-dimensional rotational test (Smolarkiewicz, 1982).

2. DESCRIPTION OF ADVECTION SCHEMES

The one-dimensional case will be considered here. In the multi-dimensional case the splitting technique can be applied. The one-dimensional advection equation in non-divergent and in divergent form is

$$\frac{\partial C}{\partial t}+u\frac{\partial C}{\partial x}=0, \qquad \frac{\partial C}{\partial t}+\frac{\partial uC}{\partial x}=0, \tag{1}$$

where $C(x,t)$ is the concentration of the tracer, $u(x,t)$ – the transport velocity, x – space coordinate, t - time. The simplest way to solve Eq. (1) is to replace the derivatives with some differential approximations. Let us introduce a homogeneous grid: $x \to x_i = i\Delta x$, $i = 1, N_x$, $t \to t_n = n\Delta t$, $n = 0, N_t$, Δx and Δt being the space and time steps. The corresponding values of speed and concentration are $u(x_i,t_n) = u_i^n$ and $C(x_i, t_n) = C_i^n$. The discreet form of Eq. (1) for cell i is

$$C_i^{n+1} = C_i^n - \left(F_{i+1/2} - F_{i-1/2}\right)\Delta t / \Delta x \quad \text{or} \quad C_i^{n+1} = C_i^n - \left(Fr_i - Fl_i\right) \tag{2}$$

where $F_{i\pm1/2} = F(u,C)$ are the mass-fluxes trough the edges of the cell, Fr and Fl being masses transported through the right and left edges of the ith cell for one time step, respectively. Fr and Fl can be positive or negative depending on the transport direction: $Fm = \text{sign}(u_m)A_m$, $(m = r,l)$, where A_m is the so-called ***flux area*** – the shadowed part of Figure 1. Usually, the problem can be normalized by introducing the Courant number $U_i^n = u_i^n \Delta t/\Delta x$ and setting $\Delta x = \Delta t = 1$.

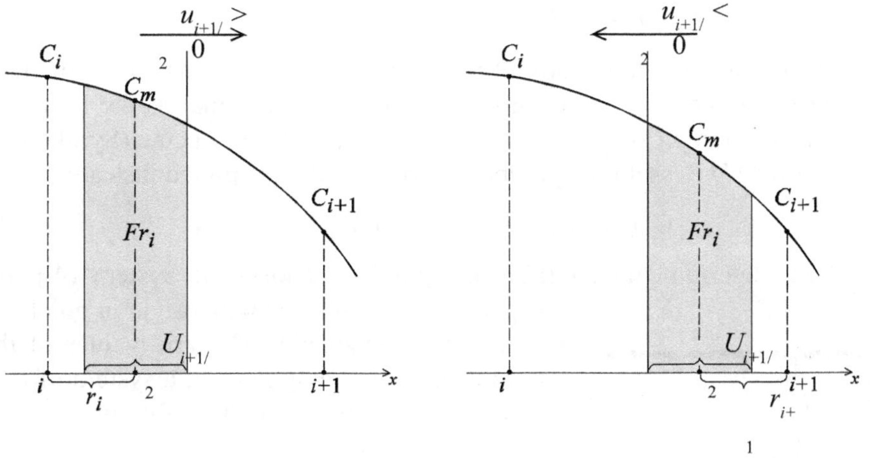

Figure 1: Flux through the right border of cell i at positive and negative transport velocity.

The schemes are *explicit* when the fluxes $F_{i\pm1/2}$ are calculated for the moment t_n. Here, such a type of schemes will be discussed; therefore the upper index n will be omitted further on. Practically, all schemes can be re-formulated and written in the form of Eq. (2). They differ in the way of determining the fluxes. A scheme is ***mass conservative*** if it is constructed in such a way that the condition $Fl_i = Fr_{i-1}$ is fulfilled during the whole integration period. Because this condition is fulfilled for all schemes here, only the right edge fluxes of each cell will be considered further.

2.1. TRAP scheme (*Tb*)

Let us suppose a staggered grid, i.e. the concentrations C_i are defined in the points i, $i = 1,N_x$ and the velocities U_i – in the points $i+1/2$, $i = 1,N_x-1$.

According to the TRAP concept, the flux area is approximated by rectangular trapezium laying on its height (see Figure 1) and its value can be determined as a product of this height (i.e., the Courant number) and the half-sum of the concentrations representing the two bases. Estimates for these bases can be obtained using the proper approximating polynomial. In TRAP, instead of calculating the two bases, a single estimate for the point in the middle of the passed distance (the trapezium height) is obtained exploiting the same polynomial. In such a way, additional acceleration of the scheme is achieved.

$$Ar_i = Cm.|U_i| \qquad C_m = C_i^p(r), \tag{3}$$

where C_m is the value of the concentration in point r, $C_i^p(r)$ - interpolation polynomial of proper order and r is the non-dimensional distance from point i to the centre of the trapezium height. According to Figure 1, this distance is

$$r = (1 - U_{i+1/2})/2, \tag{4}$$

and it is one and the same both for positive and negative transport velocities.

Bott recommended the 4th order Lagrangian polynomial as the best fit to the concentration profile. In the *Tb*-variant of the TRAP scheme (Syrakov and Galperin, 1997) Bessel type polynomial of 3rd order is applied instead:

$$C^b(r) = b_0 + b_1 r + b_2 r^2 + b_3 r^3 . \tag{5}$$

In order to determine the coefficients b_k a local coordinate system (4-pont pattern) is introduced with origin in point i as shown in Scheme 1. The grid values of the concentration being known, a system of four ordinary algebraic equations follows:

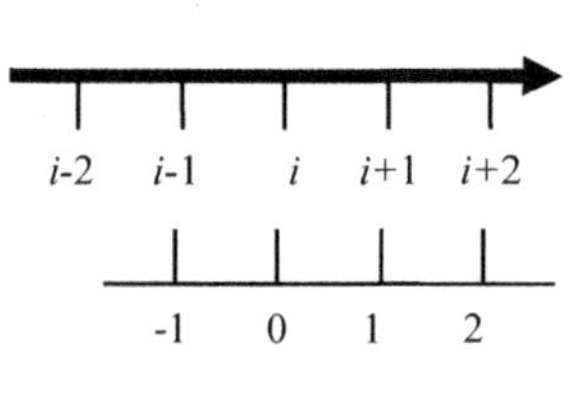

Scheme 1

$$\begin{aligned} x = -1: &\quad C_{i-1} = b_0 - b_1 + b_2 - b_3 \\ x = \ \ 0: &\quad C_i = b_0 \\ x = \ \ 1: &\quad C_{i+1} = b_0 + b_1 + b_2 + b_3 \\ x = \ \ 2: &\quad C_{i+2} = b_0 + 2b_1 + 4b_2 + 8b_3 \end{aligned} \tag{6}$$

Its solution gives the polynomial coefficients:

$$\begin{aligned} b_0 &= C_i \\ b_1 &= (-2C_{i-1} - 3C_i + 6C_{i+1} - C_{i+2})/6 \\ b_2 &= (C_{i-1} - 2C_i + C_{i+1})/2 \\ b_3 &= (-C_{i-1} + 3C_i - 3C_{i+1} + C_{i+2})/6 \end{aligned} \tag{7}$$

This local approach of polynomial fitting (separate coefficients for each cell) leads to a small numerical diffusion, but in case of strong gradients in the concentration field, the values of the polynomial can become negative or unrealistically high. That is why the next step is to introduce upper and lower limits for the flux area, as proposed by Bott

$$\begin{aligned} 0 \le Ar_i < C_i, \quad & at \quad U_{i+1/2} > 0 \\ 0 \le Ar_i < C_{i+1}, \quad & at \quad U_{i+1/2} < 0 \end{aligned}. \tag{8}$$

In the above described operations, the mass transported through the cell edge is determined using an interpolation polynomial, derived assuming that the concentration in the points i, $i\pm1$ etc. is C_i, $C_{i\pm1}$ etc. This is not fully correct, because C_i, $C_{i\pm1}$ are the average concentrations in respective cells, but not values in the specific points (the centers of the cells). In order to correct this, Bott introduced a "normalization" of the fluxes, multiplying the flux area by the factor C_i/A_i when $U_{i+1/2} > 0$ and C_{i+1}/A_{i+1} when $U_{i+1/2} < 0$, where A_i is the whole area of the cell i (hole mass in the cell) determined through the same polynomial as

$$A_i = \int_{-1/2}^{1/2} C_i^b(x)dx\,, \text{ i.e., } \quad \begin{aligned} A_i \; &= (C_{i-1} + 22C_i + C_{i+1})/24, \quad at \quad U_i > 0 \\ A_{i+1} &= (C_i + 22C_{i+1} + C_{i+2})/24, \quad at \quad U_i < 0 \end{aligned}. \tag{9}$$

Finally, the change of concentration in the cell is determined from Eq. (2). At every time step the procedure recurs for each cell consecutively starting from the first one.

The introduction of the Bessel polynomials leads to some important results. Firstly, the lower order leads to a smaller number of coefficients, so less computation is needed. Secondly, the number of the boundary points (points in which it is necessary to introduce boundary conditions) is decreased. Finally, it is well known that the Lagrangian polynomials give the best interpolation quality for the interval $|r| \le 0.25\Delta x$, i.e., close around the central point i while the highest accuracy of the Bessel polynomial is in the region $0.25 \le r \le 0.75$, i.e., around the cell edge ($r = 1/2$), where the flux area is usually located.

The results from the experiments made with the Bott scheme and some variants of the TRAP scheme, exposed in Syrakov (2003), show that Bott's scheme and the 3rd order TRAP schemes (with both Bessel and Lagrange polynomials) possess practically equal simulation abilities, but the TRAP schemes are faster. As a final conclusion from these experiments, the Bessel variant of the TRAP scheme (*Tb*) is recommended for practical use there.

2.2. Self-normalized TRAP schemes (Tn1, Tn2)

Further optimization of the TRAP scheme can be achieved using the proposed by Galperin (1998) self-normalizing. During the explanation of the *Tb* scheme it was mentioned, that a "normalizing" step is necessary, because the interpolation polynomial is built supposing that the values of the concentration in points i, $i\pm1$ etc. are equal to C_i, $C_{i\pm1}$ etc. It is clear *a priori* that C_i is the average for the whole cell concentration, but not the concentration in the centre of the cell. The idea of Galperin is to take into account this fact during the calculation of the polynomial coefficients. As a result, the necessity of this "normalizing" step will fall out. The coefficients of polynomials must be determined in such a way that the integral of the polynomial over the whole cell be equal to the mass in the respective cell. This is achieved if in Eq. (6) we put the integrals of the chosen polynomial in the respective boundaries equal to C_i, $C_{i\pm1}$ etc.

2.2.1. Description of the scheme Tn1

For obtaining this modification of the TRAP scheme, the same 4-point pattern [-1,0,1,2] with origin in point i is implemented as shown in Scheme 1. This means that a polynomial of third order will be used again

$$C_i^{Tn1}(x) = p_0 + p_1 x + p_2 x^2 + p_3 x^3. \tag{10}$$

In accordance with Galperin's approach a system of equations is derived:

$$\begin{aligned}
&at\ \ x=-1: && \int_{-3/2}^{-1/2} C_i^{Tn1}(x)dx = p_0 - p_1 + \frac{13}{12}p_2 - \frac{5}{4}p_3 = C_{i-1} \\
&at\ \ x=\ \ 0: && \int_{-1/2}^{1/2} C_i^{Tn1}(x)dx = p_0 + \frac{1}{12}p_2 = C_i \\
&at\ \ x=\ \ 1: && \int_{1/2}^{3/2} C_i^{Tn1}(x)dx = p_0 + p_1 + \frac{13}{12}p_2 + \frac{5}{4}p_3 = C_{i+1} \\
&at\ \ x=\ \ 2: && \int_{3/2}^{5/2} C_i^{Tn1}(x)dx = p_0 + 2p_1 + \frac{49}{12}p_2 + \frac{17}{2}p_3 = C_{i+2}
\end{aligned} \quad . \tag{11}$$

The solution of the system gives the interpolation polynomial coefficients

$$\begin{aligned}
p_0 &= \left(-C_{i+1} + 26C_i - C_{i-1}\right)/26 \\
p_1 &= \left(-5C_{i+2} + 27C_{i+1} - 15C_i - 7C_{i-1}\right)/24 \\
p_2 &= \left(C_{i+1} - 2C_i + C_{i-1}\right)/2 \\
p_3 &= \left(C_{i+2} - 3C_{i+1} + 3C_i - C_{i-1}\right)/6
\end{aligned} \quad . \tag{12}$$

After calculating the rightmost flux area for cell i according to

$$Ar_i = |U_{i+1/2}|.C_i^{Tn1}(r_i),\ r_i = (1 - U_{i+1/2})/2, \tag{13}$$

the Bott limiters, Eq. (8), are applied and the flux with its sign is determined. Finally, the concentration at the new time level is calculated from Eq. (2).

2.2.2. Description of the scheme Tn2

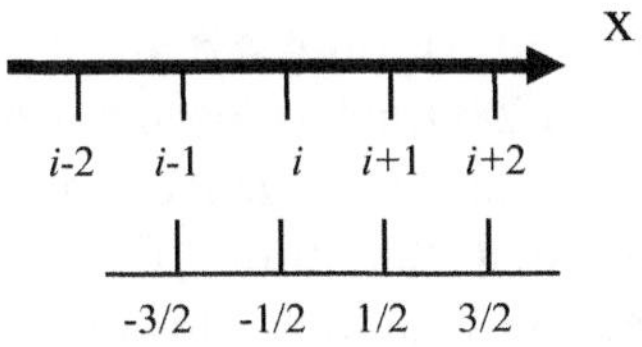

Scheme 2

The scheme *Tn2* is obtained on the 4-point pattern [-3/2,-1/2,1/2,3/2], i.e., the same pattern shifted half a step right, so its origin matches the point $i + 1/2$ (Scheme 2). Denoting the polynomial as

$$C_i^{Tn2}(x) = q_0 + q_1 x + q_2 x^2 + q_3 x^3, \quad (14)$$

and using the chosen approach and pattern, the following system for the coefficients of the interpolation polynomial is obtained:

$$\begin{aligned}
&at\ \ x=-1: && \int_{-2}^{-1} C_i^{Tn2}(x)dx = q_0 - \frac{3}{2}q_1 + \frac{7}{3}q_2 - \frac{15}{4}q_3 = C_{i-1} \\
&at\ \ x=\ 0: && \int_{-1}^{0} C_i^{Tn2}(x)dx = q_0 - \frac{1}{2}q_1 + \frac{1}{3}q_2 - \frac{1}{4}q_3 = C_i \\
&at\ \ x=\ 1: && \int_{0}^{1} C_i^{Tn2}(x)dx = q_0 + \frac{1}{2}q_1 + \frac{1}{3}q_2 + \frac{1}{4}q_3 = C_{i+1} \\
&at\ \ x=\ 2: && \int_{1}^{2} C_i^{Tn2}(x)dx = q_0 + \frac{3}{2}q_1 + \frac{7}{3}q_2 + \frac{15}{4}q_3 = C_{i+2}
\end{aligned} \quad . \quad (15)$$

The solution of Eq. (15) leads to the following estimates for q_i:

$$\begin{aligned}
q_0 &= (-C_{i+2} + 7C_{i+1} + 7C_i - C_{i-1})/12 \\
q_1 &= (-C_{i+2} + 15C_{i+1} - 15C_i + C_{i-1})/12 \\
q_2 &= (C_{i+2} - C_{i+1} - C_i + C_{i-1})/4 \\
q_3 &= (C_{i+2} - 3C_{i+1} + 3C_i - C_{i-1})/6
\end{aligned} \quad . \quad (16)$$

Next, the rightmost flux area for the ith cell is calculated from

$$Ar_i = |U_{i+1/2}|.C_i^{Tn2}(r_i), \; r_i = -U_{i+1/2}/2. \quad (17)$$

The other steps (limiting, flux calculation, etc.) are as in the case of *Tn1*.

3. NUMERICAL EXPERIMENTS

The described variants of the TRAP scheme have passed the two-dimensional rotational test (Smolarkiewicz, 1982). Instantaneous releases with different initial profiles have been rotated with constant angular velocity. The area where the tests have been made is a grid field with 101×101 points and horizontal steps $\Delta x = \Delta y = 1$.

Following Smolarkiewicz, a rotational wind field is imposed on this domain with a constant angular velocity $\omega \approx 0.1$ (628 time steps per one rotation) and centre in point (51, 51). Setting $u = -\omega y$ and $v = \omega x$ a counter-clockwise rotation is achieved. Keeping in mind that the initial field of concentration C^0_{ij} is the exact solution of the advection equation after one or several full rotations, the following criteria for estimation of the quality of the numerical simulation are established (Table 1).

Estimate	Meaning
Cmax = max(C_{ij})/max(C^o_{ij})	**Cmax**<1 – numerical diffusion presents
Cmin = min(C_{ij})/max(C^o_{ij})	**Cmin**<0 absence of positive definiteness
CM = $(\sum_{ij} C^o_{ij} - \sum_{ij} C_{ij}) / \sum_{ij} C^o_{ij}$	**CM** - normalized difference of masses, **CM**≠0 absence of conservativeness
CM2 = $(\sum_{ij} {C^o_{ij}}^2 - \sum_{ij} {C_{ij}}^2) / \sum_{ij} {C^o_{ij}}^2$	**CM2** reflects the second moment conservativeness of the scheme. The ideal advection scheme has **CM2** = 0.
DXc = $\sum_{ij} iC_{ij} / \sum_{ij} C_{ij} - \sum_{ij} iC^o_{ij} / \sum_{ij} C^o_{ij}$ **DYc** = $\sum_{ij} jC_{ij} / \sum_{ij} C_{ij} - \sum_{ij} jC^o_{ij} / \sum_{ij} C^o_{ij}$	**DXc** and **DYc** estimate the displacement of the mass centre due to numerical effects. **DXc** = **DYc** = 0 after a full rotation indicate ideal transport ability.
DD = $(\mathbf{D} - \mathbf{D}^0)/\mathbf{D}^0$, where $\mathbf{D} = \sum_{ij} C_{ij}\left[(x - \mathbf{Xc})^2 + (y - \mathbf{Yc})^2\right] / \sum_{ij} C_{ij}$	**D** - dispersion around the mass centre, **DD** shows the degree of deconcentration of masses due to the numerical diffusion
T = $\Delta T_{calc}/\Delta T^{ref}$,	relative speed of performance, $\Delta T^{ref} = \Delta T^{Tb}$

Table 1: Estimates for simulation quality of the numerical advection schemes (rotational test).

The simulation quality of the three schemes described above is examined using different types of sources. Only the best and worst cases, those of Gaussian and point sources will be presented here. After one rotation the values of the estimates are as shown in Table 2.

The analysis of these values shows that the two new schemes possess better properties than *Tb*, especially in the case of the Gaussian shaped source. They show higher **Cmax**, zero **Cmin**, better conservation of first and second moments of masses, better location of the center of masses and negligible numerical dispersion. At the same time, they are up to 20% faster.

None of the tested schemes describes satisfactorily the advection of a single disturbance (point source). No difference scheme exists capable to describe adequately this severe discontinuity in the concentration field. This can be seen in **Cmax** values, which fall down to some percents of the initial maximum. The steep gradients are squashed down immediately after the start of movement to a nearly Gauss-shaped disturbance that is transported further. In any case, *Tn1* and *Tn2* show better performance than *Tb*.

		Tb	***Tn1***	***Tn2***
Gaussian shaped source	**Cmax**	95.266900	98.300710	98.300760
	Cmin	-6.582216E-17	0.000000E+00	0.000000E+00
	CM	-9.036330E-05	0.000000E+00	-2.008073E-05
	CM2	-4.003794	-1.550007E-01	-1.549592E-01
	DXc	1.351967E-01	-7.348633E-02	-7.347107E-02
	DYc	1.831055E-04	-7.705688E-04	-7.591248E-04
	DD	6.183589	3.731381E-01	3.731808E-01
	T	4.125000 sec	89.2	82.8
Point source	**Cmax**	3.609844	4.466961	4.466957
	Cmin	-7.460410E-16	0.000000E+00	0.000000E+00
	CM	-8.008893E-05	-1.971420E-04	-1.971420E-04
	CM2	-98.140120	-97.496740	-97.496740
	DXc	2.987843	3.396282	3.396278
	DYc	8.216019E-01	9.276657E-01	9.276619E-01
	T	3.742188 sec	93.9	87.9

Table 2: Values (in %) of the simulation quality estimates for the three TRAP schemes.

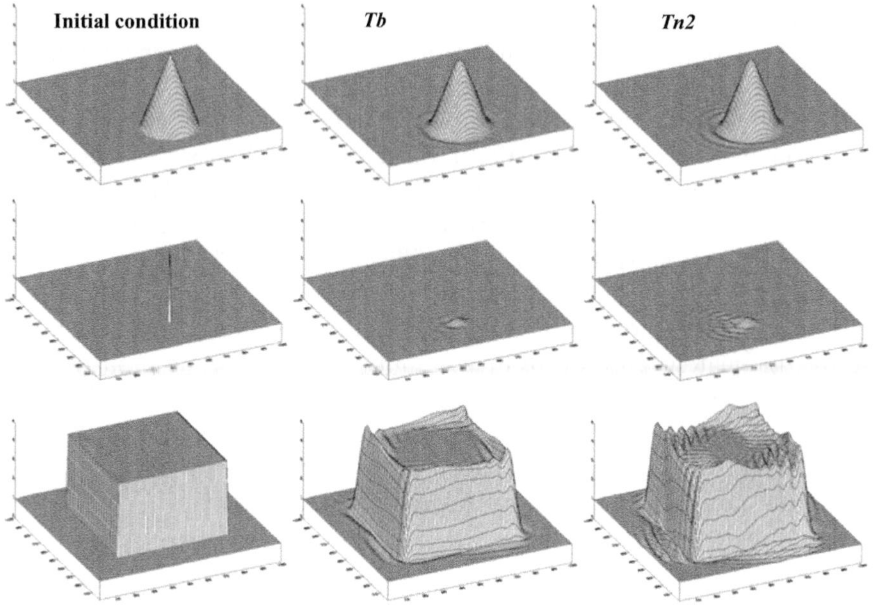

Figure 2: One rotation of cone, single disturbance (point source) and cub by ***Tb*** and ***Tn2.***

The experiments with cone-shaped source show much better results. As *Tn2* occurs to have the best quality of simulation, its performance is demonstrated in Figure 2 compared with *Tb*'s simulations.

4. DISCUSSION AND CONCLUDING REMARKS

The shapes in Figure 2 are rotated over a constant background field with height of $1\Delta x$. The tests over background are very important in proving the positive definiteness of a numerical scheme. In case of zero background, an option for checking for and neglecting the negative concentrations can keep the positive definiteness of a scheme. This limiter does not work when background tests are made. The undershooting and overshooting of the schemes usually dig "holes" in the background or produce peaks as to avoid the discontinuity. The shallower the holes, the better the scheme. *Tb* demonstrates better quality in this sense. *Tn2* transports the smooth forms better and faster than *Tb*.

It must be noted also, that in the reality discontinuities are produced by intensive point sources and the wet removal processes. Turbulent diffusion works in direction of smoothening of these discontinuities, so as a whole the concentration fields are smooth enough and *Tn2* can be used with certainty. It is up to the modeler to decide what property of a scheme is more valuable for the specified task.

Acknowledgements: The present work is supported by the NATO Collaborative Linkage Grant EST.CLG 979794, by the BULAIR project (EU 5FP Contract EVK2-CT-2002-80024), as well as by the Greek-Bulgarian bilateral project (Contract No. ES-1212-Gr/02 NSNI).

REFERENCES

Bott, A., 1989, A positive definite advection scheme obtained by nonlinear renormalization of the advective fluxes. *Mon.Wea.Rev.* **117**, pp. 1006-1015.

Galperin, M, 1997, Two numerical advection schemes for atmospheric models: development and comparison testing. in Proceeding from Swedish-Bulgarian Workshop, 1997, Sozopol, Bulgaria, Pensoft, pp 49- 61.

Smolarkiewicz, P. K., 1982, The multidimensional Crowley advection scheme. *Mon.Wea.Rev.* **113**, pp. 1109-1130.

Syrakov, D., 1995, On a PC-oriented Eulerian Multi-Level Model for Long-Term Calculations of the Regional Sulphur Deposition. In Gryning, S.-E. and Schiermeier, F. A. (eds.), Air Pollution Modelling and its Application XI, NATO/CMSS 21, Plenum Press, N.Y. and London, pp. 645-646.

Syrakov, D., 2003, Once more on the advection schemes: description of TRAP- schemes, in Proceedings of the 26th ITM, 26-30 May 2003, Istanbul, Turkey, pp 225-258.

Syrakov, D. and M. Galperin, 2000, On some explicit advection schemes for dispersion modelling applications. *International Journal of Environmental Pollution* **14**, pp. 267-277.

BULGARIAN EMERGENCY RESPONSE SYSTEM: DESCRIPTION AND *ENSEMBLE* PERFORMANCE

Dimiter Syrakov, Maria Prodanova, Kiril Slavov
National Institute of Meteorology and Hydrology 66, Tzarigradsko chaussee, Sofia 1784, Bulgaria

Abstract: A PC-oriented Emergency Response System (BERS) is developed and works at the National Institute of Meteorology and Hydrology with the Bulgarian Academy of Science. The creation and the development of BERS were highly stimulated by ETEX (European Tracer Experiment).
ERS consists of two main parts – operational and accidental ones, realized for both regions "Europe" and "Northern Hemisphere". The operational part runs automatically twice a day after new analysis and forecast meteorological data are received via GTS of WMO. It prepares the current met-files, calculates trajectories for selected NPP, visualizes the results and puts the pictures in a specialized Web-site. The accidental part is activated manually when real radioactive releases occur or during emergency exercises. The source term is specified by the users – Bulgarian emergency authorities. The EMAP model is the core of this part. This is a 3-D Eulerian dispersion model producing concentration and deposition fields. The visualized output is sent back via fax and/or specialized ftp-site.
In the paper, the BERS overall structure and modules are described and its applications in emergency management are shown. Special emphasis is made on BERS performance during the runs of the EC 5FP project ENSEMBLE.

Key words: Air pollution modeling, nuclear accident, emergency response.

1. INTRODUCTION

In spite of the fact that many safety measures are permanently taken, industrial accidents leading to release of harmful material (chemical or radioactive) are still probable. In the atmosphere, it can have consequences extending to hundreds and thousands of kilometers. In order to defend the population, the decision makers need information about the possible long-range transport of pollutants. Computer-based Emergency Response Systems

I. Farago et al. (eds.), Advances in Air Pollution Modeling for Environmental Security, 361–372.

are established in many countries, simulating and predicting the distribution of the released pollution. Such a system is developed and works operationally in the National Institute of Meteorology and Hydrology (NIMH) of Bulgaria. Its creation and development was highly stimulated by the European Tracer Experiment (Girardi et al., 1998).

NIMH took part in all activities of ETEX with the puff model LED (Syrakov et al., 1983), the results are described in Syrakov and Prodanova (1994). In the second phase of ETEX, a new model EMAP (Syrakov, 1995) was tested and found to perform better than LED (Mosca et al., 1997, Graziani et al., 1998). The main LED and EMEP simulation results connected with ETEX are given in Syrakov and Prodanova (1998a,b).

2. STRUCTURE OF BERS

Currently BERS consists of two parts, operational and accidental ones, both of them having two variants: region "Europe" and region "Northern Hemisphere". BERS overall structure and modules are presented in Figure 1.

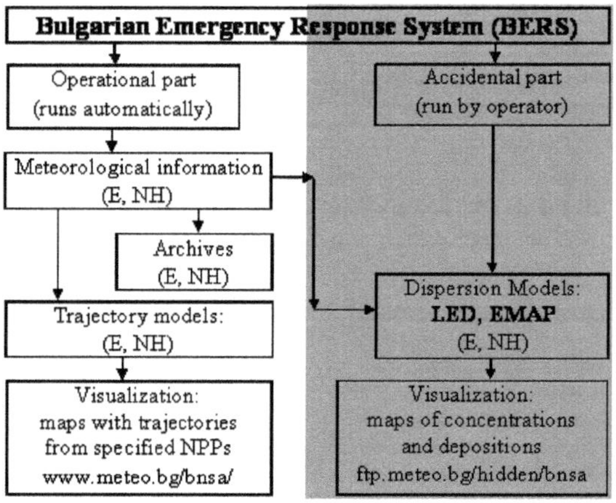

Figure 1: Structure of BERS. **E** *stands for version "Europe"*, **NH** – *for version "Northern Hemisphere"*.

The operational part runs automatically every 12 hours after fresh meteorological information comes in. As NIMH numerical forecast covers quite a limited area around the country, BERS uses the products of big centers like Bracknell, Offenbach etc. There, the numerical maps are coded in a proper way and distributed via GTS of WMO. The operational part of BERS consists of several modules preparing input meteorological files to both trajectory and dispersion models (operational data base); creating archives with analyzed met-data (archive data base); running trajec-tory models; visualizing the results and uploading the concentration and deposition maps to the specialized web-site of NIMH.

The met-input to the models is an input to the built-in PBL model. A simple barotropic model called YORDAN (Yordanov et al., 1983) is used. It is based on the similarity theory and needs only characteristics at the upper and lower boundaries of the PBL. These characteristics are the geostrophic level (850 hPa) wind components u_g and v_g, and the top-bottom potential temperature difference ($\Delta\theta = \theta_{850} - \theta_{ground}$). YORDAN calculates wind and turbulence profiles at every grid point, in such a way resconstracting 3D fields. The aim of the BERS meteorological pre-processing is to obtain u_g, v_g and $\Delta\theta$ in the grid points. Additional information, used only by dispersion models, is the precipitation intensity. All these characteristics are prepared from the respective numerical weather forecast products by a number of calculations and interpolations. For the moment, Deutscher Wetterdienst (DWD) "Global model" numerical products are used (resolution $1.5^o \times 1.5^o$).

Example of one of the trajectory maps (there are five for Europe and one for the Northern Hemisphere) is shown in Figure 2. On the picture, trajectories from the enumerated points in the list are given. The stars show the location of the plants. Three trajectories begin from each point corresponding to three starting heights - 100, 300 and 1000 m. Initial moment of all trajectories is the current synoptic term (00 UTC or 12 UTC) as indicated. Along each trajectory, the points that the emitted pollution parcels would reach after 12, 24, 36,..., 72 hours are marked.

The accidental part is activated by the operator when a real radioactive release occurs or during emergency exercise. Usually, it is initiated by an authorized user. The dispersion models LED and EMAP use to be the cores of the accidental part. Although the puff-model LED was successfully used during both ETEX stages and demonstrated good simulation abilities (see Mosca et al., 1998, and Graziani et al., 1998) it showed some shortcomings and was replaced by the recently created model EMAP.

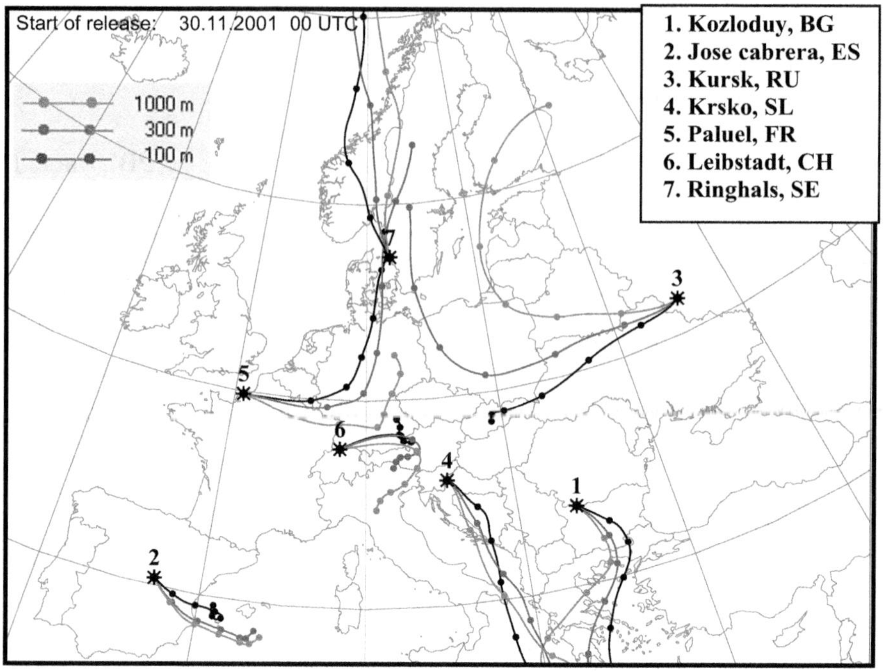

Figure 2: Forward trajectories from 7 European NPP.

EMAP requires three types of input information: permanent, operational and source one. Permanent is the land-use information: sea-land mask, relief height, roughness. Operational is the meteorological information prepared during the automated runs of BERS. The source information (source coordinates, release height and rate, release start and duration) is provided by the user. The fields of ground-level concentration and accumulated deposition are the output. The concentration and deposition maps are visualized automatically and then uploaded to a specific ftp-site accessible by the interested institutions. For demonstration, a 72-hour forecast for the concentration and deposition fields due to a “release” in Kursk NPP are shown in Figure 3. They correspond to the trajectories shown in Figure 2.

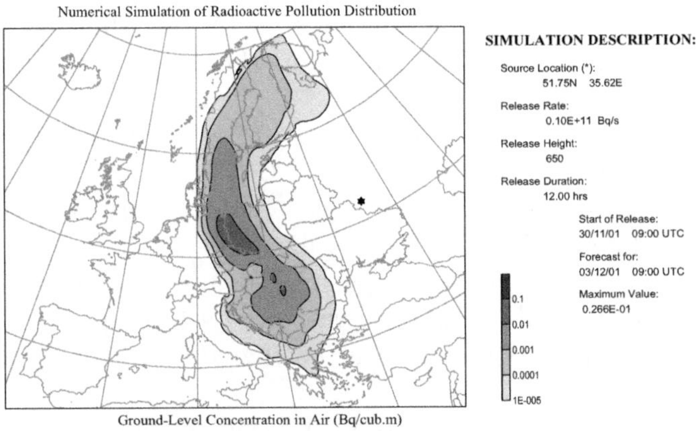

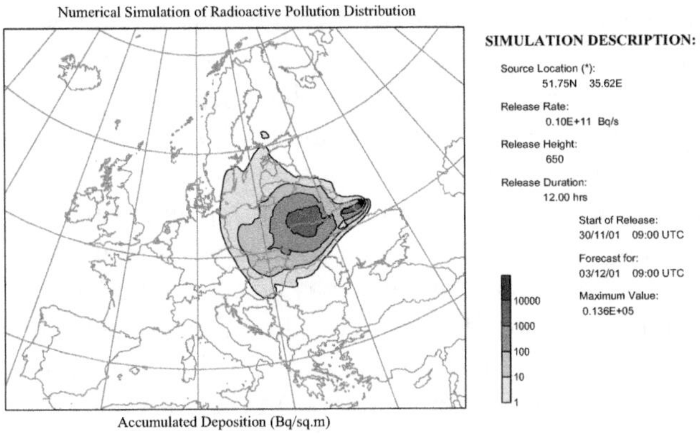

Figure 3: 72-hour forecast for the concentration and accumulated deposition fields created by a "release" in Kursk NPP.

3. SHORT DESCRIPTION OF THE DISPERSION MODEL EMAP

EMAP (**E**ulerian **M**odel for **A**ir **P**ollution) is a 3-D simulation model that allows describing the dispersion of multiple pollutants (BC-EMEP, 1994-1998; Syrakov, 1995). It takes into account the processes of horizontal and vertical advection, horizontal and vertical diffusion, dry deposition, wet removal, gravitational settling and specific chemical transformations. Within EMAP, the semi-empirical diffusion-advection equations for scalar quantities are solved numerically on an Arakawa C-type staggered grid. The horizontal resolution depends on the task to be solved. Vertically, terrain-following non-equidistant grid spacing (log-linear grid) is settled. The time splitting approach is applied for transforming the complex problem to a number of simple tasks. As to decrease the splitting error, the one-dimensional operators are applied in reverse order each time step. The temporal resolution depends on the Courant stability condition.

Advective terms are treated with the TRAP scheme (Syrakov, 1995; Syrakov and Galperin, 1997a), which is a Bott-type one. A 3rd order Bessel polynomial is used for fitting the concentration distribution in the space around any grid point. While displaying the same simulation properties as the Bott scheme (explicit, conservative, positive definite, transport ability, limited numerical dispersion), the TRAP scheme occurs to be several times faster. The boundary conditions are fixed at income flows, and "open boundary" type – at outcome ones.

Diffusion equations are discretized by means of the simplest implicit (in vertical) and explicit (in horizontal) schemes. Both schemes have 1st order accuracy in time and 2nd order accuracy in space. The horizontal diffusion coefficients are constants (defined by the user) and the lateral boundary conditions for diffusion are "open boundary" type. The vertical diffusion coefficient varies in space and time depending on PBL stability (as calculated by YORDAN). The bottom boundary condition is a flux-type one; the top boundary condition is optionally "open boundary" or "hard-lid" type.

The dry deposition is accounted for as bottom boundary condition for the vertical diffusion equation. The dry deposition flux is determined as a product of the roughness level concentration and the dry deposition velocity. The latter parameter depends on many factors. In the current version of EMAP, it is assumed to be depending only on the type of the pollutant and on the character of the land coverage. This parameter must be specified in advance. In the surface layer (SL), a parameterization that allows one to have the first computational level at the top of the SL is applied. It provides a good estimate for the roughness level concentration and accounts also for the action of continuous sources placed on the earth surface (Syrakov and Yordanov, 1997). The simplest decay approach is used for the wet removal with coefficient depending on pollutant properties and on rain intensity.

The EMAP model has been applied in many tasks: study of annual acid, lead, cadmium, mercury and benzo(a)pyrene loads over the region of Southeast Europe (BC-EMEP, 1984-1988). The model passed validation and evaluation in some international exercises: participation in the ETEX study, its results rated 9th among 34 models (Syrakov and Prodanova, 1994); EMEP/MSC-E inter-calibrations of lead and cadmium models (Syrakov and Galperin, 1987a; Gussev et al., 2000). In Figure 4, an EMAP simulation of the ETEX First Release is shown.

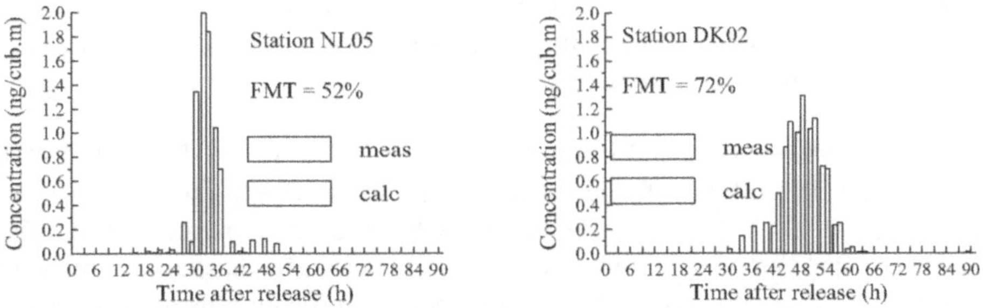

Figure 4: Time evolution of measured and calculated concentrations.

4. THE ENSEMBLE PROJECT

The Chernobyl accident definitely outlined our ill-preparedness with respect to the prediction capacity and the compelling necessity for an improvement of the modeling tools. Since then, many institutes all around the world have invested efforts in the development and implementation of long-range atmospheric transport models for real-time forecasting of atmospheric dispersion.

The approach to long-range dispersion modeling is not unique and in most of the cases it has to be based on the delicate balance that exists between having a reliable model and a tool that can be used in real time. The various approaches that have been developed and which are currently operational in meteorological services are source of an intrinsic diversity in model results. In order to estimate the effect of these differences, ad-hoc tracer experiments were designed and carried out. The most recent one is the already mentioned European Tracer Experiment (ETEX) that took place in 1994 and that allowed the comparison of several types of models with experimental evidence. One of the conclusions of the experiment was that, although modeling has greatly advanced in the last few decades, still several aspects need to be improved in order to bring the various approaches to a state of reliable maturity. Moreover, it was pointed out that model inter-comparison and the evaluation of the results should be performed on a regular basis in order to promote a constant improvement of the quality of the tools and advancements in its development. The following activity that aims at the improvement of dispersion model performance for nuclear emergency application was RTMOD (Real Time Model Evaluation, Bellasio et al., 1999). Given the difficulty in performing tracer experiments and their cost (especially on the long range), one way to test modeling approaches to atmospheric dispersion is in fact model inter-comparison. RTMOD was an internet based system that gathers and compares couples of results of models operating in different national institutions and applied to a pre-defined case. The modelers have therefore the possibility to check for the presence of systematic errors or biases in the prediction over a range of meteorological situations.

The experience obtained in the frame of ETEX, ATMES-II and RTMOD lead to crystallization of the idea of ensemble (Belassio et al., 1999; Galmarini et al., 2002). Ensembles have been used in numerical weather prediction for many years with a beneficial impact on forecast accuracy. More recently, the ensemble concept has been applied to predictions of atmospheric dispersion in the frame of the EC 5FP project ENSEMBLE.

The web-based decision support system ENSEMBLE has been created to collect forecasts produced in the various EU Member States and to provide emergency management with the tools to explore and assess these differences. It also produces a comprehensive European forecast based on the multi-model ensemble dispersion concept. ENSEMBLE enables qualitative and quantitative comparisons of the forecasts produced by several models (more than 20), and to determine their degree of consensus, essential for the decision making process. Model results are uploaded in the newly developed ENSEMBLE format via the Internet, and can be displayed in real time as single model output or grouped (ensembles) for ad-hoc statistical treatments. ENSEMBLE can be accessed by means of a simple web-browser, through which the analysis of uploaded information is calculated and displayed dynamically. The tools provided support not just for the analyses carried out by the technical advisors to the emergency management teams in the various Member States, but also provide a number of plots aimed at conveying the forecast uncertainty to the decision makers. For the moment, ENSEMBLE has no analogue in the field and the approach proposed allows novel, rapid and sound comparative analyses of available atmospheric dispersion forecasts.

Throughout the project duration 12 regular and 2 additional exercises ("dry runs") were performed, which simulated different types of nuclear accidents in different places in Europe. In each exercise, based on the information about the accident type and location, each project participant had to run his dispersion model in real time and send the results to JRC in Ispra, Italy, via Internet. Information about the accident was sent via e-mail, fax and Internet. Once the results of different models were uploaded, the Internet tools could be used for visualization and extended analysis. The same tool can be used for all tasks related, i.e., creation of different maps, scatter diagrams, frequency distributions, comparison of the results of different models and advanced statistical analysis.

5. PARTICIPANTION OF BERS IN ENSEMBLE

BERS joined ENSEMBLE at the half of the project duration. First, the five already launched exercises were simulated *off-line* and the results uploaded to the ENSEMBLE database. To the end of the project, BERS regularly provided real-time forecast for all dry runs. The comparison of BERS results with ensemble ones show that the score of BERS varies between the different dry runs. For demonstration, among the big variety of statistics and graphs the FMS statistics (Figure-of-Merit-in-Space - the percentage of overlapping of modeled and measured pollution spots) will be used together with the maps of pollution evolution as produced by ENSEMBLE web-site. In Figures 5 and 6, the best and the worst BERS performance are shown versus the averaged results from 9 European models.

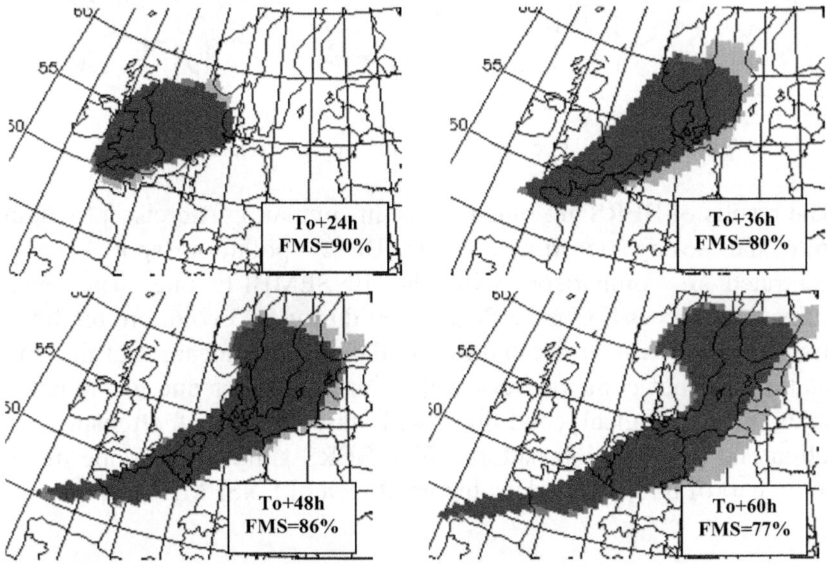

Figure 5: Simulation of ENSEMBLE dry run No.6 – release from Dublin. Light gray - 9 models average simulation; dark gray - BERS simulation; black - overlapping; concentration threshold of 0.1 Bq/m^3.

From the first moments of run 6 (Figure 5) BERS prediction coincides very well with the mean field and the percentages of overlapping are quite high. They will be even greater if the out-comers are not taken into consideration in the ENSEMBLE simulation.

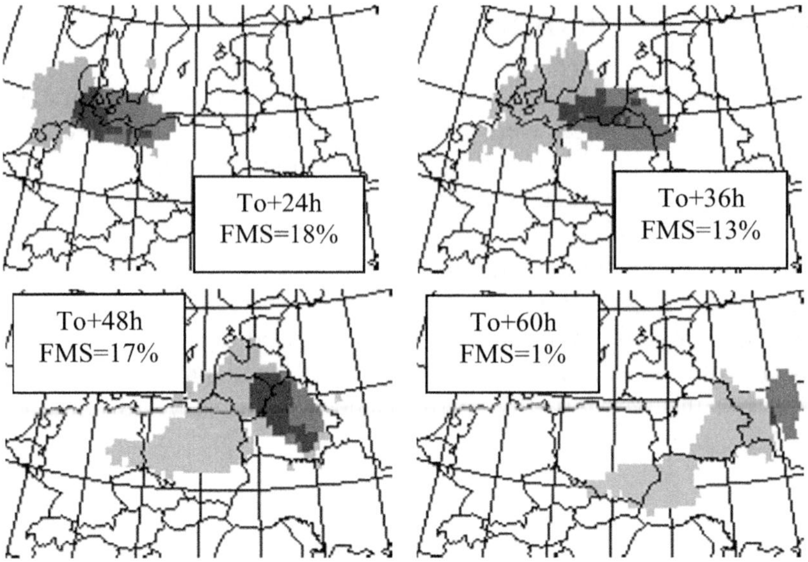

Figure 6: As in Figure 5 but for dry run No.11 – release from London.

The results of BERS are much worse in simulating exercise 11 (Figure 6). Despite the strong deformation, the BERS pollution spot keeps quite concentrated in comparison with the ENSEMBLE one. The ensemble pollution spot shows a strange disposition during its evolution, tending to be separated to two parts at the end of simulation. This means that the different simulations do not coincide very well with each other due to the extremely complex meteorological conditions with fast change of atmospheric flow direction and precipitation patterns. The BERS simulation is one of this set and only part of its spot overlaps the great area of ENSEMBLE spots.

6. CONCLUSION

The presented results prove that good emergency responses can be produced on the base of personal computers using meteorological information, distributed via GTS. This can be useful for the development of reliable warning systems in the Eastern European countries. The EMAP model demonstrates that rather sophisticated models can be run on small platforms producing relatively good forecasts.

Acknowledgement: The present work is supported by the NATO Collaborative Linkage Grant EST.CLG 979794, by the BULAIR project (EU 5FP Contract EVK2-CT-2002-80024), as well as by the Greek and Bulgarian ministries of education and science (Contract No. SE-1212-Gr/02 NSNI). The work is also based on the results obtained within the ENSEMBLE Consortium, which is acknowledged. ENSEMBLE (http://ensemble.ei.jrc.it) is a project supported by the European Commission DG-RTD Nuclear Fission Program. Contact point <stefano.galmarini@jrc.it>.

REFERENCES

BC-EMEP, 1994, 1995, 1996, 1997, 1998, Bulgarian contribution to EMEP, Annual reports for 1994, 1995, 1996, 1997, 1998, NIMH, EMEP/MSC-E, Sofia-Moscow.

Bellasio, R., Bianconi, R., Graziani, G. and Mosca, S., 1999, RTMOD: An Internet based system to analyse the predictions of long-range atmospheric dispersion models. *Computers and Geosciences* **25**(7), pp. 819-833.

Galmarini, S., Bianconi, R., Bellasio, R., Graziani, G., 2001, Forecasting the consequences of accidental releases of radionuclides in the atmosphere from ensemble dispersion modelling. *Journal of Environmental Radioactivity* **57**, pp. 203-219.

Girardi, F., Graziani, G., van Velzen, D., Galmarini, S., Mosca, S., Bianconi, R., Bellasio, R., Klug, W., and Fraser, G., 1998, ETEX - The European tracer experiment. EUR 18143 EN, pp. 1-108.

Graziani, G., Klug W. and Moksa, S. (eds), 1998, Real-Time Long-Range Dispersion Model Evaluation of the ETEX First Release. Joint Research Centre, EU, Luxemburg.

Gussev, A., Ilyin, I., Peterson, G., van Pul, A. and D. Syrakov, 2000, Long-range transport model intercomparison studies. *EMEP/MSC-E Technical Note 2/2000*, Moscow.

Mosca, S., Graziani, G., Klug, W. Bellasio, R., and Biankoni, R., 1997, ATMES-II - Evaluation of Long-range Dispersion Models using 1st ETEX release data. Volume 1, JRC-Ispra, Environmental Institute, pp. 25, 49-52, 259-261.

Syrakov, D., 1995, On a PC-oriented Eulerian Multi-Level Model for Long-Term Calculations of the Regional Sulphur Deposition. In Gryning, S.-E. and Schiermeier, F. A. (eds), *Air Pollution Modelling and its Application XI* 21, Plenum Press, N.Y. and London, pp. 645-646.

Syrakov, D. and Galperin, M., 1997a, A Model for Airborne Poli-Dispersive Particle Transport and Deposition. In: *Proceedings of the 22nd NATO/CCMS International Technical Meeting on Air Pollution Modelling and its Application*, June 2-6, 1997, Clermont-Ferrand, France, pp. 111-118.

Syrakov, D. and Galperin, M. 1997b, On a new Bott-type advection scheme and its further improvement. In Hass, H. and Ackermann, I. J. (eds.), *Proc. of the first GLOREAM Workshop, Aachen, Germany, September 1997*, Ford FZ, Aachen, pp. 103-109.

Syrakov, D., Djolov, G. and Yordanov, D, 1983, Incorporation of planetary boundary layer dynamics in a numerical model of long-range air pollution transport. *Boundary Layer Meteorology* **26**, pp. 1-13.

Syrakov, D. and Prodanova, M., 1994, European Tracer Experiment and Bulgarian Participation in its Dry Runs. *Bulgarian Journal of Meteorology & Hydrology* **5**, pp. 26-48.

Syrakov, D. and Prodanova, M., 1998a, Bulgarian Emergency Response Models - Validation against ETEX First Release. *Atmospheric Environment* **32**, pp. 4367-4375.

Syrakov, D. and Prodanova, M., 1998b, Simulation of the ETEX first release by Bulgarian emergency response models, in Gryning S.-E. and E. Batchvarova (eds.), Air Pollution

Modeling and Its Application XIII, Kluwer Academic/Plenum Publishers, New York, pp. 281-290.

Syrakov, D. and Yordanov, D., 1997, Parameterization of SL Diffusion Processes Accounting for Surface Source Action. *Proc. of 22nd NATO/CCMS International Technical Meeting on Air Pollution Modelling and its Application, 2-6 June 1997, Clermont-Ferrand, France*, pp. 111-118.

Yordanov, D., Syrakov D. and Djolov, G., 1983, A barotropic planetary boundary layer. *Boundary Layer Meteorology* **25**, pp. 363-373.

GLOBAL AND REGIONAL AEROSOL MODELLING: A PICTURE OVER EUROPE

Elisabetta Vignati[(1)], Maarten Krol[(2)], and Frank Dentener[(1)]
[(1)]European Commission, Joint Research Centre, Institute for Environment and Sustainability, TP 280, Via E. Fermi, 21020 Ispra (VA), Italy; [(2)]Institute for Marine and Atmospheric Research, Princetonplein 5, 3584 CC Utrecht, The Netherlands

Abstract: The TM5 atmospheric chemistry-transport model has been developed to study the interaction between the global and regional scales. It is an off-line model with a spatial global resolution of 6°x 4° and a two-way zooming algorithm that allows a finer resolution of 1°x 1° over selected regions (e.g. Europe, N. America, Africa and Asia). Photochemistry and aerosols are coupled in this version of the model. Aerosols, which can contain sulphate, ammonium and nitrate, are described using a bulk approach. The equilibrium model EQSAM is used to calculate the partitioning between the aerosol and gas phases. Preliminary results of the model application for the year 2000 over Europe are presented. The annual average modelled concentrations of sulphate, ammonium and nitrate are compared to the measurements from the EMEP stations. Generally the comparison with observations indicates that too low SO_4^{2-} and too high SO_2 concentrations are found. Modelled nitrate and ammonium concentrations are also high compared to EMEP measurements.

Key words: Aerosol, chemistry-transport model, European scale.

1. INTRODUCTION

Aerosols play an important role at global and regional scales. They contribute to radiative forcing of the climate system. The positive radiative forcing of the greenhouse gases is well quantified. In contrast, the quantification of aerosol radiative forcing of the climate system is less certain. Most components of aerosol scatter solar radiation, typically exerting a cooling effect on the atmosphere. However, black carbon, and to a lesser extent dust and some organic species, absorb sunlight, potentially warming the atmosphere. Moreover, aerosols have an effect on clouds, influencing their

I. Farago et al. (eds.), Advances in Air Pollution Modeling for Environmental Security, 373–381.

albedo, lifetime and precipitation efficiency. This indirect effect of aerosol on clouds and its impact on the climate system is poorly understood.

Satellite pictures of aerosol optical depth (e.g. Robles Gonzales et al. 2000) show that high aerosol concentrations are found close to the emission sources, but also that aerosols are spread over the continent. In fact they may have a long lifetime and can be transported over long distances. During transport they change their physical and chemical properties due to a variety of processes. Moreover, aerosols are formed as secondary products of gaseous precursors. Models are necessary to study aerosol transport, and the changes of their properties, the fields of their concentrations and their effects. Modelling these processes is important to asses the effects of emission reductions on particle concentrations. The TM5 atmospheric transport model has been developed with the purpose to study also the interactions between the global and regional scales.

2. AEROSOLS IN TM5

The TM5 model is an off-line global transport chemistry model (Krol et al. 2004) that uses the meteorological data calculated by the ECMWF model. It has a spatial global resolution of 6°x 4° and a two-way zooming algorithm that allows resolving regions (e.g. Europe, N. America, Africa and Asia) with a finer resolution of 1°x 1°. To smooth the transition between the global 6°x 4° region and the regional 1°x 1° domain, a domain with a 3°x 2° area resolution has been added. The algorithm gives the advantage of a local high resolution for the locations where measurements are taken and to study the effects of grid refinement on intercontinental transport of air pollution. The vertical structure has 25 layers, defined in a hybrid sigma-pressure coordinate system with a higher resolution in the boundary layer and in the free troposphere.

Transport, chemistry, deposition and emissions are solved using the operator splitting. The slopes advection scheme (Russel and Lerner, 1981) has been implemented and deep and shallow cumulus convection are parameterised according to Tiedtke (1989). All meteorological fields, including large-scale and convective cloud data are taken from the European Centre for Medium-Range Weather Forecasts (ECMWF).

The gas phase chemistry is calculated using the CBM-IV chemical mechanism (Gery et al., 1989a,b) solved by means of the EBI method (Hertel. et al., 1993). Photochemistry and aerosols are coupled in this version of the model. Aerosols are assumed internally mixed and in an accumulation mode size distribution for the calculation of scavenging and deposition processes. They can contain sulphate, ammonium and nitrate, and are described using a bulk approach. Dry deposition is calculated using the ECMWF surface characteristics. In-cloud as well as below-cloud wet removals are described in the model.

The equilibrium model EQSAM (Metzger et al., 2002a,b) is used to calculate the partitioning between the aerosol and gas phases of ammonia, nitric acid, ammonium and nitrate. The water attached to the particle is determined from the ambient relative humidity. Sulphate is assumed to be present only in the aerosol phase. It is formed by the oxidation of SO_2 (and DMS) in the gas phase by OH and in the aqueous phase by H_2O_2 and ozone.

2.1 Modelled and measured annual averages

Preliminary results of the model run using the zoom option over Europe for the year 2000 are presented. The annual averages of sulphate, ammonium and nitrate mapped over Europe are shown in Figure 1. The highest concentrations of SO_4^{2-} are found in the Po Valley, in the Balkans and in Turkey. It is evident that sulfur emissions from Eastern Europe are transported down towards the Mediterranean Sea and Africa. There is a gradient in the concentrations over the continent from the South to the North, which also appears in the measurements (Figure 2). TM5 tends to underestimate the sulphate concentrations in regions such as Spain or Denmark and Sweden; this will be discussed in the next paragraph.

High ammonium concentrations (>2 μgN m^{-3}) are predicted by the model over Southern England, Central and Eastern Europe and Italy. The levels are generally overpredicted compared to the EMEP observations. The same appears for nitrate, even though fewer measurements are available. This may be due to an overestimate of NO_3^- formation in the aerosol calculated by EQSAM for very high relative humidity. Alternatively, the nitrate overestimate can be a consequence of a too low SO_4^{2-} concentration. In that case, nitric acid is transferred to the aerosol phase if ammonia is available. These hypotheses have to be investigated further.

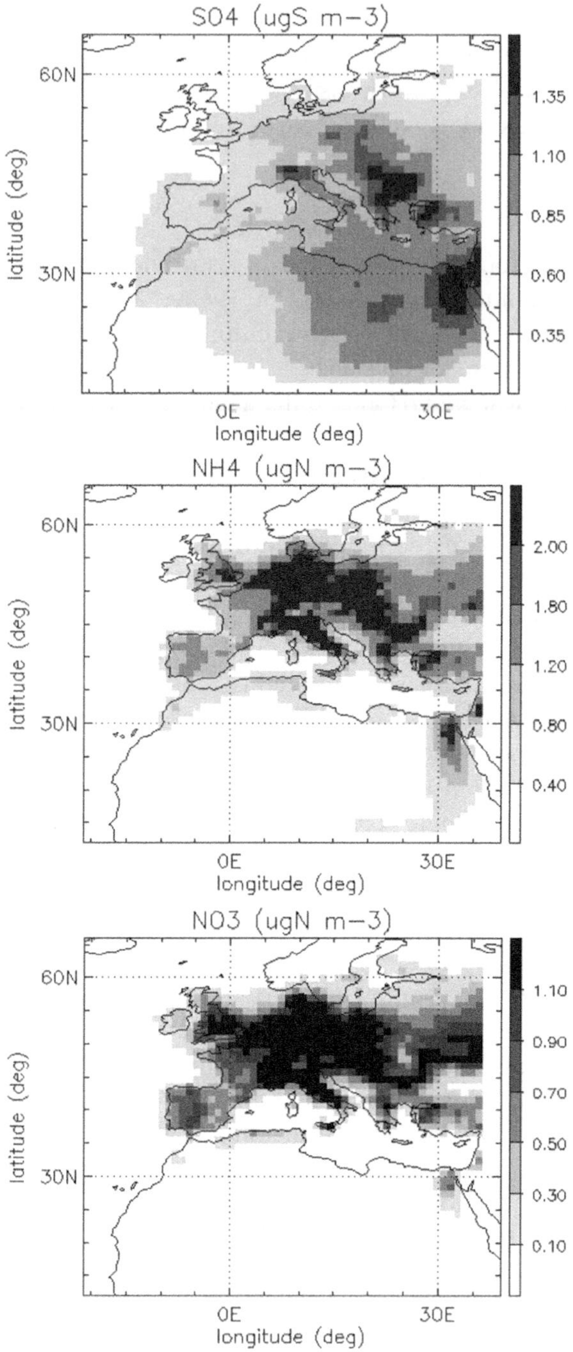

Figure 1: SO_4^{2-} (upper panel), NH_4^+ (middle) and NO_3^- (lower panel) annual average concentrations modelled with TM5.

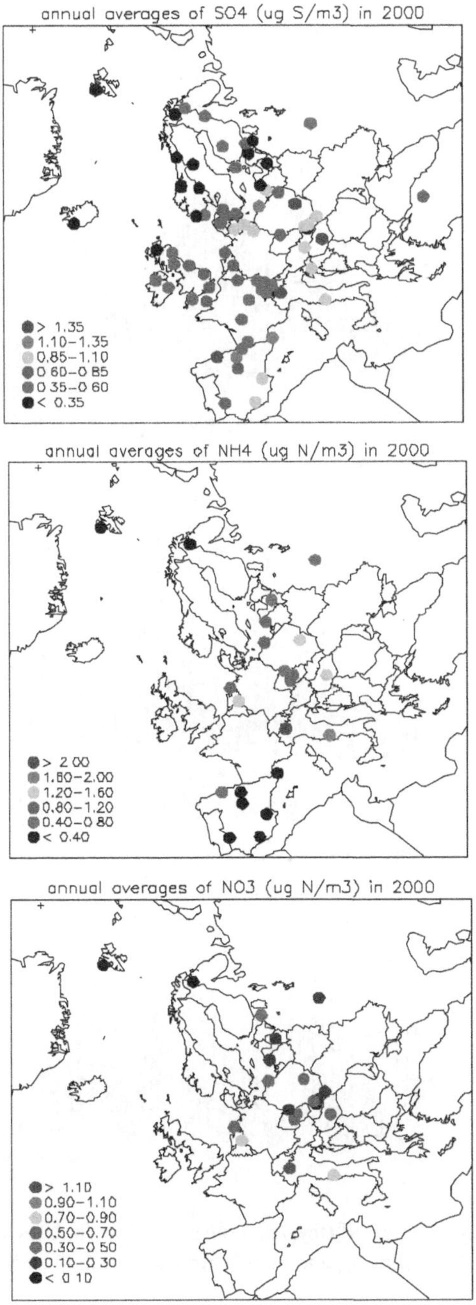

Figure 2: SO_4^{2-} (upper panel), NH_4^+ (middle) and NO_3^- (lower panel) annual averages (EMEP, 2003).

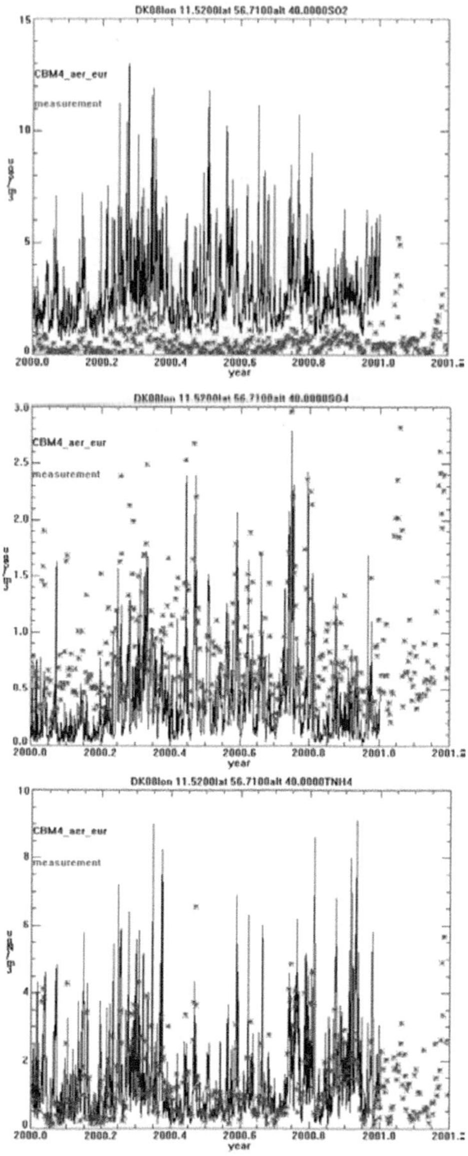

Figure 3: Time series of modeled (line) and measured (symbols) SO_2 (upper panel), SO_4^{2-} (middle) and NO_3^- (lower panel) concentrations; Anholt station (Denmark).

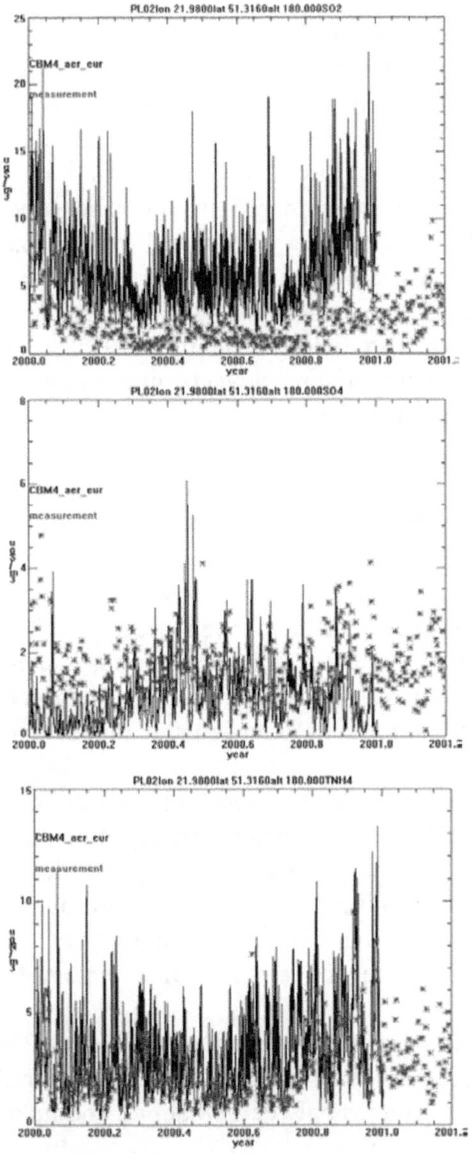

Figure 4: Time series of modeled (line) and measured (symbols) SO_2 (upper panel), SO_4^{2-} (middle) and NO_3^- (lower panel) concentrations; Jarczew station (Poland).

2.2 Time series comparison

The EMEP stations of Anholt (Denmark) and Jarczew (Poland) have been selected to represent areas with low and high SO_2 and sulphate concentrations, respectively. Figure 3 and 4 show time series of SO_2, SO_4^{2-} and ammonium modelled and observed concentrations for the two stations. In general the model underpredicts sulphate concentrations and overpredicts SO_2. This is a common problem in large-scale atmospheric aerosol models (Barrie et al., 2001). For the sites with lower SO_2 concentrations TM5 calculates much higher SO_4^{2-} concentrations compared to the observations than for areas with high SO_2. In fact, the differences in Figure 4 (middle panel) are not as large as in Figure 3 (middle panel). For ammonium the results of the comparison are opposite: at Anholt the agreement is quite good, while at Jarczew the concentrations are overpredicted by the model.

More extensive comparison between model results and measurements is necessary in order to evaluate TM5.

3. CONCLUSIONS

Preliminary results of surface aerosol fields predicted by the TM5 atmospheric-transport model have been presented. Using a two-way zoom algorithm the model can calculate sulphate, ammonium and nitrate concentrations with a resolution of 1°x 1° over the continent. As TM5 is a global model, the interactions between the global and European scales are taken into account.

The comparisons against the measurements collected at the EMEP sites reveal that the sulphate concentrations are reproduced reasonably well except some areas such as Spain, Denmark and Sweden where the levels are underestimated. At the same time SO_2 levels are overpredicted by the model suggesting that maybe the SO_2 oxidation to SO_4^{2-} is not always efficient enough. Modelled nitrate and ammonium concentrations are generally too high.

Further comparison with observations is needed to evaluate the model performance. Moreover, sensitivity tests are required to characterise the reasons of the discrepancies between modelled and measured concentrations.

Acknowledgments: We thank Julian Wilson for reading the paper and for his valuable comments.

REFERENCES

Barrie, L. A., Yi, Y., Leaitch, W. R., Lohmann, U., Kasibhatla, P., Groelofs, G.-J., Wilson, J., McGovern, F., Benkovitz, C., Meliere, M. A., Law, K., Prospero, J., Kritz, M., Bergmann, D., Bridgeman, C., Chin, M., Chistensen, J., Easter, R., Feichter, H., Land, C., Jeuken, A.,

Kjellstrom, E., Koch, D., Rasch, P., 2001, A comparison of large scale atmospheric sulphate aerosol models (COSAM): overview and highlights. *Tellus* **53B**, pp. 615-645.

EMEP, 2003, Transboundary particulate matter in Europe Status report 2003, Kahnert, M. and Tarrasón, L., ed., Joint CCC & MSC-W & CIAM Report 2003, EMEP Report 4/2003, O-98134.

Gery, M. W., Whitten, G. Z., Killus, J. P. 1989a, Development and testing of the CBM-IV for urban and regional modelling, EPA-600/3-88-012. US EPA, Research Triangle Park.

Gery, M. W., Whitten, G. Z., Killus, J. P. and Dodge, M. C. 1989b, A photochemical kinetics mechanism for urban and regional scale computer modeling. *J. Geophys. Res.* **94D**, pp. 12925-12956.

Hertel, O., Berkowicz, R., Christensen, J., 1993, Test of two numerical schemes for use in atmospheric transport-chemistry models. *Atmos. Environ.* **27A**, 16, pp. 2591-2611.

Krol, M., Houweling, S., Bregman, B., van den Broek, M., Segers, A., van Velthoven, P., Peters, W., Dentener, F., Bergamaschi, P., 2004, The two-way nested global chemistry-transport zoom model TM5: algorithm and applications. *Accepted for pub. ACPD.*

Metzger S., Dentener, F., Pandis, S., Lelieveld, J., 2002a, Gas/aerosol partitioning: 1. A computationally efficient model. *J. Geophys. Res.*, 107, D16, doi: 1-.1029/2001JD001102.

Metzger, S., Dentener, F., Lelieveld, J., Jeuken, A., Krol, M., Pandis, S., 2002b. Gas/aerosol partitioning: 2. Global Modeling results. *J. Geophys. Res.* **107**, D16, doi:10.1029, 20002.

Robles Gonzales, C., Veefkind, J. P., de Leeuw, G., 2000, Aerosol optical depth over Europe in August 1997 derived from ATSR-2 data. *Geophys. Res. Let.* **27**, pp. 955-958.

Russel, G., Lerner, J., 1981, A new finite-differencing scheme for the tracer transport equation. *J. Appl. Meteorol.* **20**, pp. 1483-1498.

Tiedtke, M. 1989, A comprehensive mass flux scheme for cumulus parameterisation in large scale models. *Mon. Wea. Rev.* **177**, pp. 1779–1800.

THE ABL MODELS YORDAN AND YORCON – TOP-DOWN AND BOTTOM-UP APPROACHES FOR AIR POLLUTION APPLICATIONS

Dimiter Yordanov[(1)], Maria Kolarova[(2)] and Dimiter Syrakov[(2)]
[(1)] Geophysical Institute, Bulgarian Academy of Sciences, Acad. G. Bonchev str. Bl. 3, Sofia 1113, Bulgaria; [(2)] National Institute of Meteorology and Hydrology, Bulgarian Academy of Sciences, Tsarigradsko chaussee 66, Sofia 1784, Bulgaria

Abstract: Two atmospheric boundary layer (ABL) models developed in agreement with the similarity theory are applied: YORDAN - for stable and neutral conditions and YORCON - for convective cases. Both models consist of a surface layer and an overlying Ekman layer. The vertical profiles of the temperature, wind velocity and the turbulent exchange coefficient in the ABL can be obtained applying the models. A proper parameterization is developed in both cases applying the resistance laws for stable, neutral and convective conditions. The relation between the external and internal to ABL parameters are given by the resistance laws numerically solved in both models. The input parameters for the models are obtained from the experimental data applying two approaches: the first one (called "top-down" approach) uses data for the geostrophic wind and the potential temperature. The second one (called "bottom-up" approach) uses data from the surface meteorological observations.

Key words: Atmospheric boundary layer, ABL models, ABL parameterization, similarity theory, resistance laws, similarity functions, surface layer.

1. INTRODUCTION

For the purposes of air pollution modeling the vertical profiles of the wind velocity, temperature and the turbulent exchange coefficient in the ABL has to be calculated at each grid point from the routine meteorological data or from the synoptic data. That is why proper ABL models have to be applied at neutral, stable and convective conditions. The ABL models YORDAN (Yordanov et al., 1983; 1998) and YORCON – for convective conditions

I. Farago et al. (eds.), Advances in Air Pollution Modeling for Environmental Security, 383–393.

(Yordanov et al., 1990; 1997) developed in agreement with the similarity theory are used in air pollution tasks accounting for two approaches – "top-down" and "bottom-up".

In the classical "top-down" approach (Yordanov et al., 1997) the internal parameters: L (the Monin-Obukhov length scale), the friction velocity u_*, and the cross isobaric angle (α) are determined from the external parameters as the geostrophic wind and the difference between the potential temperature at ABL height and at the ground. The input data are obtained from the numerical weather prediction or from radio sounding data. At the output the wind and temperature profiles are obtained together with other basic ABL characteristics applying the ABL models.

In the "bottom-up" approach the inverse problem is solved: starting from the surface meteorological measurements of the wind and atmospheric stability as input data, the external parameters are determined from the surface turbulent fluxes defined by L and u_* and finally the ABL profiles are obtained applying the ABL models as shown by Yordanov et al. (2003a, 2003b).

In the present paper a summarized description of the "bottom-up" approach applied for the ABL models YORDAN and YORCON is developed.

2. THE INTERNAL PARAMETERS L AND U*

2.1 Determination of M.-O. length *L*

The Monin-Obukhov length scale L is a basic parameter often employed in the numerical models to characterize the turbulence within the surface layer (SL). Here the M.-O. length scale L is defined as:

$$L = -u_*^3 \big/ \kappa\beta(\overline{w'\theta'})_0 \,, \tag{1}$$

where u_* is the friction velocity, κ is the von Karman constant, $\beta = g/\overline{\theta}$ is the buoyancy parameter, g is the gravity acceleration, $\overline{\theta}$ is the mean potential temperature, averaged for the whole layer, $(\overline{w'\theta'})_0$ is the normalized by $c_p\rho$ vertical turbulent heat flux at the ground, c_p is the specific heat capacity at constant pressure, and ρ is the air density.

The values of L can be computed directly from the field measurements of u_* and the heat flux if data are available, but often the M.-O. length is estimated from routine meteorological measurements.

Applying the "bottom-up" approach, L is determined using the Golder relations. To relate the M.-O. length scale L to the roughness length z_o the empirical curves of Golder are used fitted by power low functions:

$$L^{-1} = a z_0^b . \tag{2}$$

The constants a and b depend on the stability class as shown by Yordanov et al. (2003a; 2003b). Having the stability class experimentally determined from the measurements, the corresponding M.-O. length scale L can be calculated at given roughness from Eq. (2).

2.2 Wind and temperature profiles in the SL

According to the Monin-Obukhov similarity theory, the universal profiles of wind velocity and of the potential temperature in the SL can be presented as:

$$\kappa \frac{u}{u_*} = f_u(\zeta, \zeta_0) \quad \text{and} \quad \frac{(\theta - \theta_0)}{\theta_*} = \frac{(\theta - \theta_0)\kappa^2 \beta L}{u_*^2} = f_\theta(\zeta, \zeta_0) \tag{3}$$

where u and θ are the wind velocity and the potential temperature at height z, θ_0 is the potential temperature at height z_0, $\theta_* = u_*^2 / \kappa^2 \beta L$ is the temperature scale, f_u and f_θ are universal functions of the dimensionless height $\zeta = z / L$ and dimensionless roughness $\zeta_0 = z_0 / L$.

In the SL the turbulent exchange coefficients for momentum k_m and for heat k_θ can be expressed as:

$$k_m = \frac{\kappa u_* z}{\varphi_u(\zeta)}, \quad \text{and} \quad k_\theta = \frac{\kappa u_* z}{\varphi_\theta(\zeta)}, \tag{4}$$

where $\varphi_u(\zeta) = \zeta \dfrac{\partial f_u(\zeta, \zeta_0)}{\partial \zeta}$, and $\varphi_\theta(\zeta) = \zeta \dfrac{\partial f_\theta(\zeta, \zeta_0)}{\partial \zeta}$.

2.3 Determination of u*

If the universal SL profiles f_u and f_θ are known, the friction velocity u_* can be determined from Eq. (3) considering the wind velocity u_a and the anemometer height z_a as known parameters, so that:

$$u_* = \frac{\kappa u_a}{f_u(\zeta_a, \zeta_0)}, \tag{5}$$

where $\zeta_a = z_a / L$, and L is determined from Eq. (2).

Applying the ABL model YORDAN (Yordanov et al., 1983) the surface profile functions f_u and f_θ are taken as log-linear for stable and neutral stratification as shown by Yordanov et al. (2003a). The model calculates the vertical profiles of the wind and temperature as well as the vertical turbulent exchange coefficient and demonstrates good coincidence with a number of different experimental data sets as shown by Yordanov et al. (1998) and Djolov et al. (2004).

In the SL, the wind profile applied according to the model YORDAN is determined as:

$$\kappa \frac{u}{u_*} = \begin{cases} \ln(z/z_0) + 10\zeta & at \;\; \mu \geq 0 \\ \ln(z/z_0) & at \;\; -0.07 \leq \zeta \leq 0 \;\; and \;\; \mu < 0 \\ \ln(-0.07/\zeta_0) + 3\left[1 + (0.07/\zeta)^{1/3}\right] & at \;\; \zeta \leq -0.07 \;\; and \;\; \mu < 0 \end{cases} . \quad (6a)$$

Comparing Eq. (6a) with Eq. (3) an expression for the universal SL profile f_u can be obtained. Replacing it in equation (5) and taking into account that $u = u_a$ and $z = z_a$ the friction velocity can be determined.

Applying the convective ABL model YORCON the universal profiles of Businger are used, as shown by Yordanov et al. (1997) and Yordanov et al. (2003a). In this case the functions f_u and f_θ have the form:

$$f_u(\zeta, \zeta_0) = \Phi_u(\zeta) - \Phi_u(\zeta_0) \text{ and } \alpha_\theta(0) f_\theta(\zeta, \zeta_0) = \Phi_\theta(\zeta) - \Phi_\theta(\zeta_0), \quad (6b)$$

where the functions $\Phi_u(\zeta)$ and $\Phi_\theta(\zeta)$ are the following:

$$\delta\theta\, \Phi_u(\zeta) = \ln(-\zeta) + 2arcctg(\eta_u) - \ell n\left[\left(1 + \eta_u^2\right)\left(1 + \eta_u\right)^2 \eta_u^{-4}\right] \text{ and}$$
$$\Phi_\theta(\zeta) = \ell n(-\zeta) - 2\ell n\left[\left(1 + \eta_\theta\right)\eta_\theta^{-1}\right].$$

Here η_u and η_θ are defined as: $\eta_u = (1 - 15\zeta)^{-1/4}$, $\eta_\theta = (1 - 9\zeta)^{-1/2}$.

Replacing the expressions for the functions f_u and f_θ given by Eq. (6b) in Eq. (5), the friction velocity u_* can be determined in that particular case. The model YORCON was also tested on a number of experimental data as shown by Yordanov et al. (1990; 2003b).

In the first part we determined the dimensional internal parameters of ABL L and u_* using data for the surface wind and stability class as input information. In the following part we shall determine the external to ABL parameters as the geostrophic wind (V_g), the angle between the surface and the geostrophic wind (α) and the difference between the potential temperature at the top of ABL and at the ground ($\delta\theta$) applying the resistance laws and the similarity theory.

3. THE EXTERNAL PARAMETERS ($\overrightarrow{V_g}$, α, δθ)

3.1 The ABL model YORDAN

The turbulent regime in a barotropic ABL under stable and neutral conditions is parameterized on the basis of the similarity theory using the following non-dimensional external parameters:

$$S = \frac{\beta(\theta_H - \theta_0)}{f\left|\overrightarrow{V_g}\right|} \quad \text{and} \quad Ro = \left|\overrightarrow{V_g}\right| / f z_0 . \tag{7a}$$

Here S is the external to ABL non-dimensional stratification parameter and Ro is the so called Rossby number, $\delta\theta = \theta_H - \theta_0$ is the difference between the potential temperatures at the upper boundary of the ABL (θ_H) and at the ground (θ_0), V_g is the geostrophic wind vector, f is the Coriolis parameter.

The non-dimensional internal ABL parameters that determine the turbulent regime of a barotropic ABL under stable and neutral stratification are:

$$c_g = \frac{u_*}{\left|\overrightarrow{V_g}\right|}, \quad |\alpha|, \quad \mu = \frac{H}{L} = \frac{\kappa u_*}{fL}, \tag{8a}$$

where, c_g is the drag coefficient, α is the cross-isobaric angle or the surface wind deviation from the geostrophic wind (being negative in the Northern Hemisphere), and μ is the internal stratification parameter, $H = \kappa u_* / f$ is the ABL height scale.

To determine the external to ABL parameters, such as the geostrophic wind ($\overrightarrow{V_g}$), the cross-isobaric angle (α) and the difference between the potential temperature at the top and at the bottom of the ABL ($\delta\theta$), the resistance and heat exchange laws for the case of barotropic ABL are applied (Yordanov et al., 1983; 1998):

$$\frac{\kappa \cos\alpha}{c_g} = \ln\frac{u_*}{f z_0} - A(\mu), \tag{9a}$$

$$\frac{\kappa \sin|\alpha|}{c_g} = B(\mu), \tag{10a}$$

$$a_\theta(0)\frac{\kappa^3 S}{c_g \mu} = \ln\frac{u_*}{f z_0} - C(\mu). \tag{11a}$$

The absolute value of the geostrophic wind can be obtained from Eq. (9a) and Eq. (10a):

$$\left|\overrightarrow{V_g}\right| = \frac{u_*}{\kappa}\left[\left(\ln\frac{u_*}{f z_0} - A(\mu)\right)^2 + B^2(\mu)\right]^{1/2}. \tag{12a}$$

The absolute value of the cross-isobaric angle α can be obtained from Eq. (10a):

$$|\alpha| = \arcsin\left[\frac{c_g}{\kappa} B(\mu)\right]. \tag{13a}$$

The terms $A(\mu)$, $B(\mu)$, $C(\mu)$ are universal functions of the internal stratification parameter μ, deduced from the analytical solutions for the wind and temperature profiles (Yordanov, et al., 1998; 2003a).

The non-dimensional external stratification parameter S is determined from Eq. (11a) as:

$$S = \frac{c_g \mu}{a_\theta(0)\kappa^3}\left[\ln\frac{u_*}{f z_0} - C(\mu)\right]. \tag{14a}$$

For the difference between the potential temperature at the top and at the bottom of the PBL we obtain from Eq. (7a) and Eq. (14a) the following expression:

$$\delta\theta = \theta_{\mathrm{H}} - \theta_0 = \frac{u_* \mu f}{\alpha_\theta(0)\kappa^3 \beta}\left[\ln\frac{u_*}{f z_0} - C(\mu)\right]. \tag{15a}$$

Replacing the analytical form for the universal functions *A*, *B*, *C*, determined from the ABL model YORDAN in Eqs. (12a)-(15a), the final expressions for the geostrophic wind $\vec{V}_g$, the angle α, and $\delta\theta$, are obtained as shown by Yordanov, et al. (2003a).

Replacing the expressions for *Ro* and *S* from (7a) in the resistance laws (9a-11a) together with the defined universal functions *A*, *B*, *C*, we obtain a system of non-linear algebraic equations for the internal parameters c_g, α and μ. The numerical solution of this system of equations at different values of the external parameters *Ro* and *S* was found and approximated with polynomials by Yordanov et al. (1983) for barotropic ABL and by Yordanov et al. (1998) for baroclinic ABL.

3.2 The ABL model YORCON

The turbulent regime in a barotropic ABL capped by inversion is parameterized using the following non-dimensional external parameters defined under convective conditions:

$$Ri_B = \frac{\beta\,\delta\theta\, z_i}{\vec{V}_g^2}, \quad Z_0 = \frac{z_0}{z_i}, \quad Ro_i = \frac{|\vec{V}_g|}{f z_i} \quad . \tag{7b}$$

Here Ri_B is the bulk Richardson number, Ro_i is the bulk inversion Rossby number, Z_0 is the non-dimensional roughness, $\delta\theta = (\theta_{z_i} - \theta_0)$ is the difference between the potential temperatures at the inversion height z_i and at the ground (at z_o).

The turbulent regime in a convective ABL (CBL) is also determined by the following internal parameters:

$$c_g = \frac{u_*}{|\vec{V}_g|}, \quad \alpha, \quad \mu_c = \frac{z_i}{L} \quad \text{or} \quad \mu = \frac{\kappa\, u_*}{f L}. \tag{8b}$$

Here μ and μ_c are the internal stratification parameters defined in the CBL model YORCON.

The resistance laws giving the relationship between the external and internal CBL parameters are:

$$\frac{\kappa}{c_g} = \left\{\left[(\ell n Z_0 + A(\mu_c)\right]^2 + B^2(\mu_c, \mu)\right\}^{1/2} \tag{9b}$$

$$\sin|\alpha| = \frac{c_g}{\kappa} B(\mu_c, \mu) \tag{10b}$$

$$\alpha_\theta(0) Ri_B = -\frac{c_g^2}{\kappa^2}\left[\ell n Z_0 + C(\mu_c)\right] , \tag{11b}$$

where $\alpha_\theta(0) = 1.35$ and $\alpha_\theta(0) = K_\theta / K_m$ is the inverse Prandtl number (expressing the ratio between the turbulent exchange coefficients for heat and for momentum at the ground).

For the absolute value of the geostrophic wind the following expression is obtained from Eq. (9b):

$$\left|\overrightarrow{V_g}\right| = \frac{u_*}{\kappa}\left[(\ell n Z_0 - A(\mu_c))^2 + B^2(\mu_c, \mu)\right]^{1/2} . \tag{12b}$$

The absolute value of the cross-isobaric angle α is obtained from Eq. (10b):

$$|\alpha| = \arcsin\left[\frac{c_g}{\kappa} B(\mu_c, \mu)\right] . \tag{13b}$$

And $\delta\theta$ - the difference between the potential temperature at the top and at the bottom of the ABL, is obtained from Eq. (11b) as:

$$\delta\theta = \theta_{z_i} - \theta_0 = \frac{\theta_* \mu_c}{\alpha_\theta(0)}\left[-\ell n Z_0 - C(\mu_c)\right] . \tag{14b}$$

The corresponding analytical form for the universal functions *A*, *B*, *C* in the resistance laws are determined from the CBL model YORCON as shown by Yordanov and Kolarova (1988). Replacing the expressions for the universal similarity functions in the resistance laws (9b-11b) a system of non-linear algebraic equations for the internal parameters c_g , α and μ is obtained. The numerical solution of this system of equations at different values of the external parameters Ri_B , Z_o , *Ro* is given by Kolarova et al. (1989). The full procedure applying the bottom-up approach for the case of CBL is described by Yordanov et al. (2003b).

3.3 The inversion height

The evolution of the convective ABL height (mixing layer height) at conditions of horizontal homogeneity is calculated from Eq. (15b) as shown by Yordanov et al. (1990):

$$\frac{dz_i}{dt} = \frac{(1 + 2c_1)(\overline{w'\theta'})_0}{\Gamma z_i} + c_2 \frac{u_*^3}{\Gamma \beta z_i^2} \tag{15b}$$

where $\Gamma = \partial\theta / \partial z$ is the potential temperature gradient above the inversion layer, $c_1 = 0.2$, $c_2 = 2.5$ are constants. The evolution of mixing layer height depends on the surface turbulent heat and momentum fluxes:

$$(\overline{w'\theta'})_0 = -\frac{\mu f u_*^2}{\kappa^2 \beta} = -\frac{\mu_c u_*^3}{\kappa \beta z_i} \qquad u_* = c_g \left|\vec{V}_g\right|, \tag{16b}$$

i.e., on the internal CBL parameters.

Applying the "top-down" approach, the calculations are based on a known time evolution of the external parameters (7b) and the numerical solution of the system of equations for the resistance laws (9b-11b). The surface turbulent fluxes are estimated from Eq. (16b) together with the equation for the time evolution of the CBL height (15b) as it was shown by Yordanov and Kolarova (1988) and Kolarova et al. (1989).

Applying the "bottom-up" approach the surface turbulent fluxes are estimated from the surface observations by calculating u_* from Eq. (5) and using the relation $(\overline{w'\theta'})_0 = -u_*^3 / \kappa \beta L$ according to Eq. (1).

The equation (16b) is solved numerically at initial condition: $z_i = 0.1\ \kappa u_* / f$ at $t = t_{sunrise}$.

We can resume that applying the relationship between the external and internal parameters given by the resistance laws and solving numerically the corresponding equations in the cases of stable, neutral and convective ABL, the geostrophic wind $|V_g|$ the cross isobaric angle $|\alpha|$ and $\delta\theta$ can finally be determined. From these parameters applying the ABL models YORDAN and YORCON we can obtain the velocity defects and from them the wind and temperature profiles for stable, neutral and convective conditions as shown by Yordanov et al. (2003a, 2003b) and Djolov et al. (2004).

4. CONCLUSIONS

The ABL models YORDAN and YORCON are successfully applied in different practical tasks concerning air pollution modeling. The "top-down" version of the ABL model YORDAN is built in various Bulgarian dispersion models performing offline and online meteorological processing. These models are successfully used for solving many practical tasks with different time and space scales like assessment of the long range sulphur and heavy metal pollution in the region of South-Eastern Europe and some regions in Bulgaria, exchange of sulphur pollution between Bulgaria and Greece, etc. The model YORCON was applied in modeling the dispersion of passive radioactive pollutants as Kr-85 around APS Ignalina in Lithuania under daytime convective conditions.

Acknowledgements: This study was partly supported by the projects BULAIR (EVK2-CT-2002-80024), NATO CLG (EST-CLG-979863) and National Research Council Grant No. H3-1004/00.

REFERENCES

Djolov, G. D., Yordanov, D. L., and Syrakov, D. E., 2004, Baroclinic PBL model for neutral and stable stratification conditions. *Boundary Layer Meteorology* **111**, pp. 467-490.

Kolarova, M., Yordanov, D., Syrakov, D., Djolov, G., Karadjov, D., and Aleksandrov, L., 1989, Parameterization of a convective planetary boundary layer. *Izv. Acad. Sci. USSR Atmos. Ocean Phys.* **25**, pp. 659-661.

Yordanov, D. and Kolarova, M., 1988, An Analytical model of convective planetary boundary layer. *WMO/TD No 187*, March 1988, pp. 195-209.

Yordanov, D., Syrakov, D., and Djolov, G., 1983, A barotropic planetary boundary layer. *Boundary Layer Meteorology* **25**(4), pp. 363-373.

Yordanov, D., Kolarova, M., Syrakov, D., and Djolov, G., 1990, Convective boundary layer - theory and experiment. *Proc. of the 9th Symp. on Turbulence and Diffusion*, RISO, Denmark, April 1990.

Yordanov, D., Syrakov, D., and Kolarova, M., 1997, On the parameterization of the PBL of the atmosphere. *The Determination of the Mixing Height-Current Progress and Problems, EURASAP Workshop Proc.*, 1-3 Oct. 1997, RISO Nat. Lab., Roskilde, Denmark, Gryning, S.-E., Beyrich, F., Batchvarova, E. (eds.), Riso-R-997(EN), pp. 117-120.

Yordanov, D., Syrakov, D. and Djolov, G., 1998, Baroclinic PBL model: neutral and stable stratification condition. *Bulg. Geophys. J.*, **XXIV**(1-2), pp. 5-25.

Yordanov, D. L., Syrakov, D. E., and Kolarova, M. P., 2003a, Parameterization of PBL from the surface wind and stability class data. *Proc. of NATO ARW on Air Pollution Processes in Regional Scale*, Halkidiki, Greece, 13-15 June 2002, NATO Science Series, D. Melas and D. Syrakov (eds.), Kluwer Acad. Publ., Netherlands, Vol. 30, pp. 347-364.

Yordanov, D., Kolarova, M., and Syrakov, D., 2003b, Parameterization of convective PBL using surface data for the wind and stability classes. *Int. J. Environment and Pollution* **20**(1-6), pp. 165-176.

Yordanov, D., Syrakov, D., and Kolarova, M., 2003c, Parameterization of PBL using surface data for application in dispersion modeling. *Proc. of 26th NATO/CCMS ITM on Air Pollution Modeling and its Application*, Istanbul-Turkey, 26-30 May 2003, pp. 550-551.

MAJOR CONCLUSIONS FROM THE DISCUSSIONS

Zahari Zlatev [(1)], Adolf Ebel [(2)], István Faragó [(3)] and Krassimir Georgiev [(4)]
[(1)] *National Environmental Research Institute, Frederiksborgvej 399, P. O. Box 358, DK-4000 Roskilde, Denmark ;* [(2)] *Rhenish Institute for Environmental Research (RIU), EURAD-Project, University of Cologne, Cologne, Germany;* [(3)] *Eötvös Loránd University, Pázmány Péter sétány 1/C, Budapest, 1117, Hungary;* [(4)] *Institute for Parallel Processing, Bulgarian Academy of Sciences, Acad. G Bonchev str., bl. 25-A, 1113 Sofia, Bulgaria*

Abstract: Discussions were scheduled within the program of the NATO Advanced Research Workshop on "Advances in Air Pollution Modelling for Environmental Security". Three major topics were discussed: (a) the use of more powerful computers in the treatment of some large-scale environmental models, (b) the application of data assimilation techniques for improving the output from the models and (c) the need of better emission inventories. The major conclusions from these discussions well be presented in this paper.

Key words: Large-scale air pollution models, efficient utilization of powerful computers, variational data assimilation, air pollution forecasts, emission inventories, climatic changes and their influence on air pollution levels.

1. USE OF MORE POWERFUL COMPUTERS FOR THE NUMERICAL SIMULATION OF LARGE-SCALE AIR POLLUTION PROBLEMS

Computers are becoming more and more powerful. Many problems which had to be treated only two or three years ago on supercomputers can now be treated on work-stations and PCs. Nevertheless, the demand for more powerful computers is steadily increasing.

I. Farago et al. (eds.), Advances in Air Pollution Modeling for Environmental Security, 395–399.

There are several reasons for this:

- The application of fine resolution grids in the discretization of the large-scale air pollution models leads to huge computational tasks.
- The implementation of new modules in the air pollution models (as, for examples, modules for handling aerosols and modules for performing data assimilation) is also leading to an enormous increase of the computational tasks.
- The requirement for studying different trends in the development of the air pollution levels leads to runs over long time intervals (10-20 or even more years). Such studies can only be accomplished when powerful computers are available.
- If comprehensive studies are to be performed (for example, for studying the sensitivity of the pollution levels to changes of different key parameters; meteorological changes, emission variations, etc.), then the models are to be run with many scenarios. This is also increasing the demands for more powerful computers.
- The goal to treat atmospheric composition including anthropogenic changes in the context of integral earth system behaviour (changes) will only be possible with the employment of maximum computer power available.

The major conclusion was that, while many problems can now be run on work-stations and PCs, powerful computers are highly needed in this field. Moreover, even the most powerful computers available at present are not sufficient when some large air pollution problems with great time and storage requirements are to be handled. This is especially true for studies related to the influence of future climatic changes on the pollution levels in Europe and/or in different parts of Europe.

Computer grids are practically not yet used in the treatment of large-scale air pollution problems. Some of the participants expressed their belief that the situation will soon change.

As an example of possible use of computer grids in the field of air pollution modelling let us mention the task of running a **system** of models to treat a **set** of complex problems involving:

- weather forecasts on different regional scales (starting with results obtained on a global scale),
- weather forecasts on an urban scale (perhaps in parallel for several urban areas),
- air pollution forecasts on different regional scales,
- air pollution forecasts on an urban scale (perhaps in parallel for several urban areas),

- treatment of the output results in order to prepare them for the people who will use them (utilizing data mining algorithms and high-speed visualization tools),
- sending the relevant data to appropriate media (TV stations, radio stations, internet sites, GMSs, etc.) for dissemination and animation.

It is clear that computer grids will be very powerful tools in the solution of the very challenging task related to the possibility to treat efficiently the set of problems described above.

Such sets of problems are at present solved only by imposing many simplifying (and very often not physical) assumptions and IT methods. At the same time, it is also clear that a lot of difficulties must be overcome in the efforts to run such complex tasks efficiently on a computer grid. The greatest difficulties appear in the treatment of the tasks of

- achieving reliable and robust transition from one scale to another,
- communicating relevant data from one part of the computational grid to another
- preparing the final results, which should be easily understandable by the recipients.

2. APPLICATION OF DATA ASSIMILATION TECHNIQUES IN AIR POLLUTION MODELS

Observation data are becoming both more reliable and more easily available. It is evident that the best insight into a complex system will be achieved when observational and theoretical knowledge are combined by objective methods as it is done through application of advanced data assimilation techniques. Among the existing approaches Kalman filtering and 4-dimensional variational data assimilation are the most powerful ones. They are also the most demanding methods regarding computational power. Their development and application to air pollution modeling requires highest computer performance. Ongoing research and development in this field has shown that the quality of the output of air pollution models is considerably increased through advanced data assimilation. Improvements can be achieved in several different ways:

- One can increase the quality of the initial concentration data used in the treatment of the air pollution models. This is especially important when air pollution forecasts are to be produced.

- One can improve the quality of emission data. This is again important when the models are used to prepare air pollution forecasts.
- One can use data assimilation to improve the chemical part of the models by testing the relevance and interdependence of chemical reactions.
- One can explore the sensitivity of complex air quality models and contribute to their evaluation with the help of data assimilation.

Such tasks lead to very time-consuming runs. Some of these tasks cannot yet comprehensively be treated on now available computers. It will, for example, be very difficult, if possible at all, to perform a comprehensive study over a long time-period by using advanced models with a module for performing 4-dimensional variational data assimilation during the whole run. Special strategies have to be developed and applied to overcome such problems while waiting for further growth of computer performance.

The selection of representative and reliable measurements (and measurement stations) is another difficulty when data assimilation techniques are to be implemented in a large-scale air pollution model. It is clear, that the results obtained by employing data assimilation techniques essentially depend on the availability of measurements describing the treated system with sufficient complexity. Therefore, methods have to be developed which allow less accurate or even proxy-data to be implemented in data assimilation schemes. In particular, 4-dimensional variational data assimilation has the potential for such an approach.

Some of the chemical species involved in a large-scale air pollution model are not measured, while for some other species the measurements are very sparse (and not available at all in large parts of the space domain of the model). The great challenge is to find out what impact the lack of measurements (the partial lack of measurements) has on the results obtained when data assimilation modules are implemented in the model.

The conclusion was that data assimilation techniques can very successfully be used to increase the reliability of the model results. To fully exploit the advantages of data assimilation more efforts are needed in order to improve the performance of data assimilation modules in air pollution models.

3. NEED FOR BETTER EMISSION INVENTORIES

The emissions are among the most important input parameters in a large-scale air pollution model. The emission inventories, which are commonly available (in Europe, at least), are prepared on an annual basis. This means that the annual means of the emissions are given for areas and/or predefined grid-boxes (i.e. one number per area or grid-box per year). Different algorithms are to be used to obtain temporal variations of the emissions. It is clear that this situation is not very satisfactory.

Details and quality of emission estimates and the specification of volatile organic compounds may not always be sufficient for complex air pollution modelling. In the long run it will be desirable to provide scenarios with gritted source category dependent emission estimates on global, regional and local scales. Sophisticated impact studies would benefit from anthropogenic and biogenic VOC splits which are time dependent. Not enough is known about emission of aerosols from anthropogenic as well as biogenic sources. Reliable estimates of size distribution and composition are needed. Problems usually arise when emission inventories from different institutions and agencies are combined in models, for instance, for the purpose of downscaling of simulations. The need for homogenization has been stressed.

The main conclusion of the discussion was that better emission inventories are needed (this is especially true when fine resolution models are to be run). It is at least necessary to prepare some standard modules for producing temporal variations from the available annual means of the emissions.

As mentioned in the previous section data assimilation techniques can be used to improve the available emission inventories.

LIST OF PARTICIPANTS

Aloyan, Artash
Institute of Numerical Mathematics, Russian Academy of Sciences, 8 Gubkin str., 119991, Moscow, **Russia**, aloyan@inm.ras.ru

Balla, Katalin
Computer and Automation Research Institute, Hungarian Academy of Sciences, 1111 Budapest, Kende utca 13-17, **Hungary**, balla@sztaki.hu

Batchvarova, Ekaterina
National Institute of Meteorology and Hydrology, 66 Blvd. Tzarigradsko Chaussee, BG 1784 Sofia, **Bulgaria**, ekaterina.batchvarova@meteo.bg

Bozó, László
Hungarian Meteorological Service, 1675 Budapest, P.OB.39, **Hungary**, bozo.l@met.hu

Cchervenkov, Hristo
National Institute of Meteorology and hydrology, 66, Tzarigradsko chaussee Bulvd.,Sofia 1784, **Bulgaria,** hristo.tchervenkov@meteo.bg

Chock, David
Ford Research and Advanced Engineering, P.O. Box 2053, MD-3083 SRL, Dearborn, MI 48121 **USA,** dchock@ford.com

Cigizoglu, Hikmet Kerem
Civil Engineering Faculty, Istanbul Technical University, Maslak, 34469, Istanbul, **Turkey**, cigiz@itu.edu.tr

Coppalle, Alexis
UMR 6614 CORIA, CNRS-Université-INSA de Rouen
INSA - Avenue de L'Université - BP 8
F-76801 Saint Etienne du Rouvray CEDEX, **France**, Alexis.Coppalle@coria.fr

Csomós, Petra
Eötvös Loránd University, H-1117 Budapest, Pázmány Péter sétány 1/C, **Hungary**, csomos@ludens.elte.hu

Dimitriu, Gabriel
University of Medicine and Pharmacy "Gr. T. Popa", Faculty of Pharmacy, Department of Mathematics and Informatics, 16, Universitatii street, Iasi, **Romaina**, dimitriu@umfiasi.ro

Dimov, Ivan
Institute for Parallel Processing, Bulgarian Academy of Sciences, Acad. G. Bonchev str., bl. 25-A, 1113 Sofia, **Bulgaria**, ivdimov@acad.bg

Dinis, Maria de Lurdes
Department of Mining Engineering, Geo-Environment and Resources Research Center (CIGAR), Engineering Faculty of Oporto University, Rua Dr. Roberto Frias, s/n 4200-465 Oporto, **Portugal**, mldinis@fe.up.pt

Doroshenko, Anatoliy
Institute of Software Systems of the National Academy of Sciences of Ukraine, Acad. Glushkov prosp. 40, block 5, 03187 Kiev, **Ukraine**, dor@isofts.kiev.ua

Ebel, Adolf
Rhenish Institute for Environmental Research at the University of Cologne (RIU), Aachener Str. 201 – 209, 50931 Cologne, **Germany**,
eb@eurad.uni-koeln.de

Faragó, István
Dept. of Applied Analysis, Eotvos Lorand University, Pázmány P. sétány 1/C, 1117, Budapest, **Hungary**,
faragois@cs.elte.hu

Galabov, Vasko
National Institute of Meteorology and hydrology, 66, Tzarigradsko chaussee Bulvd.,Sofia 1784, **Bulgaria**,
Vasko.Galabov@meteo.bg

Galatchi, Liviu-Daniel
Ovidius University, Mamaia Street 124, 900241 Constanta, **Romania**,
galatchi@univ-ovidius.ro

Ganev, Kostadin
Institute of Geophysics, Bulgarian Academy of SciencesAcad. G. Bonchev Str., block.3, Sofia 1113, **Bulgaria**,
kganev@geophys.bas.bg

Garcia, Raul
Environmental Software and Modelling Group, Computer Science School, Technical University of Madrid (UPM), Boadilla del Monte, 28660 Madrid, **Spain**,
ragar@sol.lma.fi.upm.es

Geels, Camilla
National Environmental Research Institute, Department of Atmospheric Environment, Frederiksborgvej 399, DK- 4000 Roskilde, **Denmark**,
cag@dmu.dk

Genikhovich Eugene
Voeikov Main Geophysical Observatory, Karbyshev Street 7, 194021 St. Petersburg, **Russia**,
ego@main.mgo.rssi.ru

Georgiev, Krassimir
Institute for Parallel Processing, Bulgarian Academy of Sciences, Acad. G. Bonchev str. Bl. 25-A, 1113 Sofia, **Bulgaria**,
georgiev@parallel.bas.bg

Gnandt, Boglárka
Hungary Eötvös Loránd University, Department of Meteorology, Pázmány Péter sétány 1/A, 1117 Budapest, **Hungary**,
g_boglarka@ludens.elte.hu

Gryning, Sven-Erik
Wind Energy Department, Risø National Laboratory, Frederiksborgvej 399, Build. 125, P.O. 49, DK-4000 Roskilde, **Denmark**,
sven-erik.gryning@risoe.dk

Havasi, Ágnes
Eötvös Loránd University, Department of Meteorology, Pázmány Péter sétány 1/A, 1117 Budapest, **Hungary**,
hagi@nimbus.elte.hu

Karaivanova, Aneta
Institute for Parallel Processing, Bulgarian Academy of Sciences, Acad. G Bonchev str., bl. 25-A, 1113 Sofia, **Bulgaria**,
anet@parallel.bas.bg

Karatzas, Kostas
Aristotle Uuniversity, Dept. of Mechanical Engineering, Box 483, 54124 Thessaloniki, **Greece**,
kostas@aix.meng.auth.gr

Kharytonov, Mykola
State Agrarian University, Voroshilov st. 25, 49027, Dnepropetrovsk, **Ukraine**,
mykola_kh@yahoo.com

Kirova, Hristina
National Institute of Meteorology and Hydrology 66, Tzarigradsko chaussee, Sofia 1784, **Bulgaria**,
Hristina.Kirova@meteo.bg

Kolarova, Maria
National Institute of Meteorology and hydrology, 66, Tzarigradsko chaussee Bulvd.,Sofia 1784, **Bulgaria**,
Maria.Kolarova@meteo.bg

Krysta, Monika
CEREA (Centre d'Enseignement et de Recherche en Environnement Atmosphérique), Joint Research Laboratory ENPC/EDF R&D (Ecole Nationale des Ponts et des Chaussées/Electricité de France), Cité Descartes, Champs sur Marne, 77455 Marne la Vallée Cedex, **France**,
krysta@cerea.enpc.fr

Ladics, Tamás
Budapest University of Technology and Economics, Department of Analysis, H-1111 Budapest, Egry J. u. 1., **Hungary**,
tladics@math.bme.hu

Lefebre, Filip
Vito - Centre for Integrated Environmental Studies, Boeretang 200, B-2400 Mol, **Belgium**,
filip.lefebre@vito.be

Lysaridis, Iraklis
Laboratory for atmospheric Physics, Aristotle University, Campus Box 149, 54006 Thessaloniki, **Greece**,
hralys@auth.gr

Margenov, Svetozar
Institute for Parallel Processing, Bulgarian Academy of Sciences, Acad. G. Bonchev str., bl. 25-A, 1113 Sofia, **Bulgaria**,
margenov@parallel.bas.bg

Markakis, Kostas
Laboratory for atmospheric Physics, Aristotle University, Campus Box 149, 54006 Thessaloniki, **Greece**,
Kostasmarkakis79@yahoo.gr

Melas, Dimitrios
Laboratory of Atmospheric Physics, Box 149, Aristotle University of Thessaloniki, 54124, Thessaloniki, **Greece**,
melas@auth.gr

Mensink, Clemens
Vito - Centre for Integrated Environmental Studies, Boeretang 200, B-2400 Mol, **Belgium**,
clemens.mensink@vito.be

Ostromsky, Tzvetan
Institute for Parallel Processing, Bulgarian Academy of Sciences, Acad. G Bonchev str., bl. 25-A, 1113 Sofia, **Bulgaria**,
ceco@parallel.bas.bg

Perez, Juan L.
Environmental Software and Modelling Group, Computer Science School, Technical University of Madrid (UPM), Boadilla del Monte, 28660 Madrid, **Spain**,
jupe@sol.lma.fi.upm.es

Perkauskas, Donatas
Air Quality Management Department, Environmental Protection Agency, A.Juozapaviciaus 9, LT-09311 Vilnius, **Lithuania**,
d.perkauskas@gamta.lt

Petrova, Silvia
National Institute of Meteorology and hydrology, 66, Tzarigradsko chaussee Bulvd.,Sofia 1784, **Bulgaria**,
silviaspetrova@icqmail.com

Planinsek, Anton
Environmental Agency of the Republic of Slovenia, Vojkova 1 b, SI-1000 Ljubljana, **Slovenia**,
anton.planinsek@gov.si

Práger, Tamás
Hungarian Meteorological Service, 1525 Budapest, Pf. 38, **Hungary**,
prager.t@met.hu

Prodanova, Maria
National Institute of Meteorology and Hydrology 66, Tzarigradsko chaussee, Sofia 1784, **Bulgaria**,
Maria.Prodanova@meteo.bg

San Jose, Roberto
Environmental Software and Modelling Group, Computer Science School, Technical University of Madrid (UPM), Boadilla del Monte, 28660 Madrid, **Spain**,
roberto@fi.upm.es

Sandu, Ion
National Intitute of Meteorology and Hydrology,Sos bucuresti Ploiesti, No. 97, Sector1, Bucharest, **Romania**,
sandu@meteo.inmh.ro

Schaap, Martijn
Martijn Schaap, TNO, Institute of Environmental Science P.O. Box 342, 7300 AH Apeldoorn,
The Netherlands,
M.Schaap@mep.tno.nl

Steib, Roland
Hungary Hungarian Meteorological Service, 1675 Budapest, P.O.B. 39,
Hungary,
steib.r@met.hu

Syrakov, Dimiter
National Institute of Meteorology and Hydrology 66, Tzarigradsko chaussee, Sofia 1784, **Bulgaria**,
dimiter.syrakov@meteo.bg

Tenchova, Anna
National Institute of Meteorology and hydrology, 66, Tzarigradsko chaussee Bulvd.,Sofia 1784, **Bulgaria**,
silviaspetrova@icqmail.com

Vavalis, Manolis
Dept. of Mathematics, University of Crete, P.O. Box 2208, GR-71409 Heraklion, Crete, **Greece**,
mav@math.uoc.gr

Veliov, Vladimir M.
BAS Institute of Mathematics and Informatics, Bulgarian Academy of Sciences, Acad. G. Bonchev str., Bl. 8, 1113 Sofia, **Bulgaria** and Vienna University of Technology, Argentinierstr. 5/119, 1040 Vienna, **Austria**,
vveliov@eos.tuwien.ac.at

Vignati, Elisabetta
European Commission, Joint Research Centre, Institute for Environment and Sustainability, TP 280, Via E. Fermi, 21020 Ispra (VA), **Italy**,
elisabetta.vignati@jrc.it

Yordanov, Dimiter
Institute of Geophysics, Bulgarian academy of Sciences, 1, acad.G.Bonchev Str., Bl.3, Sofia 1113, **Bulgaria**,
Maria.Kolarova@meteo.bg

Zlatev, Zahari
National Environmental Research Institute, Frederiksborgvej 399, P. O. Box 358, DK-4000 Roskilde, **Denmark**,
zz@dmu.dk

SUBJECT INDEX

GPSR Compliance

The European Union's (EU) General Product Safety Regulation (GPSR) is a set of rules that requires consumer products to be safe and our obligations to ensure this.

If you have any concerns about our products, you can contact us on ProductSafety@springernature.com

In case Publisher is established outside the EU, the EU authorized representative is:

Springer Nature Customer Service Center GmbH
Europaplatz 3
69115 Heidelberg, Germany

Batch number: 10823957

Printed by Printforce, the Netherlands